WASTEWATER SLUDGE PROCESSING

WASTEWATER SLUDGE PROCESSING

IZRAIL S. TUROVSKIY
Wastewater Sludge Treatment
Jacksonville, Florida

P. K. MATHAI
Jacobs Civil Inc.
St. Louis, Missouri

WILEY

Wastewater Sludge Processing

First Indian Reprint: 2014

Authorized reprint by Wiley India Pvt. Ltd., 4435-36/7, Ansari Road, Daryaganj, New Delhi – 110002.

Published by John Wiley & Sons, Inc.

Printed at: *Sanat Printers, Kundli, Haryana*

ISBN: 978-81-265-4634-3

CONTENTS

PREFACE

The processing of wastewater sludge for use or disposal has been a continuing challenge for many municipal agencies. The federal, state, and local regulations coupled with the public awareness of the beneficial use of this valuable resource are forcing a closer look at how to process sludge effectively to reap its benefits, yet protect public health.

The objective of this book is to bring together a wide body of knowledge from the field of wastewater sludge processing and present it in a format that is useful as a textbook for graduate students in environmental engineering and as a reference book for practicing engineers. We discuss unit operations used for processing sludge and the methods available for final disposition of the processed product. The book can be used for planning, designing, and implementing municipal wastewater sludge management projects.

This book could not have been written without the assistance of many staff members of Jacobs Civil Inc. We wish to express our thanks to Jeff Westbrook and Bruce Thomas-Benke, who were the motivating forces behind the preparation of this book. Mr. Thomas-Benke reviewed the entire manuscript and made valuable suggestions that make the book truly comprehensive. We also wish to thank Natalie Preston for her expert preparation of the graphics for all the figures in the book and JoAnn Null for her word-processing skills in the preparation of the manuscript.

Finally, our eternal gratitude to Emilia Turovskiy and Elizabeth Mathai for their support and encouragement throughout the writing of the book.

Izrail S. Turovskiy
P. K. Mathai

ABOUT THE AUTHORS

Izrail S. Turovskiy is a wastewater and biosolids consultant in Jacksonville, Florida. He received a B.S. degree in civil engineering from the Civil Engineering Institute, St. Petersburg, Russia; an M.S. degree in sanitary engineering from the Civil Engineering Institute, Moscow, Russia; and a D.Sc. degree in environmental engineering from the Municipal Academy, Moscow, Russia. He has more than 50 years of experience in environmental engineering, including as head of the All-Union Research Institute of Water Supply, Sewage Systems and Hydrotechnical Structures in Moscow, Russia. Izrail invented and developed almost 50 unit processes and new technologies and has patents in Russia, France, Germany, Italy, Finland, and the United States. Izrail has also authored or coauthored more than 200 technical publications, including eight books. His recommendations for municipal and industrial wastewater and sludge treatment have been implemented in many cities and factories in Russia, Bulgaria, Germany, Hungary, Finland, and Poland.

P. K. Mathai is a senior project manager and associate fellow in environmental engineering with Jacobs Civil Inc. in St. Louis, Missouri. He received a B.S. degree in civil engineering from the University of Kerala, India; and an M.S. degree in environmental engineering from the University of Dayton, Ohio. He has over 30 years of wide-ranging experience in environmental engineering and has been involved in the planning, design, and technical review of over 60 wastewater treatment plants and sludge processing facilities. He is a registered professional engineer in Missouri, Illinois, Ohio, Kentucky, and Maryland.

1

INTRODUCTION

Wastewater Sludge Processing, By Izrail S. Turovskiy and P. K. Mathai

1.1 INTRODUCTION

Water is the most useful and important resource of life—life depends on it. When wastewater is treated to return this resource to the environment, a semisolid, nutrient-rich by-product called *wastewater sludge* is produced. It typically contains 0.25 to 7% solids by weight. When processed properly, the resulting product, known as *biosolids*, has several beneficial uses, including (1) applying to cropland to improve soil quality and productivity because of the nutrients and organic matter it contains, (2) using as a soil amendment in landscaping, and (3) using as a daily cover or part of a final landfill cover.

The U.S. Environmental Protection Agency (EPA) and the Water Environment Federation (WEF) promote the use of the term *biosolids* to emphasize the beneficial use of wastewater sludge solids. However, in this book the term *biosolids* is used only to refer to sludge that has been processed for beneficial use, and the term *sludge* is used exclusively before beneficial use criteria have been achieved and in conjunction with a process descriptor, such as *primary sludge, secondary sludge, return activated sludge*, and *waste activated sludge*. The term *solids* is also used both before and after the beneficial use criteria have been met. Note that screenings and grit are also wastewater solids but are not sludge solids.

Approximately 6.3 million metric tons (6.9 million U.S. tons) (dry solid weight) of municipal wastewater sludge was produced in the United States in 1998 (U.S. EPA, 1999). This is projected to increase to 6.9 million and 7.4 million metric dry tons (7.6 million and 8.2 million U.S. dry tons) in the years 2005 and 2010, respectively. Figure 1.1 shows the estimates of biosolids use and disposal in 1998.

The quantity of sludge produced in a wastewater treatment plant is appro–ximately 1% of the quantity of treated wastewater. While treatment of wastewater takes several hours, processing of the sludge generated and preparing it for disposal or beneficial use takes several days or even several weeks and necessitates the use of more complex equipment. That is why sludge management costs 40 to 50% of the total wastewater treatment costs. Therefore, the processing, reuse, and disposal of wastewater sludge must be managed by municipalities in a cost-effective way and at the same time taking into account prevailing local, state, and federal regulations. The purpose of this book is to describe the prevailing methods for the planning, design, and implementation of wastewater sludge management programs. A discussion of the U.S. EPA's 40 CFR Part 503 regulations, *Standards for the Use or Disposal of Sewage Sludge*, is given in the rest of this chapter.

1.2 40 CFR PART 503 REGULATION

Under the authorities of the Clean Water Act as amended, the U.S. EPA promulgated in 1993 at 40 Code of Federal Regulation (CFR) Part 503, *Stan-*

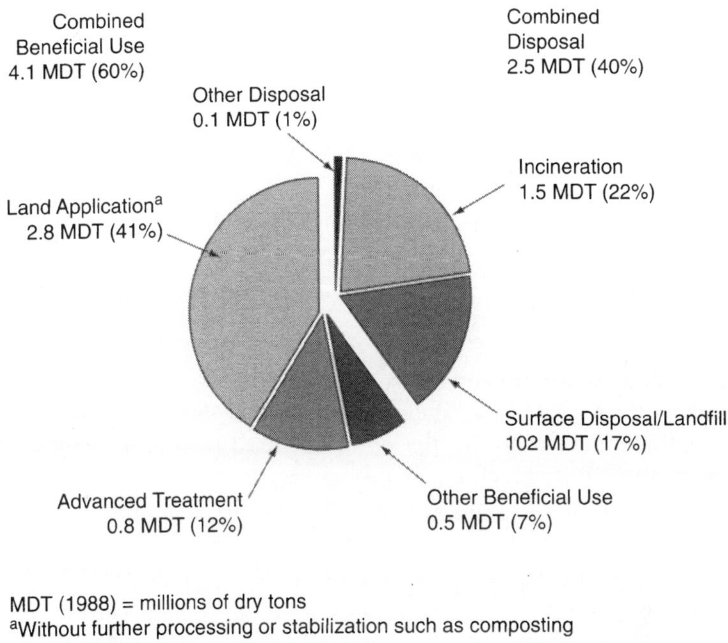

Figure 1.1 Estimates of biosolids use and disposal.

dards for the Use or Disposal of Sewage Sludge. The intent of this regulation is to ensure that sewage sludge is used or disposed of in a way that protects both human health and the environment. The regulation is divided into five subparts: (1) general provisions, (2) land application, (3) surface disposal, (4) pathogens and vector attraction reduction, and (5) incineration. The subparts are described in the following sections.

1.2.1 General Provisions

Part 503 establishes the general requirements, pollutant limits, operational standards, and management practices as well as frequency of monitoring, recordkeeping, and reporting requirements that have to be met when biosolids are applied to the land, placed on a surface disposal site, placed on a municipal solids waste landfill unit, or fired in a sewage sludge incinerator. In addition to establishing requirements for the quality of biosolids that are used or disposed and for sites on which biosolids are used or disposed, Part 503 indicates who has to ensure that the requirements are met. These requirements include:

- The person who prepares the biosolids
- The person who applies the biosolids

- The owner or operator of a surface disposal site
- The person who fires sludge in a sludge incinerator
- Class I sludge management facilities (wastewater treatment plants that are required to have an approved pretreatment program)
- Publicly owned treatment works (POTWs) with a flow rate equal to or greater than 3800 m^3/day (1 mgd)
- POTWs that serve a population of 10,000 people or more

The following activities and types of sludge and other wastewater solids are not subject to Part 503 regulation:

- Requirements for processes used to treat wastewater or for processes used to treat wastewater sludge prior to final use or disposal, with the exceptions that pertain to the processes used to reduce pathogens and vector attraction reduction.
- The selection of a use or particular disposal practice of biosolids. Determination of the manner in which the biosolids are used or disposed of are the responsibility of the local authority.
- Co-firing of sewage sludge in an incinerator with other wastes or for the incinerator in which sewage sludge and other wastes are co-fired. If the auxiliary fuel is municipal solid waste, it cannot exceed 30% of the dry weight of the sewage sludge and auxiliary fuel combined. The requirements of 40 CFR Parts 60 and 61 have to be met in cases where total mass fired contains more than 30% municipal solid waste.
- Sludge generated during treatment of industrial process wastewater at an industrial facility, including sewage sludge generated at an industrial facility during the treatment of industrial wastewater combined with domestic sewage. If the sludge is applied to or placed on land, requirements in 40 CFR Part 257 have to be met. If the sludge is placed in a municipal solid waste landfill facility, the requirements of 40 CFR Part 258 apply.
- Hazardous sludge, including hazardous sewage sludge, which are subject to the requirements in 40 CFR Parts 261 to 268 when they are used or disposed of.
- Incinerator ash.
- Grit and screenings generated during the treatment of domestic sewage. If used or disposed of, requirements of 40 CFR Part 257 apply.
- Drinking water treatment sludge. Requirements of 40 CFR Part 257 apply if used or disposed of.
- Commercial septage, industrial septage, a mixture of domestic septage and commercial septage, or a mixture of domestic septage and industrial septage. Part 503 applies if the device in which the septage is generated, such as a septic tank, receives only domestic sewage.

When the EPA designed Part 503, the regulation was written to be self-implementary. This means that even without the terms and conditions of a biosolids permit, regulated users and disposers of biosolids are required to meet the requirements of the regulation. Part 503 is a national standard. When writing a permit, the permitting authority can impose stringent or additional requirements as long as the authority can show that they are necessary to protect the public health and the environment from any adverse effect of a pollutant in the biosolids. The permitting authority can use best professional judgment to develop more stringent conditions or to add limits for pollutants that are not regulated by Part 503.

1.2.2 Land Application

Land application is the spreading or spraying of biosolids onto the surface of the land or the injection of biosolids beneath the surface of the land to condition the soil or to fertilize crops grown in the soil. Land application has been practiced for decades and continues to be the most common method of using biosolids.

Biosolids that meet certain requirements are commonly referred to as being of *exceptional quality*. Although the phrase is not found in Part 503, exceptional quality (EQ) biosolids are biosolids or materials derived from biosolids that meet Part 503 requirements of the pollutant ceiling concentrations, one of the class A pathogen reduction alternatives and one of the vector attraction reduction alternatives. If a generator or other preparer is able to demonstrate that biosolids or material derived from biosolids meet the criteria of EQ biosolids, the biosolids are not subject to the general requirements and management practices in the land application subpart of Part 503.

General Requirements The application of biosolids to the land may involve several parties. Although not all of these parties need to receive a permit, all of them must comply with the appropriate land application subpart requirements. The first general requirement for the applier of bulk biosolids is that no person may apply biosolids to the land without meeting the requirements in Subpart B of Part 503. These requirements include, but are not limited to, management practices, certain vector attraction reduction options, and the class B site restrictions. The land applier should request information from the preparer of biosolids to ensure that treatment-related requirements have been met. This includes pollutant concentrations in biosolids, the class of pathogen reduction, and whether vector attraction reduction is achieved through treatment. The person who applies biosolids to the land must also provide the owner or leaseholder of the land on which the biosolids are applied notice and necessary information to comply with the land application requirements. Requirements that may apply to the owner or leaseholder of the land include the site restrictions for class B biosolids, such as prohibitions on harvesting certain types of crops, grazing animals for certain periods, and restricting public access for certain periods.

Pollutant Limits The first of three parameters that must be assessed to determine the overall quality of biosolids is the level of pollutants. To allow land application of biosolids of variable quality, Part 503 provides four sets of pollutant limits: pollutant ceiling concentration limits, pollutant concentration limits, cumulative pollutant loading rates, and annual pollutant loading rates. These pollutant limits are listed in Table 1.1.

Ceiling Concentration Limits All biosolids that are land applied, and both bulk biosolids and biosolids sold or given away in a bag or other container, have to meet the ceiling concentration limits. These limits are absolute values, which means that all samples of biosolids have to meet the limits.

Pollutant Concentration Limits Biosolids meeting the pollutant concentration limits (also known as exceptional quality pollutant concentration limits)

TABLE 1.1 Land Application Pollutant Limits

Pollutant	Ceiling Concentration Limits[a,b] (mg/kg)	Pollutant Concentration Limits[a,c,d] (mg/kg)	Cumulative Pollutant Loading Rates (kg/ha)	Annual Pollutant Loading Rates (kg/ha/365-day period)
Arsenic	75	41	41	2.0
Cadmium	85	39	39	1.9
Copper	4300	1500	1500	75
Lead	840	300	300	15
Mercury	57	17	17	0.85
Molybdenum	75	—	—	—
Nickel	420	420	420	21
Selenium	100	36	100	5
Zinc	7500	2800	2800	140
Applies to:	All biosolids that are land applied	Bulk biosolids and biosolids sold or given away in a bag or other container	Bulk biosolids	Biosolids sold orgiven away in a bag or other container

Source: U.S. EPA, 1994.

[a] Dry weight basis.

[b] Absolute values.

[c] Monthly averages.

[d] Exceptional quality biosolids.

achieve one of three levels of quality necessary for exceptional quality status, hence can be sold or given away in bags or other containers. These limits must also be met when bulk biosolids are land applied. The limits are monthly averages (i.e., the arithmetic average of all measurements taken during the month).

Cumulative Pollutant Loading Rates The cumulative pollutant loading rates apply to bulk biosolids that meet ceiling concentration limits but do not meet pollutant concentration limits. These rates limit the amount of a pollutant that can be applied to an area of land in bulk biosolids for the life of the application site.

Annual Pollutant Loading Rates The annual pollutant loading rates apply to biosolids that meet ceiling concentration limits but do not meet pollutant concentration limits, and are to be sold or given away in a bag or other container for application to the land. Annual pollutant loading rates rather than cumulative pollutant loading rates are applied to these biosolids because biosolids sold or given away in a bag or other container are commonly used by homeowners, and it would be impractical to expect homeowners to track cumulative pollutant loadings. Design application rates of biosolids for land application based on nitrogen and pollutant concentration are described in Chapter 10.

Pathogen Reduction The second parameter in determining biosolids quality is the presence or absence of pathogens. Pathogens are disease-causing organisms such as *Salmonella* bacteria, enteric viruses, and viable helminth ova. The preparer of biosolids is responsible for monitoring and certifying that measures have been taken to reduce these types of pathogens in the biosolids they produce.

Part 503 contains two classes of pathogen reduction: classes A and B. Class A pathogen reduction alternatives render the biosolids virtually pathogen-free after treatment. Class B pathogen reduction alternatives reduce pathogens significantly. Land appliers who apply biosolids that are certified as class A have no restrictions relative to pathogens. Land application of class B biosolids comes with it the following site restrictions, which are necessary to provide the same level of protection to public health and the environment as is provided by class A biosolids:

- Public access to land with a high potential for public exposure shall be restricted for one year after biosolids application.
- Public access to land with a low potential for public exposure shall be restricted for 30 days after biosolids application.
- Food crops, feed crops, or fiber crops shall not be harvested for 30 days after biosolids are applied.

- Food crops with harvested parts that touch the biosolids–soil mixture and are totally above the land surface (e.g., melons, cucumbers, squash) shall not be harvested for 14 months after application of biosolids.
- Food crops with harvested parts below the surface of the land (e.g., potatoes, carrots, radishes) shall not be harvested for 20 months after application when biosolids are not incorporated into the soil or remain on the soil surface for four or more months prior to incorporation into the soil.
- Food crops with harvested parts below the surface of the land (e.g., potatoes, carrots, radishes) shall not be harvested for 38 months if biosolids are incorporated into the soil within four months after application.
- Animals shall not be grazed on the site for 30 days after biosolids application.
- Turf shall not be harvested for one year after biosolids application if the turf is placed on land with a high potential for public exposure, or on a lawn, unless otherwise specified by the permitting authority.

Vector Attraction Reduction Attractiveness of biosolids to vectors is the third parameter of biosolids quality. Vectors are animals and insects, such as rodents, flies, and birds, that might be attracted to biosolids and therefore could transmit pathogenic organisms to humans or to domestic animals or livestock. One of the 10 options provided in Subpart D of Part 503 to reduce vector attractiveness must be met if biosolids are to be applied to the land. The 10 vector attraction reduction options, which are discussed in more detail at the end of this chapter, fall into two categories: *treatment options* (options 1 through 8) and *barrier options* (options 9 and 10). Treatment options are undertaken by the biosolids preparer. If any one of the treatment options is performed by the preparer, the land applier has no requirements relative to vector attraction reduction. If one of the treatment options is not performed by the preparer, the land applier is responsible for implementing and certifying compliance with either one of the barrier options, which use soil as a barrier between the biosolids and any vectors that might be present.

Management Practices Four management practices are specified for bulk biosolids that are applied to the land. The practices help ensure that biosolids are applied in a manner that is protective to human health and environment. The four management practices applicable to the land application of bulk nonexceptional quality biosolids are:

- Endangered species or critical habitat protection
- Application to flooded, frozen, or snow-covered land
- Distance to surface waters
- Agronomic application rate

Endangered Species or Critical Habitat Protection Application of bulk non-EQ biosolids is prohibited if it is likely to affect threatened or endangered species or their designated critical habitat. Any direct or indirect action that reduces the likelihood of survival and recovery of a threatened or endangered species is considered an adverse effect. *Critical habitat* is any place where a threatened or endangered species lives and grows during any stage of its life cycle. The U.S. Department of Interior, Fish and Wildlife Service (FWS) publishes a list of threatened or endangered species. To comply with this management practice, the land applier should consult with FWS to determine whether any threatened or endangered species or their designated critical habitats are present at the site.

Application to Flooded, Frozen, or Snow-Covered Land Part 503 does not prohibit the application of bulk non-EQ biosolids to flooded, frozen, or snow-covered land. It does allow that the permitting authority can withhold this permission. Part 503 does state that biosolids applied to these lands may not enter wetlands or other waters of the United States. Prior to applying biosolids to these lands, the land applier should ensure that proper runoff control measures, such as slope restrictions, berms, dikes, silt fences, and sediment basins, are in place.

Distance to Surface Waters Application of bulk non EQ biosolids on agricultural land, forest, or a reclamation site that is within 10 m of any waters of the United States is prohibited unless otherwise specified by the permitting authority. The 10-m buffer zone serves as a barrier against biosolids entering water bodies.

Agronomic Application Rate Bulk non-EQ biosolids should be applied to a site only at a rate equal to or less than the agronomic rate for the site. The *agronomic rate* is the application rate that is designed to provide the amount of nitrogen needed by the crop or vegetation and thus to minimize the amount of nitrogen that passes to the groundwater and contaminates it. In some instances, the permitting authority may specifically authorize application to a reclamation site at a rate in excess of the agronomic rate, where there is no potential for nitrate to leach down to the groundwater. However, once a vegetation cover has been established, future application of biosolids should be limited to the agronomic rate of the vegetation grown.

Only one management practice applies to biosolids sold or given away in a bag or other container. A label or information sheet must be provided with the biosolids indicating the appropriate application rate for the quality of the biosolids.

Monitoring The person who prepares biosolids must monitor biosolids quality. As mentioned earlier, the land applier must obtain information from the preparer on the pollutant concentrations in the bulk non-EQ biosolids.

TABLE 1.2 Parameters To Be Monitored in Land-Applied Biosolids

Pollutants[a]	Pathogens	Vector Attraction Reduction
Arsenic	Fecal coliform or *Salmonella*	Percent volatile solids reduction[c]
Cadmium	Enteric viruses[b]	Specific oxygen uptake rate[d]
Copper	Helminth ova[b]	pH[e]
Lead		Percent solids[f]
Mercury		
Molybdenum		
Nickel		
Selenium		
Zinc		

[a] Dry weight basis.
[b] Class options 3 and 4.
[c] Vector attraction reduction options 1, 2, and 3.
[d] Vection attraction reduction option 4.
[e] Vection attraction reduction option 6.
[f] Vector attraction reduction options 7 and 8.

TABLE 1.3 Monitoring Frequency for Land Application

Biosolids Quantity[a] (dry metric tons/365-day period)	Frequency[b]
0 to <290	Once per year
290 to <1500	Once per quarter (4 times/year)
1500 to <15,000	Once per 60 days (6 times/year)
15,000 or greater	Once per month (12 times/year)

[a] Bulk biosolids or biosolids sold or given away in a bag or other container.
[b] After two years the permiting authority may reduce the frequency for pollutant concentrations and for the pathogen density rerquirements.

Parameters that must be monitored are listed in Table 1.2. The frequency of monitoring is typically established by the permitting authority through individual permits on a case-by-case basis. However, to enhance the self-implementation of the regulation, Part 503 has established the monitoring frequency shown in Table 1.3.

Record Keeping Records must be kept to show compliance with pollutant concentration and loadings, pathogen reduction requirements, and management practices. These requirements are divided into responsibilities for the biosolids preparer and for the land applier. All preparers must keep records of biosolids quality regardless of whether the biosolids are EQ or non-EQ.

Record-keeping requirements for land appliers vary depending on biosolids quality. However, all appliers of bulk non-EQ biosolids must document implementation of applicable management practices. The land applier must also keep records tracking the cumulative pollutant loadings. When the site reaches 90% of its allowable cumulative pollutant loading, the land applier should notify the preparer annually until the site reaches 100%, and then no additional biosolids that are non-EQ for pollutants can be applied. When the bulk biosolids are non-EQ for pathogens, the applier must keep records on the implementation of class B site restrictions. When the bulk biosolids are non-EQ for vector attraction reduction, the applier must keep records documenting the implementation of the vector attraction reduction options 9 and 10 (discussed later in this chapter). Unless otherwise noted by the permitting authority, records must be kept for a minimum of five years.

Reporting Reporting requirements apply to wastewater treatment plants with a design flow of 4000 m^3/day (1 mgd) and to plants with a service population of 10,000 people or more. Reporting requirements also apply to class I sludge management facilities, which are treatment plants that are required to have an approved pretreatment program.

1.2.3 Surface Disposal

A *surface disposal site* is an area of land on which wastewater solids are placed for disposal. Examples of surface disposal practices include the following:

- Sludge-only landfills (monofills)
- Sludge piles or mounts
- Sludge lagoon used for final disposal
- Surface application sites where wastewater solids are applied at rates in excess of the agronomic rate

Surface disposal of wastewater solids differs from land application in that it principally uses the land for final disposal instead of using the solids to enhance the productivity of the land. However, some surface disposal practices, where the solids are applied on the surface of the land, may be very similar to land application practices. For example, a surface disposal site, where solids are applied at rates in excess of the agronomic rate needed by vegetation grown on the site and a food, feed, or fiber crop is grown or animals are grazed, may appear to be a land application site, but the site is a surface disposal site if it is for the final disposal of solids. In this situation, management practices must be implemented to control activities such as the growing of crops or animal grazing.

The storage or treatment of sludge, other than treatment to reduce pathogen levels and vector attraction characteristics, is not regulated by Part 503. Lagoons, in particular, are frequently used to dewater or stabilize wastewater sludge as well as for storage. The most obvious indicator that the land-based activity is treatment or storage is whether the wastewater treatment plant has designated a subsequent sludge use or disposal plan. Unless a final use or disposal plan has been identified for the sludge, the land-based activity is considered as final disposal.

Storage is placement of biosolids on land on which it remains for two years or less. Therefore, a factor used to distinguish between storage and final disposal is the length of time that the biosolids remain on the land. Biosolids remaining on the land for longer than two years is considered final disposal unless the person who prepares the biosolids establishes a basis for leaving the biosolids on the land for longer than two years prior to final disposal. Wastewater treatment plants that generate small quantities of biosolids may stockpile biosolids for a reasonable period of time that is longer than two years before use or disposal. Even some large treatment plants may stockpile biosolids for a lengthy period when they unexpectedly lose access to their disposal practice.

Some provisions of the surface disposal subpart apply to biosolids, some to the location or siting of the surface disposal site, and others to operation of the surface disposal site. Not all of the management practices, frequency of monitoring, and record-keeping requirements apply to every surface disposal site. For example, some apply only to sites with a liner and leachate collection system; others apply only to sites on which a cover is placed over the biosolids.

General Requirements The four general requirements for placing wastewater solids on an active site (*surface unit* is the phrase used in Part 503) are the following:

- Wastewater solids cannot be placed on an active sludge disposal unit unless the requirements of Subpart C of Part 503 are met.
- An active unit cannot be located within 60 m of a fault with displacement in Holocene time, in an unstable area, or in a wetland.
- The owner or operator of an active unit must submit a written closure and postclosure plan 180 days before the unit is due to close.
- The owner of a surface disposal unit must provide written notification to the subsequent owner of the site that wastewater solids were placed on the land.

The first general requirement places the responsibility on the person who places wastewater solids on an active sewage sludge unit to ensure that the requirements related to the surface disposal are met when the solids are actually placed on the unit. The person who places the solids on the land could be either the preparer or the owner/operator of the surface disposal site. If the person who places the solids is not the preparer, that person should request the information from the preparer that indicates that the treatment-related Part 503 surface disposal requirements have been met. These include pollutant concentrations in the sludge and whether pathogen reduction and vector attraction reduction have been achieved through treatment.

If an active sewage sludge disposal unit falls into one of the three categories of the second general requirement listed above, the owner or operator must develop a compliance schedule that addresses closure activities and time frames. The closure plan, at a minimum, must address the following three specific items:

- Operation and maintenance of the leachate collection system if the unit has a liner and leachate collection system
- Methane gas monitoring if a final cover is placed on the unit
- Public access restriction to the site

Leachate Collection System Leachate can contaminate groundwater, surface water, and soil if it is not controlled. If the active land disposal unit has a liner and a leachate collection system, the closure and postclosure plan must describe how the leachate collection system will be operated and maintained for three years after closure. The owner or operator must comply with all local, state, and federal requirements for leachate collection and disposal.

Methane Gas Monitoring System Since methane gas, a by-product of anaerobic decomposition of organic matter, is explosive within a certain concentration range in air, Part 503 requires continuous monitoring of air for methane gas at sites where the sewage sludge is covered. Additionally, air must be monitored for methane for a period of three years after closure if a final cover is placed on an active sewage sludge unit at closure.

Public Access Restriction Part 503 restricts public access to a closed surface disposal unit to prevent:

- Possible exposure to methane
- Direct contact with, or ingestion of, the sewage sludge or sewage sludge–soil mixture
- Traffic that could damage the final cover

Final Cover If a final cover is placed on an active sewage sludge disposal unit that is to be closed, the cover should be designed to:

- Control volatilization of pollutants
- Account for settling or subsidence in the unit

- Resist erosion
- Control runoff and prevent other damage to the cover

Notification to Subsequent Owner The owner of a surface disposal site must provide a subsequent owner of the site with written notification stating that the land has been used for surface disposal of wastewater solids. The notice must describe the wastewater solids disposal activities as well as provide details about the design and operation of the site.

Pollutant Limits If a sewage sludge disposal unit is equipped with a liner and a leachate collection system, there are no pollutant limits because the liner retards the movement of pollutants in sewage sludge into the groundwater. A liner is clay or synthetic material that has a hydraulic conductivity of 1×10^{-7} cm/s or less. If the liner does not meet this specification, the unit is considered unlined for the purpose of assigning pollutant limits to the sewage sludge placed in the unit.

Sludge placed in an active unit without a liner, and a leachate collection system must meet pollutant limits for arsenic, chromium, and nickel. The specific pollutant limits to apply depend on the distance from the active sewage sludge disposal unit boundary to the surface disposal site property line. The limits allowed are listed in Table 1.4.

The permitting authority may determine that the conditions of the active sewage sludge disposal unit warrant the development of site-specific limits, in which case the authority may use site-specific data and develop sewage sludge pollutant limits different from the limits in Table 1.4. In addition, if a surface disposal site has several different types of active sewage sludge units, the sewage sludge placed on various units will be subject to different pollutant limits, depending on the active sludge unit in which it is placed.

TABLE 1.4 Pollutant Concentration Limits in Active Sewage Sludge Disposal Unit Without a Liner and Leachate Collection System

Distance from Boundary to Property Line (m)	Pollutant Concentration[a]		
	Arsenic (mg/kg)	Chromium (mg/kg)	Nickel (mg/kg)
0 to <25	30	200	210
25 to <50	34	220	240
50 to <75	39	260	270
75 to <100	46	300	320
100 to <125	53	360	390
125 to <150	62	450	420
150 or greater	73	600	420

Source: U.S. EPA, 1994.

[a] Dry weight basis.

Management Practices Management practices dealing with the location, design, and operation of wastewater sludge surface disposal facilities include the following:

- Sludge shall not be placed in an active sewage sludge disposal unit if threatened or endangered species of plant, fish, or wildlife are identified within or near the unit during any stage of their life cycles. However, if the unit is located within the migratory path of an endangered species, the permitting authority may prohibit the disposal of sludge only during the migration period.
- An active sewage sludge disposal unit shall not restrict the flow of a base flood (100-year flood). If the owner or operator of the unit can demonstrate that the unit will not pose unacceptable threats of higher flood levels and flood velocity, the requirements of this provision are met.
- When an active sewage sludge disposal unit is located in a seismic impact zone, the unit shall withstand the maximum recorded horizontal ground-level acceleration. A seismic impact zone is an area that has a 10% or greater probability that the horizontal ground-level acceleration of the rock in the area exceeds 0.10 gravity once in 250 years.
- An active sewage sludge disposal unit shall not be located within 60 m of a fault that has a displacement in Holocene time, which is approximately the last 11,000 years.
- An active sewage sludge disposal unit shall not be located in an unstable area, such as landslide-prone areas, karst terrains, and surface areas weakened by underground mining or oil, water, or water withdrawals.
- An active sewage sludge disposal unit shall not be located in a wetland unless the owner or operator applies and receives a Section 404 permit from the U.S Army Corps of Engineers.
- The runoff collection system for an active sewage sludge disposal unit shall have the capacity to handle runoff from a 24-hour 25-year event, and the runoff shall be disposed in accordance with a National Pollutant Discharge Elimination System (NPDES) permit.
- The leachate collection system for an active sewage sludge disposal unit that has a liner and leachate collection system shall be operated and maintained, and the leachate shall be disposed of in accordance with the applicable requirements during the period the unit is active and for three years after the unit closes.
- If sewage sludge is covered with soil or other material, proper equipment shall be installed to monitor methane continuously in air in structures and at the site property line. Methane levels cannot exceed 25% of the lower explosive limit (LEL) in on-site structures, and cannot exceed the LEL at the site property line. The monitoring shall continue and the LEL limit requirements shall be met for three years after the sewage sludge unit closes.

- Growing of food, feed, or fiber crops or the grazing of animals on an active sewage sludge disposal unit is prohibited unless the owner or operator of the site demonstrates to the permitting authority through management practices that public health and the environment are protected from adverse effects when crops are grown or when animals are grazed.
- Public access to a surface disposal site is restricted when the unit is active and for three years after the site closes.
- The owner or operator of a sewage sludge disposal unit shall demonstrate to the permitting authority that the unit does not contaminate an aquifer (1) by providing a certification by a qualified groundwater scientist that the unit will not contaminate the aquifer, or (2) by performing groundwater monitoring.

Pathogen and Vector Attraction Reduction Requirements Sewage sludge to be placed in an active sewage sludge disposal unit must meet one of the class A or B pathogen reduction alternatives by the preparer unless the sludge is covered with soil or other material by the owner or operator at the end of each operating day. One of the first eight vector attraction reduction alternatives, or an equivalent alternative as determined by the permitting authority, is required to be met by the preparer of the sludge. Also, one of three vector attraction reduction alternatives (discussed further later in the chapter) is required to be met by the owner or operator of the sewage sludge unit.

Monitoring Sewage sludge placed on unlined units must be monitored for the three regulated pollutants (arsenic, chromium, and nickel), pathogen reduction (fecal coliform or *Salmonella*, enteric viruses, and helminth ova), and vector attraction reduction (volatile solids reduction, specific oxygen uptake rate, pH, and percent solids). Sludge placed on units equipped with liners and leachate collection systems is subject only to pathogen and vector attraction reduction requirements.

The frequency of monitoring is typically established through permits on a case-by-case basis. However, to enhance the self-implementation of regulation, monitoring frequencies have been established in Part 503. The frequencies established are shown in Table 1.5, but the permitting authority has the discretion to require more frequent monitoring.

Record Keeping Records must be kept to demonstrate that permit conditions that implement all applicable regulatory requirements of Part 503 are being met. Specific information must be kept to show compliance with pollutant concentrations, pathogen reduction, vector attraction reduction, and management practices. These records must be retained for at least five years.

Reporting Only the following types of facilities are required to report to the permitting authority under the Part 503 regulation:

TABLE 1.5 Frequency of Monitoring: Surface Disposal

Amount of Sludge[a] (metric tons/365-day period)	Frequency
0 to <290	Once/year
290 to <1500	Once/quarter (4 times/year)
1500 to <15,000	Once/60 days (6 times/year)
15,000 or grater	Once/month (12 times/year)

[a] Dry weight basis.

- Class I sludge management facilities
- Publicly owned treatment works with a flow rate equal to or greater than 3800 m^3/day (1 mgd)
- Publicly owned treatment works serving a population of 10,000 or greater

Surface Disposal of Domestic Septage The regulatory requirements for the surface disposal of domestic septage are not as extensive as sewage sludge. The requirements include meeting the same management practices that are required for the surface disposal of sewage sludge. There are no specific pathogen reduction requirements for domestic septage applied on surface disposal sites. However, to meet the vector attraction reduction requirements, it must be (1) injected into the subsurface of the land, (2) applied to the surface and incorporated into the soil, (3) covered with soil or other material at the end of each operating day, or (4) treated with an alkaline material to raise the pH to 12 or higher for at least 30 minutes.

Placement of Sludge in a Municipal Solid Waste Landfill There are no specific requirements for disposal of sewage sludge in a municipal solids waste landfill (MSWLF) in Part 503. Instead, Part 503 requires compliance with 40 CFR Part 258. Thus, the design standards and operating standards for MSWLFs established in Part 258 serve as alternative standards for protection of public health and the environment. To meet the Part 258 requirements, the preparer of the sewage sludge must ensure that (1) the sewage sludge is non-hazardous, and (2) it does not contain free liquids as defined by the *paint filter test*. The owner or operator of a MSWLF unit must ensure that the material used to cover the unit is capable of controlling disease vectors, fires, odors, blowing litter, and scavenging without a threat to human health and the environment. Therefore, if sewage sludge is used as an alternative cover material, it may have to be treated for vector attraction reduction prior to its use. Use of sewage sludge as a cover material must also be approved by the state agency regulating MSWLFs.

1.2.4 Pathogen and Vector Attraction Reduction

Pathogens are disease-causing organisms such as certain bacteria, fungi, viruses, protozoans and their cysts, and intestinal parasites and their ova. Vector attraction is the characteristic of sewage sludge that attracts rodents, flies, mosquitoes, or other organisms capable of transporting infectious agents. Subpart D of Part 503 covers alternatives for reducing pathogens in sewage sludge and domestic septage, as well as options for reducing the characteristic of sewage sludge that attracts vectors.

The pathogen and vector attraction reduction requirements apply to biosolids and their application to or placement on the land for beneficial use or disposal. Depending on how biosolids are used or disposed of and which pathogen reduction alternative and vector attraction reduction options are relied on, compliance with pathogen and vector attraction requirements is the responsibility of:

- The generator of biosolids that are either land applied or surface disposed
- The person who derives a material from biosolids that are either land applied or surface disposed
- The person who applies biosolids to land or places biosolids on a surface disposal site
- The owner or operator of a surface disposal site

Pathogen Reduction Alternatives Biosoilds are classified as class A or B based on the level of pathogens present in biosolids that are used or disposed of. Biosolids meet class A designation if pathogens are below detectable levels. All biosolids that are sold or given away in a bag or other container for application to land, lawns, or home gardens must meet class A pathogen requirements. Biosolids are designated class B if pathogens are detectable but have been reduced to levels that do not pose a threat to public health and the environment as long as actions are taken to prevent exposure to the biosolids after their use or disposal. All biosolids that are land applied or placed on surface disposal sites (except sludge placed on a surface disposal site that is covered with soil or other material) must meet class B pathogen requirements. In general, class A corresponds to Part 257, *process to further reduce pathogens* (PFRP), designation, and class B corresponds roughly to Part 257, *process to significantly reduce pathogens* (PSRP), designation.

Class A Pathogen Requirements Part 503 lists six alternatives for treating wastewater sludge to meet class A requirements with respect to pathogens. These alternatives are summarized in Table 1.6, and a short description of each alternative follows. All six alternatives must meet one of the two following criteria at the time biosolids are used or disposed of, prepared for sale or

TABLE 1.6 Summary of Alternatives for Meeting Class A Pathogen Requirements

Alternative 1: Thermally Treated Biosolids
Biosolids must be subjected to one of four time–temperature regimes.

Alternative 2: Biosolids Treated in a High-pH/High-Temperature Process
Biosolids must meet specific pH, temperature, and air drying requirements.

Alternative 3: Biosolids Treated in Other Known Processes
It must be shown that the process can reduce enteric viruses and viable helminth ova. Operating conditions used in the demonstration should be maintained after pathogen reduction demonstration has been completed.

Alternative 4: Biosolids Treated in Unknown Processes
Biosolids must be treated for pathogens—*Salmonella* sp. or fecal coliform bacteria, enteric viruses, and viable helminth ova—at the time the biosolids are used or disposed of, or in ceratain situations, prepared for use or disposal.

Alternative 5: Biosolids Treated in a PFRP
Biosolids must be treated by one of the *processes to further reduce pathogens* (PFRPs) (see Table 1-8).

Alternative 6: Biosolids Treated in a Process Equivalent to a PFRP
Biosolids must be treated by a process equivalent to one of the PFRPs, as determined by the permitting authority.

Source: U.S. EPA, 1994.

giveaway in a bag or other container for land application, or prepared to meet EQ requirements:

- The density of fecal coliform in the biosolids must be less than 1000 MPN (most probable number) per gram of total solids (dry weight basis); or
- The density of *Salmonella* sp. bacteria in the biosolids must be less than 3 MPN per 4 g of total solids (dry weight basis).

Alternative 1: Thermally Treated Biosolids This alternative applies when specific thermal heating procedures are used to reduce pathogens. The length of heating time at a given temperature needed to obtain class A pathogen reduction is determined by equations for each of the four thermal heating regimes, which are shown in Table 1.7. Any one of the four regimes may be used. Frequency of monitoring for pathogens is same as in Table 1.5.

Alternative 2: Biosolids Treated in a High-pH/High-Temperature Process This alternative describes conditions of a specific temperature–pH process that is effective (usually by alkaline treatment) in reducing pathogens to below detectable levels. The frequency of monitoring for pathogens is same as in Table 1.5. The process conditions required are:

TABLE 1.7 Thermal Heating Regimes for Class A Pathogen Reduction Under Alternative 1

Regime	Applies to:	Requirement	Time–Temperature Relationship[a]
A	Biosolids with 7% solids or greater (except those covered by Regime B)	Temperature of biosolids must be 50°C or higher for 20 minutes or longer	$D = 131{,}700{,}000/10^{0.14t}$
B	Biosolids with 7% solids or greater in the form of small particles and heated by contact with either warmed gases or an immiscible liquid	Temperature of biosolids must be 50°C or higher for 15 seconds or longer	$D = 131{,}700{,}000/10^{0.14t}$
C	Biosolids with less than 7% solids	Heated for at least 15 seconds but less than 30 minutes	$D = 131{,}700{,}000/10^{0.14t}$
D	Biosolids with less than 7% solids	Temperature of sludge is 50°C or higher with at least 30 minutes or longer contact time	$D = 50{,}070{,}000/10^{0.14t}$

Source: U.S. EPA, 1994.

[a] D, time in days; t, temperature in °C.

- Elevating the pH of the biosolids to greater than 12 (measured at 25°C) for a minimum of 72 hours
- Maintaining the temperature above 52°C for a minimum of 12 hours during the period when the pH is greater than 12
- Air drying to at least 50% solids after the 72-hour period of elevated pH

Alternative 3: Biosolids Treated in Other Known Processes The purpose of this alternative is to demonstrate that a new treatment process fully meets the class A pathogen requirements. The presence of enteric viruses and viable helminth ova has to be shown in the sludge prior to the treatment (as sludge in some treatment plants may not contain enteric viruses or helminth ova) to document the effectiveness of the treatment process. Subsequent testing for enteric viruses and helminth ova is not required whenever the tested set of operating parameters has been met. The values for enteric viruses and helminth ova that have to be achieved during the demonstration are:

- *Enteric viruses:* less than 1 plaque-forming unit per 4 g of total solids (dry weight basis)
- *Helminth ova:* less than 1 viable ovum per 4 g of total solids (dry weight basis)

If no enteric viruses or viable helminth ova are present in the biosolids before treatment, it is class A with respect to enteric viruses or viable helminth ova, or both, until the next sampling episode, at which time another sample of the biosolids has to be tested for those organisms. The frequency of monitoring for both enteric viruses and viable helminth ova is once per year, quarterly, bimonthly, or monthly until it is demonstrated that the limits stated have been achieved for some period of time, as determined by the permitting authority. Monitoring for fecal coliform or *Salmonella* sp. bacteria is always required at the frequency in Table 1.5.

Alternative 4: Biosolids Treated in Unknown Processes This alternative is used when the wastewater solids have been treated using a newly developed or innovative treatment process. The biosolids must meet the same pathogen test results as in alternative 3. If the biosolids meet those requirements at the time of disposal, the biosolids meet class A requirements. To continue to be class A, the pathogen requirements have to be met in every sample of biosolids that is collected. The frequency of monitoring for the pathogens is once per year, quarterly, bimonthly, or monthly, as determined by the permitting authority.

Alternative 5: Biosolids Treated in a PFRP This alternative states that biosolids are considered class A if they are treated in one of the PFRPs listed in Table 1.8 and the class A fecal coliform or *Salmonella* sp. requirement is met.

TABLE 1.8 Process to Further Reduce Pathogens

1. Composting

Using either the within-vessel composting method or the static aerated composting method, the temperature of the sludge is maintained at 55°C or higher for 3 days.

Using the windrow composting method, the temperature of the sludge is maintained at 55°C or higher for 15 days or longer. During the period the compost is maintained at 55°C or higher, there shall be a minimum of five turnings of the windrow.

2. Heat Drying

Sludge is dried by direct or indirect contact with hot gases to 90% solids or greater. Either the temperature of the sludge particles exceeds 80°C or the wet bulb temperature of the gas in contact with the sludge as the sludge leaves the dryer exceeds 80°C.

3. Heat Treatment

Liquid sludge heated to a temperature of 180°C or higher for 30 minutes.

4. Thermophilic Aerobic Digestion

Liguid sludge is agitated with air or oxygen to maintain aerobic conditions, and the mean cell residence time of the sludge is 10 days at 55 to 60°C.

5. Beta-Ray Irradiation

Sludge is irradiated with beta rays from an accelecometer at dosages of at least 1.0 Mrad at room temperature (about 20°C).

6. Gamma-Ray Irradiation

Sludge is iradiated with gama rays from certain isotopes, such as cobalt 60 and cesium 137, at room temperature (about 20°C).

7. Pasteurization

The temperature of the sludge is maintained at 70°C or higher for 30 minutes or longer.

Source: Appendix B of 40 CFR Part 503.

To meet these requirements, the treatment process must be operated according to the conditions listed in the table.

Alternative 6: Biosolids Treated in a Process Equivalent to a PFRP Under this alternative, biosolids are considered class A if they are treated by any process determined by the permitting authority to be equivalent to a PFRP. To be equivalent, the treatment process must be able to meet class A status consistently with respect to enteric viruses and viable helminth ova. The EPA's Pathogen Equivalency Committee is available as a resource to provide recommendations to the permitting authority on equivalency determinations.

TABLE 1.9 Summary of Alternatives for Meeting Class B Pathogen Requirements

Alternative 1: Monitoring of Indicator Organisms
Test for fecal coliform density as an indicator for all pathogens. Seven representative samples of the biosolids that is used or disposed shall be collected. The geometric mean of the seven samples shall be less than 2 million MPN per gram of total solids (dry weight basis), or less than 2 million colony-forming units (CFU) per gram of total solids (dry weight basis).

Alternative 2: Biosolids Treated in a PSRP
Sludge must be treated by one of the PSRPs listed in Table 1.10. Unlike the comparable class A requirement, this alternative does not require microbial monitoring for regrowth of fecal coliform or *Salmonella* sp. bacteria.

Alternative 3: Biosolids Treated in a Process Equivalent to a PSRP
Sludge must be treated in a process equivalent to one of the PSRPs, as determined by the permitting authority. The EPA Pathogen Equivalency Committee is available as a resource to provide recommendations to the permitting authorities on equivalency determinations.

Source: U.S. EPA, 1994.

Class B Pathogen Requirements For biosolids to be classified as class B with respect to pathogens, the requirements in one of the three alternatives listed in Table 1.9 must be met. Class B biosolids may contain some pathogens. Therefore, requirements for land application of class B biosolids also include site restrictions that prevent crop harvesting, animal grazing, and public access for a certain time until environmental conditions have further reduced pathogens. Site restrictions for land application of class B biosolids are summarized in the land application subpart of Part 503. The process to significantly reduce pathogens is summarized in Table 1.10.

Vector Attraction Reduction Options Vectors, which include flies, mosquitoes, and rodents, can transmit pathogens to humans physically through contact or biologically by playing a specific role in the life cycle of the pathogens. Reducing the attractiveness of biosolids to vectors or preventing the biosolids from coming into contact with vectors reduces the potential for transmitting diseases from pathogens in biosolids. The 12 options in Part 503 for reducing vector attraction are summarized in Table 1.11, and the applicability of the options is presented in Table 1.12. Option 12 applies only to domestic septage. Options 1 through 8 and 12 are designed to reduce the attractiveness of the material to vectors, and options 9 through 11 are designed to prevent vectors from coming into contact with the material.

TABLE 1.10 Process to Significantly Reduce Pathogens

1. **Aerobic Digestion**
   Sludge is agitated with air or oxygen to maintain aerobic conditions for a specific mean cell residence time at a specific temperature. Values for the mean cell residence time and temperature shall be between 40 days at 20°C and 60 days at 15°C.

2. **Air Drying**
   Sludge is dried on sand beds or on paved or unpaved basins. The sludge dries for a minimum of three months. During two of the three months, the ambient average daily temperature is above 0°C.

3. **Anaeobic Digestion**
   Sludge is treated in the absence of air for a specific mean cell residence time at a specific temperature. Values for the mean cell residence time and temperature shall be between 15 days at 35 to 55°C and 60 days at 20°C.

4. **Composting**
   Using either the within-vessel, static aerated pile, or windrow composting method, the temperature of the sludge is raised to 40°C or higher and remains at 40°C or higher for 5 days. For 4 hours during the 5 days, the temperature in the compost pile exceeds 55°C.

5. **Lime Stabilization**
   Sufficient lime is added to the sludge to raise the pH of the sludge to 12 after 2 hours of contact.

Source: Appendix B of 40 CFR Part 503.

TABLE 1.11 Summary of Options for Meeting Vector Attraction Reduction

Option 1: Meet 38% reduction in volatile solids content. This percentage is the amount of volatile solids reduction attained by anaerobic or aerobic digestion.

Option 2: Demonstrate vector attraction reduction with additional anaerobic digestion in a bench-scale unit. Frequently, sludge that has been recycled through the biological treatment unit of a wastewater treatment plant or that has resided for long periods of time in the wastewater collection system undergoes substantial biological degradation. Under these circumstances, the 38% reduction required by option 1 might not be possible. This option allows the operator to demonstrate vector attraction reduction by testing a portion of the previously digested sludge in a bench-scale unit in the laboratory. Vector attraction reduction is achieved if after anaerobic digestion for an additional 40 days at a temperature of 30 to 37°C, the volatile solids in the sludge are reduced by less than 17% from the beginning to the end of the bench test.

Option 3: Demonstrate vector attraction reduction with additional aerobic digestion in a bench-scale unit. This option is appropriate for aerobically digested sludge that cannot meet the 38% volatile solids reduction required by option 1. This includes sludge from extended aeration plants, where the minimum residence time of waste activated sludge solids generally exceeds 20 days. Under this option, aerobically digested sludge with 2% or less solids has achieved vector attraction reduction if after 30 days of aerobic digestion at 20°C in a bench test, volatile solids are reduced by less than 15%.

TABLE 1.11 *Continued*

Option 4: Meet a specific oxygen uptake rate (SOUR) for aerobically digested sludge. SOUR is the mass of oxygen consumed per unit time per unit mass of total solids in the sludge on a dry weight basis. Reduction in vector attraction can be demonstrated if the SOUR of the sludge, determined at 20°C, is equal to or less than 1.5 mg of oxygen per hour per gram of total solids. This test is only applicable to liquid sludge withdrawn from an aerobic process.

Option 5: Use an aerobic process at greater than 40°C for 14 days or longer. This option applies primarily to composted biosolids that also contain partially decomposed organic bulking agents.

Option 6: Add alkali under the conditions specified. Vector attraction reduction is achieved by:

- Raising the pH of the sludge to 12 or higher by adding alkali
- Maintaining pH at 12 or higher for at least 2 hours without the addition of more alkali
- Maintaining the pH at 11.5 or higher for another 22 hours without the addition of more alkali

Option 7: Dry sludge with no unstabilized solids to at least 75% solids. Under this option, vector attraction reduction is achieved if the sludge does not contain unstabilized solids and if the solids content of the sludge is at least 75% before it is mixed with other materials. Thus, the reduction must be achieved by removing water, not by adding inert materials.

Option 8: Dry sludge with unstabilized solids to at least 90% solids. If the sludge contains unstabilized solids, increasing the solids content by 90% or greater adequately reduces vector attraction. The solids increase should be achieved by removal of water, not by adding inert solids.

Option 9: Inject biosolids beneath the soil surface. Injection beneath the soil surface places a barrier of earth between the biosolids and the vectors. Under this option, vector attraction reduction is achieved:

- If no significant amount of biosolids remain on the land surface 1 hour after the injection of the biosolids
- If the biosolids is class A with respect to pathogens, they are injected within 8 hours after they are discharged from the pathogen reduction process.

Option 10: Incorporate biosolids into the soil within 6 hours of application to or placement on the land. If the biosolids are class A with respect to pathogens, the time between processing and application to or placement on the land must not exceed 8 hours.

Option 11: Cover biosolids placed on a surface disposal site with soil or other material at the end of each operating day. In addition to creating a physical barrier between the biosolids and the vectors, covering also helps meet pathogen requirements by allowing environmental conditions to reduce pathogens.

Option 12: Treat domestic septage with alkali to pH 12 or above for 30 minutes without adding more alkaline material.

Source: U.S. EPA, 1994.

TABLE 1.12 Vector Attraction Reduction Options for Each Use or Disposal Practice

Use or Disposal Practice	Vector Attraction Reduction Option											
	1	2	3	4	5	6	7	8	9	10	11	12
Bulk biosolids applied to agricultural land, forest, public contact sites, or reclamation sites	X	X	X	X	X	X	X	X	X	X		
Bulk biosolids applied to lawns or home gardens	X	X	X	X	X	X	X	X				
Biosolids sold or given away in a bag or other container for application to the land.	X	X	X	X	X	X	X	X				
Surface disposal	X	X	X	X	X	X	X	X	X	X	X	X

Source: U.S. EPA, 1995a.

1.2.5 Incineration

Incineration of wastewater sludge is the firing of sludge at high temperatures in an enclosed device called an *incinerator* (furnace). The most commonly used incinerators are multiple-hearth, fluidized-bed, and electric infrared furnaces. An incinerator system consists of an incinerator and one or more air pollution control devices, which are used either to remove small particles and the adhering metals in the exhaust gas from the incinerator or to further decompose organics. Wet scrubbers, dry and wet electrostatic precipitators, and fabric filters are metal-removing air pollution control devices. Afterburners, another type of air pollution control device, are used to burn organics in exhaust gases more completely.

Subpart E of Part 503 covers requirements for wastewater sludge incineration, including pollutant limits for metals, limits for total hydrocarbons, general requirements and management practices, and monitoring, recordkeeping, and reporting requirements. Anyone who fires sewage sludge in an incinerator, except as listed below, must meet the requirements in Subpart E:

- Nonhazardous ash generated from the incineration of sewage sludge is not covered by Part 503. Instead, it must be disposed in accordance with the solids waste disposal regulations in 40 CFR Part 258. If the ash is applied to land or placed on land other than a municipal solid waste landfill, the regulations in 40 CFR Part 257 govern.
- Auxiliary fuel is often used to enhance the burning of sewage sludge. Auxiliary fuels include natural gas, fuel oil, grit, screening, scum, coal, and municipal solid waste. If municipal solid waste account for more than 30% on a dry weight basis of the mixture of sludge and auxiliary fuel, the municipal solid waste is not considered an auxiliary fuel under Part 503. Instead, 40 CFR Parts 60 and 61 would regulate the process.
- Hazardous wastes are not considered auxiliary fuels. Such a process is covered by 40 CFR Parts 261 through 268.

Pollutant Limits Subpart E regulates five metals in sewage sludge before incineration: lead, arsenic, cadmium, chromium, and nickel. Subpart E contains equations for calculating pollutant limits for these five metals based on site-specific conditions. These conditions include dispersion factor, incinerator control efficiency, and sludge feed rate. In addition to the five metals noted above, emission of beryllium and mercury are regulated by the National Emission Standards for Hazardous Air Pollutants (NESHAP) per 40 CFR Part 61. The NESHAP limit for the emission of beryllium is 10 g during any 24-hour period from each incinerator. The limit for mercury emitted into the atmosphere from all incinerators at a given site is 3200 g during any 24-hour period.

Organic compounds that are emitted as a result of incomplete combustion or the generation of combustion by-products such as benzene, phenol, and vinyl chloride can be present in incinerator emissions. Part 503 regulates the emission of organic pollutants through an operational standard that limits the amount of total hydrocarbons (THCs) allowed in the incinerator stack gas. The stack gas must meet a monthly average limit of 100 ppm of THCs. The monthly average concentration is the arithmetic mean of the hourly averages. The hourly averages must be calculated based on at least two readings taken each hour that the incinerator operates. The THC concentration must also be corrected for 0% moisture and 7% oxygen.

Management Practices The management practices for sludge incineration are as follows:

- Instruments must be used that continuously measure and record:
  - THC concentration
  - Oxygen levels
  - Information needed to calculate moisture content in the stack exit gas
  - Combustion temperature in the furnace
- Instruments must be installed, calibrated, operated, and maintained according to the guidance provided by the permitting authority. The instrument used for THC measurement must:
  - Use a flame ionization detector
  - Have a sampling line heated to 150°C or higher at all times
  - Be calibrated at least once every 24-hour operating period using propane
- The incinerator must not be operated above the maximum combustion temperature set by the permitting authority based on performance test conditions.
- Conditions for operating the air pollution control devices, which are also set by the permitting authority based on performance test conditions, must be followed.
- Sludge may not be incinerated if it is likely to affect a threatened species or its critical habitat negatively, as listed in the Endangered Species Act.

Frequency of Monitoring Table 1.13 shows the monitoring frequency for sludge incinerators. Monitoring frequencies for arsenic, cadmium, chromium, lead, and nickel are the same as those for land application.

Record Keeping and Reporting Operators of sludge incinerators must develop and keep certain records for a minimum of five years. Records should include those related to pollution limits for metals; THC limit; and manage-

TABLE 1.13 Monitoring Frequency for Sewage Sludge Incinerators

Pollutant/Parameter	Amount of Sludge Fired[a]	Must Monitor at Least:
Arsenic, cadmium, chromium, lead, and nickel in sludge	0 to <290	Once per year
	290 to <1500	Once per quarter (4 times/year)
	1500 to <15,000	Once per 60 days (6 times/year)
	15,000 or greater	Once per month (12 times/year)
Beryllium and mercury in sludge or stack exit gas	NA	As often as permitting authority requires
THC concentration in stack exit gas	NA	Continuously
Oxygen concentration in stack exit gas	NA	Continuously
Information needed to determine moisture content in stack exit gas	NA	Continuously
Combustion temperature in furnace	NA	Continuously
Air pollution control device condition	NA	As often as permitting authority requires

Source: U.S. EPA, 1994.

[a] Metric tons per 365-day period, dry weight basis.

ment practices and monitoring requirements, such as THC levels, oxygen levels, and moisture content in stack exit gas, and combustion temperature in furnace. Treatment works serving a population of 10,000 or more, and treatment works with a design flow of 4000 m^3/day (1 mgd) must report these records to the permitting authority by every February 19.

REFERENCES

Federal Register (1993), FR 58 No. 32, February 19, pp. 9248–9415.

U.S. EPA (1992), *Control of Pathogen and Vector Attraction in Sewage Sludge*, EPA 625/R-92/013.

——— (1994), *A Plain English Guide to the EPA Part 503 Biosolids Rule*, EPA 832/R-93/003.

——— (1995a), *Part 503 Implementation*, EPA 833/R-95/001.

——— (1995b), *Process Design Manual: Land Application of Sewage Sludge and Domestic Septage*, EPA 625/R-95/001.

——— (1999), *Biosolids Generation, Use, and Disposal in the United States*, EPA 530/R-99/009.

2

SLUDGE QUANTITIES AND CHARACTERISTICS

Municipal wastewater is generated not only from the domestic use of water, but also from commercial establishments and from industries such as chemical or petrochemical, food and beverage processing, metal processing, pulp and paper, textile, automobile, and pharmaceutical. Therefore, the characteristics of wastewater vary from one municipality to another based on the unique mix of domestic, commercial, and industrial users. When wastewater is treated using various mechanical, biological, and physiochemical methods to remove organic and inorganic pollutants to levels required by the permitting authority, the sludge produced will also vary in quantity and character-

Wastewater Sludge Processing, By Izrail S. Turovskiy and P. K. Mathai

istics from one treatment plant to another. In this chapter we discuss primarily the quantities and characteristics of sludge produced by primary, biological, and chemical treatment of wastewater in order to determine the accurate design basis for selecting, sizing, and designing suitable sludge management systems, discussed later in the book.

2.1 TYPES OF SLUDGE

Types of sludge and other solids, such as screenings, grit, and scum, in a wastewater treatment plant vary according to the type of plant and its method of operation. The sources and types of solids generated in a treatment plant with primary, biological, and chemical treatment facilities are illustrated in Figure 2.1.

Wastewater sludge can be classified generally as *primary*, *secondary* (also called *biological*), and *chemical*. Sludge contains settleable solids such as

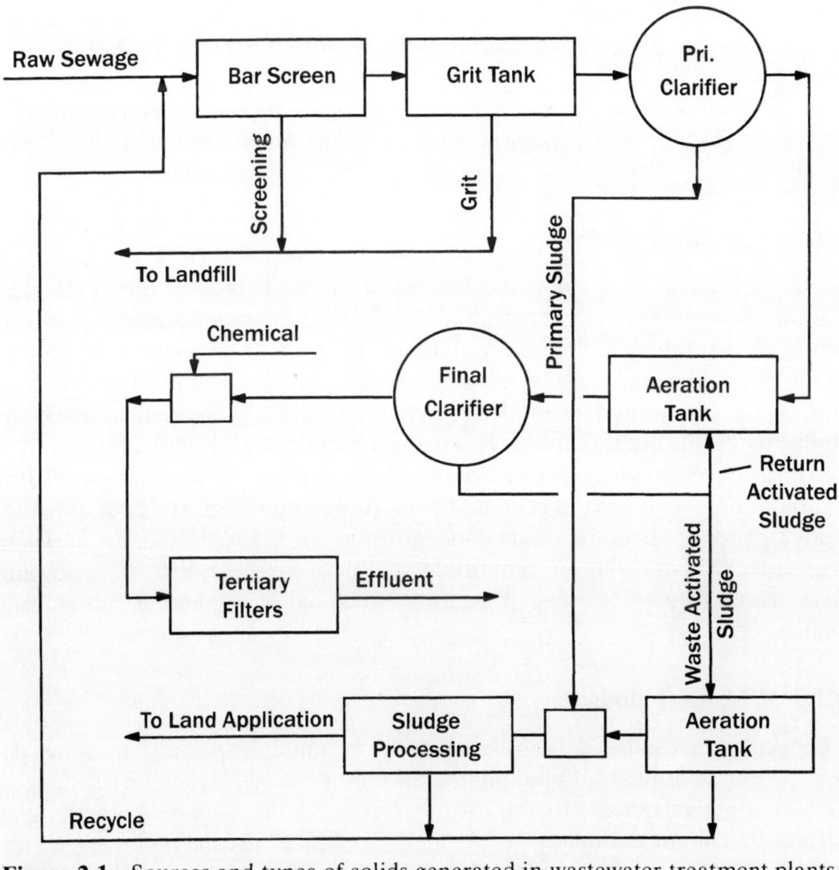

Figure 2.1 Sources and types of solids generated in wastewater treatment plants.

(depending on the source) fecal material, fibers, silt, food wastes, biological flocs, organic chemical compounds, and inorganics, including heavy metals and trace minerals. The sludge is raw sludge when it is not treated biologically or chemically for volatile solids or pathogen reduction. When the sludge is treated, the resulting biosolids can be classified by the treatment, such as aerobically digested (mesophilic and thermophilic), anaerobically digested (mesophilic and thermophilic), alkaline stabilized, composted, and thermally dried. The treated sludge can be only primary, secondary, or chemical, or a mixture of any two or three of the sludges.

2.1.1 Primary Sludge

Most wastewater treatment plants use the physical process of primary settling to remove settleable solids from raw wastewater. In a typical plant with primary settling and a conventional activated sludge secondary treatment process, the dry weight of the primary sludge solids is about 50% of that for the total sludge solids. The total solids concentration in raw primary sludge can vary between 2 and 7%. Compared to biological and chemical sludges, primary sludge can be dewatered rapidly because it is comprised of discrete particles and debris and will produce a drier cake and give better solids capture with low conditioning requirements. However, primary sludge is highly putrescible and generates an unpleasant odor if it is stored without treatment.

2.1.2 Secondary Sludge

Secondary sludge, also known as *biological sludge*, is produced by biological treatment processes such as activated sludge, membrane bioreactors, trickling filters, and rotating biological contactors. Plants with primary settling normally produce a fairly pure biological sludge as a result of the bacteria consuming the soluble and insoluble organics in secondary treatment system. The sludge will also contain those solids that were not readily removed by primary clarification. Secondary sludge generated in plants that lack primary settling may contain debris such as grit and fibers. Activated sludge and trickling filter sludge generally contain solids concentrations of 0.4 to 1.5% and 1 to 4%, respectively, in dry solids weight. Biological sludge is more difficult to dewater than primary sludge because of the light biological flocs inherent in biological sludge.

2.1.3 Chemical Sludge

Chemicals are used widely in wastewater treatment, especially in industrial wastewater treatment, to precipitate and remove hard-to-remove substances, and in some instances, to improve suspended solids removal. In all such instances, chemical sludges are formed. A typical use in removing a substance from wastewater is the chemical precipitation of phosphorus. The chemicals used for phosphorus removal include lime, alum, and "pickle liquors" such as ferrous chloride, ferric chloride, ferrous sulfate, and ferric sulfate. Some treatment plants add the chemicals to the biological process; thus, chemical precipitates are mixed with the biological sludge. Most plants apply chemicals to secondary effluent and use tertiary clarifiers or tertiary filters to remove the chemical precipitates. Some chemicals can create unwanted side effects, such as depression of pH and alkalinity of the wastewater, which may require the addition of alkaline chemicals to adjust these parameters.

2.1.4 Other Wastewater Residuals

In addition to sludge, three other residuals are removed in wastewater treatment process: screenings, grit, and scum. Although their quantities are significantly less than those of sludge in volume and weight, their removal and disposal are very important.

Screenings include relatively large debris, such as rags, plastics, cans, leaves, and similar items that are typically removed by bar screens. Quantities of screenings vary from 4 to 40 mL/m^3 (0.5 to 5 ft^3/MG) of wastewater. The higher quantities are attributable to wastes from correctional institutions, restaurants, and some food-processing industries. Screenings are normally hauled to a landfill. Some treatment plants return the screenings to the liquid stream after marcerating or comminuting. This is not recommended because many of the downstream pieces of equipment, such as mixers, air diffusers, and electronic probes, are subject to fouling from reconstituted rags and strings.

Grit consists of heavy and coarse materials, such as sand, cinders, and similar inorganic matter. It also contains organic materials, such as corn, seeds, and coffee grinds. If not removed from wastewater, grit can wear out pump impellers and piping. Grit is typically removed in grit chambers. In some treatment plants, grit is settled in primary clarifiers along with primary sludge and then separated from the sludge in vortex-type grit separators. The volume of grit removed varies from 4 to 200 mL/m^3 (0.5 to 27 ft^3/MG) of wastewater. The higher quantities are typical of municipalities with combined sewer systems and sewers that contribute excessive infiltration and inflow. Grit is almost always landfilled.

Scum is the product that is skimmed from clarifiers. Primary scum consists of fats, oils, grease, and floating debris such as plastic and rubber products. It can build up in piping, thereby restricting flow and increasing pumping costs, and can foul probes, flow elements, and other instruments in the waste stream. Secondary scum tends to be mostly floating activated sludge or biofilm, depending on the type of secondary treatment used. The quantity and moisture content of scum typically are not measured. It may be disposed of by pumping to sludge digesters, concentrating, and then incinerating with other residuals, or drying and then landfilling.

2.2 SLUDGE QUANTITY

Determining the quantity of sludge produced in the treatment of wastewater is required for the sizing of sludge processing units and equipment such as sludge pumps, storage tanks, thickeners, digesters, and incinerators. The best approach in estimating solids production is to base the estimate on historical data from similar facilities and the anticipated influent strength. Generally, solids production rates range between 0.2 and 0.3 kg/m^3 (0.8 to 1.2 dry tons/MG) of wastewater treated. In the absence of historic or plant-specific data, a rule-of-thumb approximation for solids produced in a typical wastewater treatment plant is 0.24 kg/m^3 (1 dry ton/MG) of wastewater treated (WEF, 1998). It is also important to take into account the effects of industrial contribution, stormwater flows, and temperature on the influent wastewater characteristics and therefore on solids production.

2.2.1 Primary Sludge

Primary sludge solids production can vary typically from 0.1 to 0.3 kg/m^3 (800 to 2500 lb/MG) of wastewater. A rule-of-thumb approximation is 0.05 kg/capita (0.12 lb/capita) per day of primary sludge solids production. The most common approach in estimating primary sludge production is by computing the quantity of suspended solids entering the treatment plant and assuming a removal rate. The removal rate is usually in the range 50 to 65%. A removal rate of 60% is commonly used for estimating purposes, provided that the effects of industrial contribution are minimal and no major sidestreams from the sludge processing units are discharged to the primary clarifier influent.

Total solids in wastewater comprise dissolved solids + suspended solids (settleable and colloidal) + floatable solids (scum). The solids removal rate can be correlated to either the hydraulic detention time or the surface overflow rate of the primary clarifier. The following equations (Koch et al., 1990) demonstrate how solids production relates to detention time:

$$\text{solids production} = \text{plant flow} \times \text{influent suspended solids} \times \text{removal rate} \tag{2.1}$$

$$\text{removal rate} = \frac{T}{(a + bT)} \tag{2.2}$$

where

removal rate = removal of suspended solids, %
T = detention time, minutes
a = constant, 0.406 minute
b = constant, 0.0152

The relation of the solids removal rate to the surface overflow rate is presented in Figure 2.2. The figure also shows the BOD removal rate. The typical BOD removal rate is 50% of suspended solids removed. Several factors

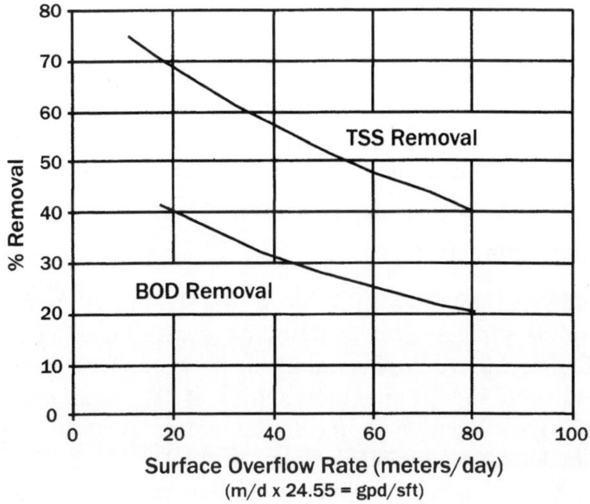

Figure 2.2 Relationship between surface overflow rate and solids removal in primary treatment.

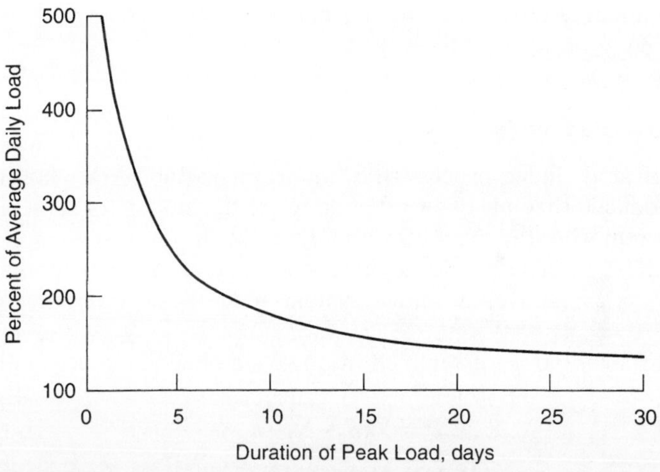

Figure 2.3 Relationship between peak solids loading and duration of peak load. (From U.S. EPA, 1979.)

can affect suspended solids removal efficiency, chief among them being industrial contribution, sludge treatment process sidestreams, and mechanical factors of primary settling tanks, such as poor flow distribution and density currents.

Daily variations in primary sludge production can occur, usually in proportion to the quantity of solids entering a wastewater treatment plant. Peak rates of sludge production can be several times the average. Figure 2.3 shows the relationship between peak suspended solids entering a plant and the duration of time that the peak persists, which is based on a study of several large wastewater treatment plants. This variation in daily quantity is appropriate for large cities that have large areas with combined sewers on flat grades. The peaks occur when the solids deposited in the combined sewers are resuspended and carried to the treatment plant by storm flows.

2.2.2 Secondary Sludge

Secondary sludge is the biomass produced by a secondary treatment process, such as activated sludge with its numerous variations, such as extended aeration, oxidation ditch, complete mix, plug flow, step feed, sequencing batch reactor (SBR), and membrane bioreactor (MBR); and attached growth systems, such as trickling filters and rotating biological contactors. Secondary sludge also includes nonbiodegradable inorganic particulate matter not removed by primary settling. The solids become incorporated into the biomass. Since the amount of organic loading to the secondary treatment process is the most important factor in the production of biological solids, the rate of removal of biological or chemical oxygen demand (BOD or COD) in primary settling is very important. As mentioned earlier, BOD removal is approximately 50% of the suspended solids removal in primary clarification.

2.2.3 Activated Sludge

In an activated sludge process, the important variables in quantifying the sludge produced that must be wasted daily are the amount of substrate (BOD or COD) removed, the mass of microorganisms in the system, and the non-biodegradable inert suspended solids in the influent to the system. Figure 2.4 shows a typical activated sludge system with these and other variables noted.

The variables can be assembled into two simple equations as follows:

$$P_x = Y(S_0 - S) - k_d X \tag{2.3}$$

$$\text{WAS} = P_x + I_0 - E_t \tag{2.4}$$

where

P_x = net growth of biomass expressed as volatile suspended solids (VSS), kg/d or lb/d
Y = gross yield coefficient, kg/kg or lb/lb
S_0 = influent substrate (BOD or COD), kg/d or lb/d
S = effluent substrate (BOD or COD), kg/d or lb/d
k_d = endogenous decay coefficient, d^{-1}
X = biomass in aeration tank (MLVSS), kg or lb

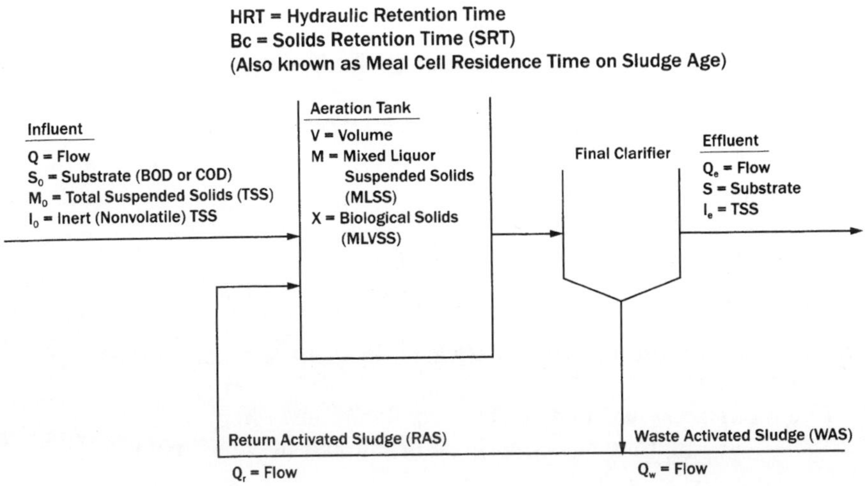

Figure 2.4 Schematic of a typical activated sludge system.

TABLE 2.1 Typical Kinetic Coefficients for the Activated Sludge Process

Coefficient	Unit	Range[a]	Typical[a]
Y	g VSS/g BOD	0.4–0.8	0.6
Y	g VSS/g COD	0.3–0.6	0.4
k_d	g VSS/g VSS·d	0.04–0.14	0.1

[a] Values listed are for 20°C.

WAS = total waste activated sludge solids, kg/d or lb/d
I_0 = influent nonvolatile suspended solids, kg/d or lb/d
E_t = effluent suspended solids, kg/d or lb/d

Equation (2.3) dates back to 1951 (U.S. EPA, 1979). To use equation (2.3), the values of Y, the gross yield coefficient, and k_d, the endogenous decay coefficient, need to be known. Table 2.1 shows typical values for these coefficients.

Solids Retention Time *Solids retention time* (SRT), also known as *mean cell residence time*, or *sludge age*, is the average time the sludge solids stay in the system, expressed in days. It is an important design and operating parameter for the activated sludge process. As the solids inventory in the clarifier and the solids lost in the effluent are small, SRT can be defined as the solids in the system divided by the mass of solids removed per day. It can be expressed as

$$\text{SRT (days)} = \frac{X}{P_x} = \frac{\text{biomass (VSS) under aeration, kg or lb}}{\text{biomass (VSS) removed per day, kg/d or lb/d}} \tag{2.5}$$

For wastewater with no nonbiodegradable volatile suspended solids (VSS) in the influent to the aeration tank, secondary biomass production and SRT have the following kinetic relationship:

$$Y_{\text{obs}} = \frac{Y}{1 + k_d\,(\text{SRT})} \tag{2.6}$$

where Y_{obs} is the observed yield coefficient (kg biomass produced/kg substrate removed).

Equations (2.3) and (2.4) can be combined and simplified by using the observed yield (net biomass yield) coefficient and the flow rate as follows to quantify the sludge to be wasted daily:

$$\text{WAS} = \frac{Q[Y_{\text{obs}}\,(S_0 - S) + I_0]}{10^3\ \text{g/kg}} \tag{2.7}$$

In U.S. customary units, the equation is

$$\text{WAS} = Q[Y_{\text{obs}}\,(S_0 - S) + I_0] \times 8.34\ \text{lb/mgd} \cdot \text{mg/L} \tag{2.8}$$

where

Y_{obs} = observed yield, g biomass/g (lb biomass/lb) substrate removed
Q = influent flow, m^3/d (mgd)
S_0 = influent substrate concentration, g/m^3 (mg/L)
S = effluent substrate concentration, g/m^3 (mg/L)
I_0 = influent nonvolatile suspended solids concentration, mg/L

Observed yield coefficients are often reported in the literature. Figure 2.5 graphically illustrates the observed yield coefficients (VSS production) versus SRTs in an activated sludge system. As SRT increases, Y_{obs} decreases due to biomass loss by endogenous respiration, and thus the biomass to be wasted from the system decreases. Therefore, the costs of sludge handling can be reduced by using a higher value of SRT in the design of the activated sludge system. However, the lower costs might be offset by higher costs for the increase in the aeration tank volume needed.

Nitrification Nitrogen in municipal wastewater occurs predominantly in the form of ammonia nitrogen and organic nitrogen. Approximately 60% of the total is in ammonia form and approximately 40% is in organic form; less than 1% will be inorganic nitrogen in the form of nitrate or nitrite. *Nitrification* is the biooxidation of ammonia nitrogen and organic nitrogen to nitrate by the

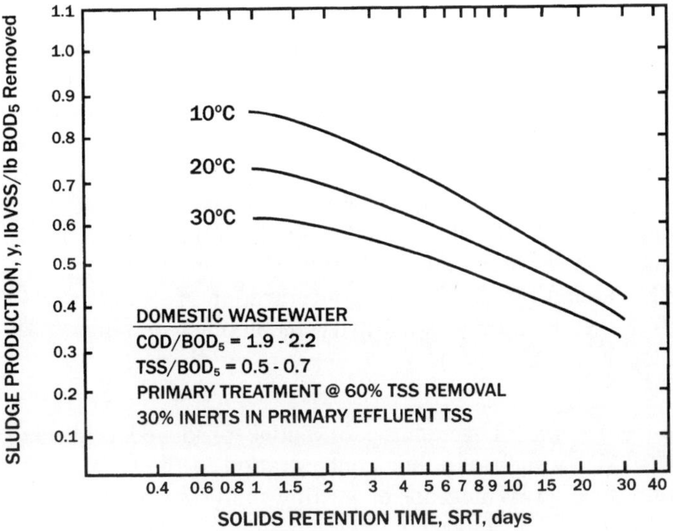

(a) With Primary Treatment

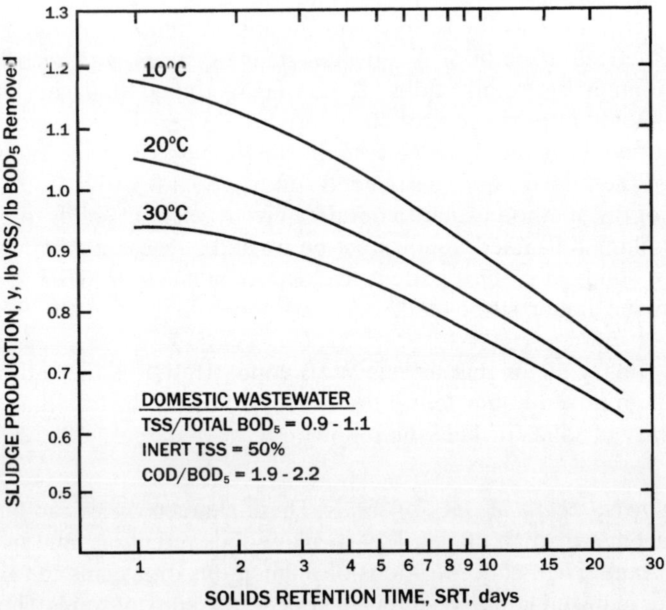

(b) Without Primary Treatment

Figure 2.5 Sludge production in activated sludge system. (Reprinted with permission from WEF, 1998.)

staged activities of the autotrophic species *Nitrosomas* and *Nitrobacter*. Compared with processes designed for carbonaceous oxidation only, nitrification processes operate at long solids retention time. As a result, an activated sludge system in nitrification mode produces fewer waste solids than a conventional system produces. However, there is an additional component in a nitrification system, which is the net yield of the nitrifying biomass. This is estimated to be 0.15 kg of biomass per kilogram of total Kjeldahl nitrogen (ammonia plus organic) removed. This is only a small fraction of the total biomass produced. For example, wastewater that contains an ammonia nitrogen concentration of 20 mg/L and an organic nitrogen concentration of 10 mg/L would add only $(20 + 10) \times 0.15 = 4.5$ mg/L of nitrifying biomass per liter (38 lb/MG) of wastewater.

WAS Concentration The volume of sludge produced is directly proportional to the dry weight of solids concentration in the waste sludge stream. There are basically two methods of wasting solids in a waste activated sludge treatment system. The most common method of wasting solids is wasting from the secondary clarifier underflow. Such waste activated sludge can vary, in practice, across a range from 0.4 to 1.5% total suspended solids. The second method is wasting solids from the mixed liquor. Although not commonly practiced in traditional activated sludge systems, it is becoming more prevalent in MBR applications. It has been argued that this method of wasting should improve control of the process. When mixed liquor is wasted in a traditional activated sludge system, waste sludge is at the same concentration of the mixed liquor suspended solids: about 0.1 to 0.4%. The disadvantage of this method is that to obtain a given weight of solids to be wasted, a large volume of mixed liquor must be wasted because of the low concentration of solids in it. This issue is avoided or reduced in MBR applications where mixed liquor suspended solids are much higher: in the 0.8 to 1.5% range.

The primary factor that affects WAS concentration is the settleability of the solids in mixed liquor fed to the clarifier. Various factors that affect the settleability of solids include the following:

- *Biological characteristics of solids*. These characteristics can be partially controlled by maintaining a particular solids retention time in the aeration tank. High concentrations of filamentous organisms can sometimes occur in mixed liquor, resulting in poor settleability of solids. These organisms can be reduced in number through control of oxygen, control of solids retention time, or sometimes by periodic chlorination of the return activated sludge to destroy the filamentous organisms.
- *Sludge volume index* (SVI). The SVI is defined as the volume of 1 g of sludge solids after 30 minutes of settling. The SVI is determined by placing a sample of mixed liquor of known solids concentration in a 1-L cylinder

and measuring the settled volume after 30 minutes. Then the SVI is computed using the equation

$$\text{SVI} = \frac{[\text{settled volume (mL/L)}](10^3\ \text{mg/g})}{\text{mixed liquor suspended solids (mg/L)}} = \text{mL/g} \qquad (2.9)$$

An SVI value of 100 or lower is considered a good settling sludge, which is sometimes called *old sludge*. SVI values above 100 represent *young sludge*. Mixed liquor with an SVI value above 150 settles poorly, possibly due to filamentous growth.

- *Sludge density index* (SDI). The SDI is determined in the same way as the SVI and is computed using the equation

$$\text{SDI} = \text{MLSS (mg/L per mL of sludge settled in 30 min)} = 100/\text{SVI} \qquad (2.10)$$

An SDI value of 1.0 is ideal.

- *Surface overflow rate.* The surface overflow rate of the clarifier is related to the zone settling velocity, which is the settling velocity of the sludge–water interface at the beginning of the sludge settleability test that is described in standard methods. The surface overflow rate is then determined using the equation

$$\text{OR} = \frac{V_i(24)}{\text{SF}} \qquad (2.11)$$

where

OR = surface overflow rate, m/d
V_i = zone settling velocity, m/h
24 = conversion factor from m/h to m/d
SF = safety factor, typically 1.75 to 2.5

This method of determining the surface overflow rate takes into account the effect of MLSS concentrations. The value of V_i will decrease, resulting in a higher clarifier surface area and better settling velocity.

- *Solids flux.* The solids flux is the solids in the mixed liquor divided by the clarifier area (kg/m^2·d). High rates of solids flux require that clarifiers be operated at lower solids concentrations (the same effect as that for the lower zone settling velocity described above).
- *Limits of flow distribution and sludge collection equipment.* Poor velocity dissipation of the flow distributed to the clarifiers can result in solids carryover and poor settling of solids. Also, because of the pseudoplastic and viscous nature of WAS, some sludge collectors are not capable of reliable operation at higher underflow sludge concentrations.

- *Higher sludge concentration with raw wastewater.* If raw wastewater is fed to the activated sludge process instead of primary clarifier effluent, a higher sludge concentration usually results. Chemicals added to the mixed liquor for removal of phosphorus and suspended solids will also result in heavy suspended solids and higher sludge concentration.

2.2.4 Attached Growth System Sludge

The microorganisms in attached growth secondary treatment systems such as trickling filters and rotating biological contactors (RBCs) are biochemically similar to microorganisms that predominate in activated sludge systems. Consequently, the biomass production from attached growth systems and activated sludge systems is roughly similar when compared on the basis of kilogram of biomass produced per kilogram of substrate removed. Therefore, equations that predict solids production in an attached growth system are similar in form to those used in predicting solids production in an activated sludge system. However, there are differences between the two systems. The main difference lies in the term used to define the quantity of biomass in the system. Figure 2.6 is a schematic of a trickling filter treatment system. The following equations for predicting solids production in a trickling filter system (the same equations can also be applicable for RBCs) are similar to equations (2.3) and (2.4), but are based on the assumption that the total mass of microorganisms present in the system is proportional to the media surface area:

$$P_x = Y'(S_0 - S_e) - k_d'(A_m) \tag{2.12}$$

$$\text{WTFS} = P_x + I_0 - E_t \tag{2.13}$$

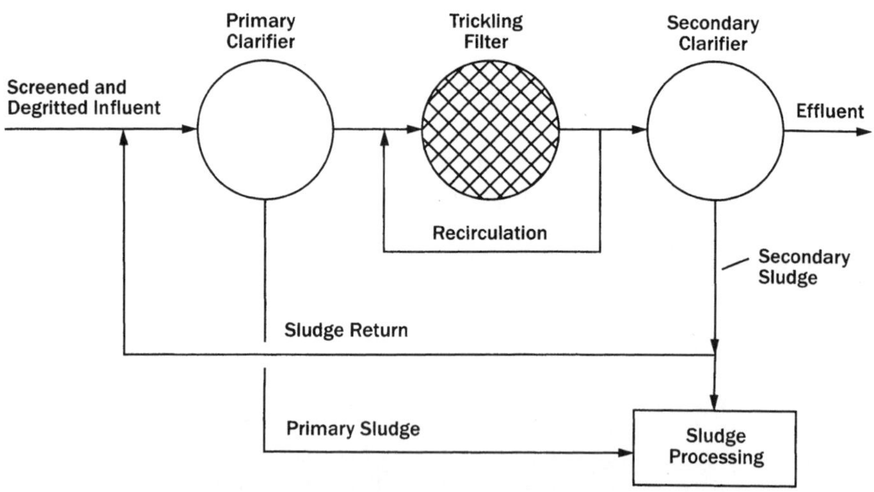

Figure 2.6 Schematic of a typical trickling filter system.

where

P_x = net growth of biomass (VSS), kg/d or lb/d
Y = gross yield coefficient, kg/kg or lb/lb
S_0 = influent substrate (BOD), kg/d or lb/lb
S_e = effluent substrate (BOD), kg/d or lb/lb
k'_d = decay coefficient, d^{-1}
A_m = total media surface area, m^2 or ft^2
WTFS = waste trickling filter solids, kg/d or lb/d
I_0 = influent nonvolatile suspended solids, kg/d or lb/d
E_t = effluent suspended solids, kg/d or lb/d

The yield coefficient values for attached growth systems are similar to the values for activated sludge systems. However, the decay coefficient values are higher for attached growth systems because of the apparently long time the solids stay attached to the media (longer effective SRT). Typical values of decay coefficients range from 0.03 to 0.3 d^{-1}.

Nitrification can occur in attached growth systems. The resulting quantity of biomass, similar to that in activated sludge systems, is small: 3 mg/L (25 lb/MG) may be used for design purposes.

Attached growth solids have better settling and thickening properties than those for activated sludge. Consequently, attached growth sludge tends to be of higher concentration, typically 1 to 4%.

2.2.5 Chemical Sludge

Chemicals are widely used in wastewater treatment to precipitate and remove phosphorus and in some cases to improve the efficiency of suspended solids removal. Chemicals can be added to raw wastewater, to a secondary biological process, or to secondary effluent, in which case tertiary filters or tertiary clarifiers are used to remove the chemical precipitates.

Although theoretical rates of chemical sludge production can be estimated from the anticipated chemical reactions, competing reactions can make the estimation difficult. For example, ferric chloride will form ferric hydroxide, which in turn will react with phosphate to form ferric phosphate. Classical jar tests are favored as a means of estimating chemical sludge quantities. Quantities of precipitates in chemical sludge are influenced by such conditions as pH, mixing, reaction time, and opportunity for flocculation. Following are some of the types of precipitates that must be considered in measuring the total sludge production (U.S. EPA, 1979):

- *Phosphate precipitates.* Examples are $AlPO_4$ or $Al(H_2PO_4)(OH)_2$ with aluminum salts, $FePO_4$ with iron salts, and $Ca_3(PO_4)_2$ with lime.
- *Carbon precipitates.* This is significant with lime, which forms calcium carbonate, $CaCO_3$. If two-stage recarbonation is used, a recarbonation sludge of nearly pure $CaCO_3$ is formed.

- *Hydroxide precipitates.* With iron and aluminum salts, excess salt forms a hydroxide, $Fe(OH)_3$ or $Al(OH)_3$. With lime, magnesium hydroxide, $Mg(OH)_2$, may form; the magnesium comes from the influent wastewater, from lime, or from magnesium salts.
- *Inert solids from chemicals.* This item is most significant with lime. If a quicklime is 92% CaO, the remaining 8% may be mostly inert solids that appear in the sludge. Many chemicals supplied in dry form may contain significant amounts of inert solids.
- *Polymer solids.* Polymers may be used as primary coagulants and to improve the performance of other coagulants. The polymers themselves contribute little to the total mass, but they can greatly improve clarifier efficiency, with a concomitant increase in sludge production.
- *Suspended solids from the wastewater.* Addition of any chemical to wastewater treatment process affects process efficiency. Therefore, the change in sludge production must be considered.

2.3 SLUDGE CHARACTERISTICS

Knowledge of the characteristics of sludge, along with the quantities produced, is important for the design of a sludge processing system in a wastewater treatment plant. Several factors influence the characteristics of sludge. These include the amount and type of industrial waste contribution, ground garbage, storm flow, sidestreams from sludge processing units, and the treatment process used for the wastewater.

2.3.1 Primary Sludge

Fresh primary sludge is a gray or light brown suspension with solids of different sizes and composition. Because of the high organic content of primary sludge, it decays quickly and becomes septic, which can be identified by its change to a dark gray or black color and an objectionable sour odor. Sludge characteristics vary widely from one treatment plant to another. Table 2.2 lists compositions of sludge typically seen in European and Russian wastewater treatment plants. These can be compared to the constituents of sludge in North America that are shown in Table 2.3. A range of values and in most cases typical values are given.

Concentration is affected by the type of solids in the raw wastewater and the frequency of sludge withdrawn from the primary settling tank. Some plants withdraw the sludge less frequently and allow the sludge to thicken further in the primary settling tanks, thereby increasing the solids concentration of sludge. However, such sludge, because of its long detention time in the tank, can generate unpleasant odors. The quantity of raw primary sludge can be approximately 0.4 to 0.5% by volume of the plant influent flow, or approximately $1.1\,m^3$ ($39\,ft^3$) per 1000 people.

TABLE 2.2 Wastewater Sludge Characteristics in Europe and Russia

Item	Raw Primary Sludge	Digested Primary Sludge	Unthickened Activated Sludge	Digested Mixture of Primary and Thickened Activated Sludge
Total dry solids (TS) (%)	4.5	6.0	0.5	3.0
Volatile solids (% of TS)	70	50	75	60
Grease and fats	18	11	6	5
Protein (% of TS)	25	18	37	22
Ammonia nitrogen (% of TS)	3.0	2.0	5.0	3.5
Phosphoric acid (% of TS)	1.4	2.0	4.0	3.0
Potash (% of TS)	0.5	0.4	0.4	0.4
pH	6	7	7	7

TABLE 2.3 Wastewater Sludge Characteristics in North America

	Primary Sludge		Activated Sludge	
Item	Range	Typical	Range	Typical
Total dry solids (TS) (%)	2–7	5	0.4–1.5	1
Volatile solids (% of TS)	60–80	65	60–80	75
Specific gravity		1.02		1.01
Grease and fats				
Ether soluble (% of TS)	6–30			
Ether extract (% of TS)	7–35		5–12	
Protein (% of TS)	20–30	25	32–41	
Nitrogen (N, % of TS)	1.5–4.0	2.5	2.4–5.0	
Phosphorus (P_2O_5, % of TS)	0.8–2.8	1.6	2.8–11.0	
Potash (K_2O, % of TS)	0–1	0.4	0.5–0.7	
Cellulose (% of TS)	9–13	10		7
Iron (not as sulfide, % of TS)	2–4	2.5		
Silica (SiO_2, % of TS)	15–20			8
pH	5–8	6	6.5–8.0	7
Alkalinity (mg/L as $CaCO_3$)	500–1,500	600	580–1,100	
Organic acids (mg/L as HAc)	200–2,000	500	1,100–1,700	
Energy content:				
kJ/kg	23,300	18,600	23,300	
Btu/lb	10,000	8,000	10,000	

Source: Adapted in part from U.S. EPA, 1979.

TABLE 2.4 Major Mineral Constituents of Sludge[a]

Content	Raw Primary Sludge	Raw Activated Sludge	Digested Mixture of Primary and Activated Sludges
SiO_2	21.5–55.9	17.6–33.8	27.3–35.7
Al_2O_3	0.3–18.9	7.3–26.9	8.7–9.3
Fe_3O_4	4.9–13.9	7.2–18.7	11.4–13.6
CaO	11.8–35.9	8.9–16.7	12.5–15.6
MgO	2.1–4.3	1.4–11.4	1.5–3.6
K_2O	0.7–3.4	0.8–3.9	1.8–2.8
Na_2O	0.8–4.2	1.9–8.3	2.6–4.7
SO_3	20–7.5	1.5–6.8	3.0–7.2
ZnO	0.1–0.2	0.2–0.3	0.1–0.3
CuO	0.1–0.8	0.1–0.2	0.2–0.3
NiO	0.2–2.9	0.2–3.4	0.2–1.0
Cr_2O_3	0.8–3.1	0.0–2.4	1.3–1.9

[a] Values shown are a percentage of total mineral constituents.

The concentration of volatile solids in sludge will approximately match the concentration of volatile solids in the influent wastewater unless large amounts of sludge-processing sidestreams are returned to the primary tank influent flow. A volatile solids content below 70% of total solids is usually influenced by the presence of groundwater infiltration, stormwater inflow, sludge-processing sidestreams, a large amount of grit, industrial waste with a low volatile solids content, or wastewater solids that have a long detention time in the sewers. Volatile solids will also be at the lower end of the range if chemicals are used to precipitate phosphorus in the primary settling tank.

Major chemical constituents of raw sludge include grease and fats, protein, nitrogen, phosphorus, potash, cellulose, iron, and silica. The major mineral constituents of sludge are shown in Table 2.4. Wastewater sludge may also contain heavy metals, such as cadmium, chromium, cobalt, copper, lead, mercury, nickel, and zinc.

Alkalinity and pH are the most important of the easily measured chemical parameters affecting sludge conditioning. Raw primary sludge has a pH range of 5.0 to 8.0 and an alkalinity of 500 to 1500 mg/L as $CaCO_3$. The organic acid content is 200 to 2000 mg/L as HAc.

2.3.2 Activated Sludge

Tables 2.2 through 2.4 also list the characteristics of raw activated sludge. Activated sludge contains mostly bacterial cells that are viscous and difficult to dewater. The sludge is light gray or dark brown in color. Suspended solids concentration is 0.4 to 1.5%. Factors affecting the concentration of solids have been discussed in Section 2.2.2. Thickening of activated sludge is a very important process because of its low solids concentration and high volume. Mechanical thickening such as gravity belt thickening can increase solids concentration to about 5% and reduce the volume to about one-fifth. For approximate computations, the quantity of the mixture of primary sludge and thickened activated sludge at the average solids concentration of about 4% can be assumed to be 0.6 to 1.0% of the volume of wastewater treated.

As can be seen from the tables, compared to primary sludge, activated sludge contains lower amounts of grease and fats and cellulose, but higher amounts of nitrogen, phosphorus, and protein. The phosphorus content will be at the higher end of the range in plants using a biological process for phosphorus removal. The alkalinity of activated sludge is 580 to 1100 mg/L as $CaCO_3$ and the pH is 6.5 to 8.0. Organic acids range from 1100 to 1700 mg/L as HAc.

2.3.3 Physical and Biological Properties

Raw primary sludge particle size distribution is: greater than 7 mm (5 to 20%), 1 to 7 mm (9 to 33%), and smaller than 1 mm (50 to 88%), of which about 45% is less than 0.2 mm. In activated sludge, the approximate distribution is: 90% below 0.2 mm, 8% between 0.2 and 1 mm, 1.6% between 1 and 3 mm, and 0.4% over 3 mm. The organic part of the sludge decays more rapidly, with an increase in the quantity of finely dispersed and colloidal particles and bound water resulting in a decrease in the separation of water from the sludge and poor dewaterability.

The density of primary sludge is 1.0 to 1.03 g/cm^3, and the density of activated sludge is about 1.0 g/cm^3. The density of dry sludge solids is 1.2 to 1.4 g/cm^3. Primary sludge at solids concentrations above 5% and activated sludge at solids concentrations above 3% are non-Newtonian, which means that head losses in piping are not proportional to the velocity and viscosity. They are also thixotropic, which means that they become less viscous when mixed.

Thermophysical characteristics of sludge are shown in Table 2.5. The specific heat of a mixture of primary and thickened activated sludge is 3.5 to 4.7×10^3/kg·K. The heat value of combustion of sludge dry solids equals 16.7

TABLE 2.5 Thermophysical Characteristics of Sludge

Type of Sludge	Temperature Conductivity (10^8 m^2/s)	Thermal Conductivity (W/m·K)	Specific Heat (kJ/kg·K)
Raw primary and waste activated sludge	—	0.4–0.6	3.5–4.7
Vacuum filter dewatered	10.9–14.3	0.2–0.5	2.1–3.0
Centrifuge dewatered	8.5–12.1	0.1–0.3	2.0–2.4
Thermally dried	14.0–21.6	0.1–0.3	1.7–2.2

to 18.4 MJ/kg. Sludge burns at a temperature of 430 to 500°C (800 to 930°F); however, to eliminate odors, the temperature needs to be raised to 800 to 850°C (1470 to 1560°F). In the process of thickened activated sludge digestion, 15 MJ of heat is produced per kilogram of volatile suspended solids.

Dewatering is the process of natural or mechanical removal of water from sludge. Water may be present in sludge as free water or bound to the particles physically or chemically. The greater the bound water present in sludge, the more the energy or reagents it takes to condition the sludge for removal of the bound water. The separation of water from sludge depends on the size of solid particles; the smaller the particles, the poorer the water separation from sludge. Therefore, any sludge treatment process that reduces the size of suspended solids particles has a negative effect on the conditioning and dewatering of sludge. The chemical composition of sludge also exerts a significance influence on its treatment and dewaterability. Compounds of iron, aluminum, chromium, and copper, as well as acids and alkalis, improve the process of precipitation and dewatering and reduce the consumption of chemical reagents for conditioning of sludge before dewatering. Oils, fats, and nitrogen compounds intensify the anaerobic sludge digestion but interfere with the conditioning and dewatering processes. The dewaterability of sludge can be evaluated by measuring its specific resistance, which is determined in a Buchner funnel test by measuring the volume of filtrate collected from sludge and the time it takes to filter. Specific resistance varies depending on the type of sludge and its characteristics; the values are shown in Table 2.6 for various types of sludge.

TABLE 2.6 Specific Resistance of Sludge

Type of Sludge	Moisture (%)	Specific Resistance (10^{10} cm/g)
Primary sludge from municipal wastewater treatment plants with substantial industrial wastewater contributions from:		
Machine and metallurgical plants	91–95	50–300
Synthetic rubber plants	92–95	200–400
Textile plants	95–97	300–700
Various industrial plants	93–96	300–1,000
Thickened activated sludge	96–98	400–8,000
Digested primary sludge	93–96	400–2,000
Digested mixture of primary and thickened activated sludges under:		
Mesophilic conditions	96–98	800–6,800
Thermophilic conditions	96–98	4,000–10,000
Aerobically digested mixture of primary and activated sludges, thickened	96–97	2,400–4,000

TABLE 2.7 Levels of Indicator Bacteria and Pathogens[a]

Agent	Range of Levels Reported (number/g)[b,c]	Average Level Reported (number/g)[b,c]
Total coliform	1.1×10^1–3.4×10^9	6.4×10^8
Fecal coliform	ND–6.8×10^8	9.5×10^6
Fecal streptococci	1.4×10^4–4.8×10^8	2.1×10^6
Salmonella sp.	ND–1.7×10^7	7.9×10^2
Shingella sp.	ND	ND
Pseudomonas aeruginosa	1.5×10^1–9.4×10^4	5.7×10^3
Enteric virus	5.9–9.0×10^3	3.6×10^2
Parasite ova/cysts	ND–1.4×10^3	1.3×10^2

Source: Lue-Hing et al., 1998.

[a] Values are for raw primary, secondary, and mixed sludge.

[b] Dry weight basis.

[c] ND, none detected.

Levels of indicator bacteria and pathogens in raw primary, activated, and mixed sludges are shown in Table 2.7. The diversity of microflora makes it difficult to enumerate the total population. Primary sedimentation and activated sludge treatment of wastewater are very efficient in removing microorganisms from wastewater and transporting them to the sludge. Primary sedimentation reduces microorganisms in sewage by 30 to 70%. After activated sludge treatment, the reduction of microorganisms reaches 90 to 99%.

The types of bacteria in activated sludge are mostly floc-forming, but the sludge also contains filamentous microorganisms. An excessive amount of filamentous organisms can cause sludge bulking in secondary clarifiers. *Sludge bulking* is a condition in the secondary clarifier where the flocs do not compact or settle well, causing large amounts of flocs to discharge with the clarifier effluent.

2.4 MASS BALANCE

A good approach to estimating solids production is to prepare a material mass balance for the entire wastewater treatment plant. A material balance is prepared for the key components of flow, BOD, and TSS, and in facilities where nutrients are removed, nitrogen and phosphorus should be included. A mass balance is typically computed for average dry weather flow and concentrations. However, if higher flows and concentrations are likely to be sustained for a long period of time, such as in communities with seasonal fluctuations in population, it is important to compute the balance for maximum conditions (usually, for maximum monthly average conditions) and to design the sludge-handling facilities for higher values to avoid shock loading to those facilities. Recycle streams from sludge-processing facilities, such as thickeners, digesters, and dewatering systems, must also be included in the mass balance.

Example 2.1 demonstrates the computation of mass balance for a wastewater treatment plant. Table 2.8 shows the solids concentrations and solids capture efficiencies for the most commonly used sludge-processing units. Table 2.9 shows the BOD and TSS concentrations in the recycle flows from the various processes. There are wide variations in some of these values. Values for mass balance computations should be chosen based on the data from treatment plants with similar wastewater concentrations and treatment systems.

Recycle streams are important in the preparation of solids mass balance. The typical approach to the computation is first to assume a fixed percentage of the influent BOD and TSS in the total recycle flow, based on typical plant data. Then an iterative computational procedure is used until the incremental change is less than 5%. However, if a spreadsheet program is used for computation, incremental changes can be made, as little as 1% or less. A variety of simulation software is available that provides mass balance calculations.

TABLE 2.8 Typical Solids Concentration and Capture Efficiencies for Various Processes

	Solids Concentration (%)		Solids Capture Efficiency (%)	
Unit Operation	Range	Typical	Range	Typical
Gravity thickening				
Primary sludge	4–12	6	85–92	90
Waste activated sludge (WAS)	2–4	3	75–90	85
Combined primary and WAS	2–6	4	80–90	85
Flotation thickening of WAS				
With chemicals	4–6	5	90–97	95
Without chemicals	2–5	4	80–95	90
Gravity belt thickening of WAS				
With chemicals	4–6	5	90–98	95
Centrifuge thickening of WAS				
With chemicals	4–8	5	90–98	95
Without chemicals	3–6	4	80–90	85
Belt filter press dewatering				
With chemicals, raw sludge	18–30	23	90–98	95
With chemicals, digested sludge	12–25	18	90–98	95
Centrifuge dewatering				
With chemicals	15–35	24	85–98	92
Filter press dewatering				
With chemicals	20–45	38	90–98	95

Source: Adapted in part from Metcalf & Eddy, 2003.

TABLE 2.9 BOD and TSS Concentrations in Recycle Flows

	BOD (g/m^3)		TSS (g/m^3)	
Unit Operation	Range	Typical	Range	Typical
Gravity thickening supernatant				
Primary sludge	100–400	250	100–400	200
Waste activated sludge (WAS)	100–500	300	100–400	300
Combined primary and WAS	80–400	300	100–400	250
Flotation thickening subnatant	100–1000	250	100–2000	300
Gravity belt thickening filtrate	100–2500	800	100–2000	1000
Centrifuge thickening centrate	200–3000	1000	500–3000	1000
Belt filter press dewatering filtrate	50–600	300	100–2000	1000
Centrifuge dewatering centrate	50–300	1000	200–9000	5000
Filter press dewatering filtrate	50–300	200	50–1000	600
Aerobic digestion supernatnat	100–1800	500	100–9000	3500
Anaerobic digestion supernatant	500–5000	1000	800–9000	4500
Sludge lagoon supernatant	100–250	200	10–200	100

Source: Adapted in part from Metcalf & Eddy, 2003.

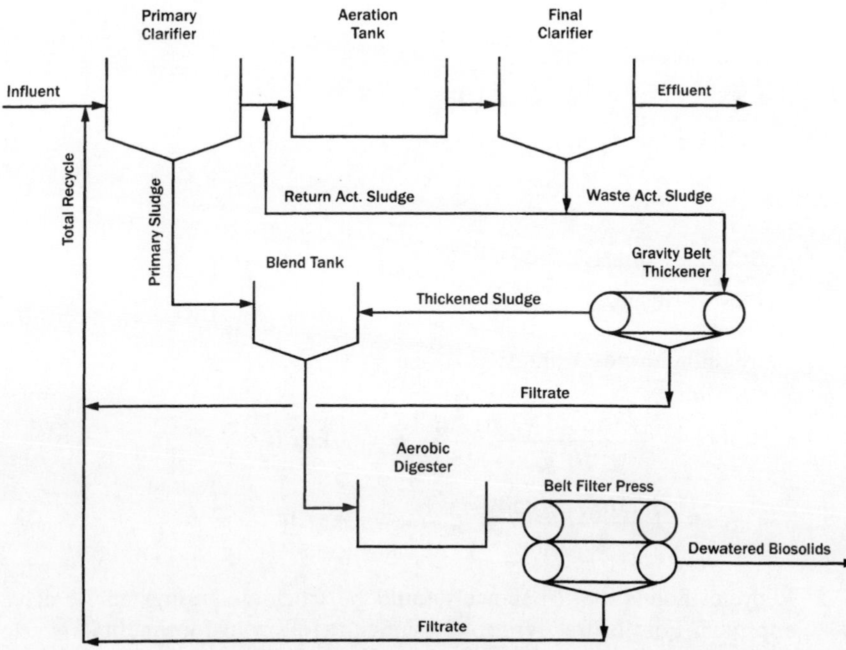

Figure 2.7 Flow schematic of WWTP for Example 2.1.

Example 2.1: Computation of Solids Mass Balance This example illustrates the preparation of solids mass balance for a hypothetical wastewater treatment plant. Figure 2.7 is a flow schematic of the plant. Some of the units for sludge processing shown in the figure are typical equipment and processes, and similar equipment and processes may be substituted depending on the selection and design of the sludge handling system. For example, a gravity thickener, floatation thickener, or centrifuge thickener may be substituted for the gravity belt thickener. However, the appropriate characteristics of the processed sludge and solids capture efficiencies, shown in Tables 2.8 and 2.9, should be used in the mass balance computations.

Following is the information required for the mass balance calculations:

average daily flow = 15,000 m^3/d (4 mgd)

Influent characteristics:

BOD: 200 g/m^3

TSS: 250 g/m^3

Note: g/m^3 and mg/L are numerically the same.

Primary clarifier:

BOD removal: 30%

TSS removal: 60%

VSS: 65% of TSS

Sludge concentration: 6%

Effluent:

BOD: 10 mg/L

TSS: 10 mg/L

1. Daily influent mass values:

a. $\text{BOD} = \dfrac{(15{,}000\,\text{m}^3/\text{d})(200\,\text{g}/\text{m}^3)}{10^3\,\text{g}/\text{kg}} = 3000\,\text{kg}/\text{d}$

b. $\text{TSS} = \dfrac{(15{,}000\,\text{m}^3/\text{d})(250\,\text{g}/\text{m}^3)}{10^3\,\text{g}/\text{kg}} = 3750\,\text{kg}/\text{d}$

2. Recycle: Solids mass balance should be calculated using an iterative approach. For the first iteration, assume the following for the total recycle flow:

Flow: 1% of influent flow

BOD: 2% of influent BOD

TSS: 4% of influent TSS

a. Flow = $(15{,}000\,\text{m}^3/\text{d})(0.01) = 150\,\text{m}^3/\text{d}$

b. BOD = $(3000\,\text{kg/d})(0.02) = 60\,\text{kg/d}$

c. TSS = $(3750\,\text{kg/d})(0.04) = 150\,\text{kg/d}$

3. Primary clarifier:

a. Influent flow = $(15{,}000 + 150)\,\text{m}^3/\text{d} = 15{,}150\,\text{m}^3/\text{d}$

b. Influent BOD = $(3000 + 60)\,\text{kg/d} = 3060\,\text{kg/d}$

c. Influent TSS = $(3750 + 150)\,\text{kg/d} = 3900\,\text{kg/d}$

d. BOD removed = $(3060\,\text{kg/d})(0.30) = 918\,\text{kg/d}$

e. TSS removed = $(3900\,\text{kg/d})(0.60) = 2340\,\text{kg/d}$

f. Effluent BOD = $(3060 - 918)\,\text{kg/d} = 2142\,\text{kg/d}$

g. Effluent TSS = $(3900 - 2340)\,\text{kg/d} = 1560\,\text{kg/d}$

h. VSS removed = $(2340\,\text{kg/d})(0.65) = 1521\,\text{kg/d}$

i. Effluent VSS = $(2340 - 1521)\,\text{kg/d} = 819\,\text{kg/d}$

j. At 6% concentration, sludge flow $= \dfrac{2340\,\text{kg/d}}{0.06 \times 10^3\,\text{kg/m}^3} = 39\,\text{m}^3/\text{d}$

k. Effluent flow = $(15{,}150 - 39)\,\text{m}^3/\text{d} = 15{,}111\,\text{m}^3/\text{d}$

4. Plant effluent: In the first iteration, assume the plant effluent flow to be the same as the plant influent flow, although this may vary depending on the recycle flows and primary sludge and WAS flows discharged to the sludge processing system.

a. $\text{BOD} = \dfrac{(15{,}000\,\text{m}^3/\text{d})(10\,\text{g/m}^3)}{10^3\,\text{g/kg}} = 150\,\text{kg/d}$

b. $\text{TSS} = \dfrac{(15{,}000\,\text{m}^3/\text{d})(10\,\text{g/m}^3)}{10^3\,\text{g/kg}} = 150\,\text{kg/d}$

5. Secondary process:

a. Operating parameters:

RAS and WAS concentration = 0.8% ($8000\,\text{g/m}^3$)

MLSS = 80% of MLSS

SRT = 6 days

$Y_{obs} = 0.56$ (from Fig. 2.5)

b. Influent flow $= 15{,}111\,\text{m}^3/\text{d}$ (see 3k)

c. Influent BOD concentration $= \dfrac{(2142\,\text{kg/d})(10^3\,\text{g/kg})}{15{,}111\,\text{m}^3/\text{d}}$

$= 142\,\text{g/m}^3$

d. Biomass produced that must be wasted: Using the biomass portion of equation (2.5), we have

$$P_x = \frac{Y_{\text{obs}}Q(S_0 - S_e)}{10^3\,\text{g/kg}}$$

$$= \frac{0.56(15{,}111\,\text{m}^3/\text{d})(142 - 10)\,\text{g/m}^3}{10^3\,\text{g/kg}}$$

$$= 1117\,\text{kg/d}$$

Note: In reality, the substrate in the effluent (influent soluble BOD escaping treatment) should be determined based on the BOD of the biodegradable effluent TSS. However, the error in using the effluent BOD as the effluent substrate is negligible. In sizing the aeration basin, some designers ignore the effluent BOD altogether, as it is usually a small amount.

e. Determine the solids to be wasted based on the fact that the MLVSS is 80% of MLSS:

$$\text{WAS} = \left(\frac{1117\,\text{kg/d}}{0.80}\right) - 150\,\text{kg/d} = 1246\,\text{kg/d}$$

f. Determine the WAS flow at a solids concentration of 0.8%:

$$\text{WAS} = \frac{1246\,\text{kg/d}}{0.008 \times 10^3\,\text{kg/m}^3} = 156\,\text{m}^3/\text{d}$$

g. Return activated sludge: Assuming an MLSS of $3500\,\text{g/m}^3$, compute the RAS ratio:

$$3500(Q + Q_r) = 8000(Q_r)$$

$$3500Q = Q_r$$

$$Q_r/Q = \text{RAS} = 0.78\ (78\%\ \text{return rate})$$

$$\text{RAS flow} = (15{,}111\,\text{m}^3/\text{d})(0.78) = 11{,}787\,\text{m}^3/\text{d}$$

h. Total mixed liquor (ML) flow $= (15{,}111 + 11{,}787)\,\text{m}^3/\text{d}$

$= 26{,}898\,\text{m}^3/\text{d}$

i. TSS in RAS $= \dfrac{(11{,}787\,\text{m}^3/\text{d})(0.80)(10{,}000\,\text{g/m}^3)}{10^3\,\text{g/kg}}$

$= 94{,}296\ \text{kg/d}$

j. TSS in ML to the aeration tank $= (1560 + 94{,}296)\,\text{kg/d}$ (3.g + 5.i)

$= 95{,}856\,\text{kg/d}$

k. TSS in ML from the aeration tank $= (95{,}856 + 1117)\,\text{kg/d}$

$= 96{,}973\,\text{kg/d}$

l. TSS concentration in the mixed liquor $= \dfrac{(96{,}973\,\text{kg/d})(10^3\,\text{g/kg})}{26{,}898\,\text{m}^3/\text{d}}$

$= 3605\,\text{g/m}^3$

Note: If solids are wasted from mixed liquor, the volume to be wasted based on an MLSS concentration of $3368\,\text{g/m}^3$ is

$$\frac{(1246\,\text{kg/d})(10^3\,\text{g/kg})}{3605\,\text{g/m}^3} = 346\,\text{m}^3/\text{d}$$

6. Gravity belt thickening:

a. Operating parameters:

WAS flow $= 156\,\text{m}^3/\text{d}$ (see 5.f)

WAS solids $= 1246\,\text{kg/d}$ (see 5.e)

WAS concentration = 0.8%

Solids capture efficiency = 95%

Thickened sludge concentration = 5%

Belt wash water flow and the weight of polymer are not considered in the mass balance calcwulations.

b. Solids in thickened sludge $= (1246\,\text{kg/d})(0.95) = 1184\,\text{kg/d}$

c. Thickened sludge flow $= \dfrac{1184\,\text{kg/d}}{0.05 \times 10^3\,\text{kg/m}^3} = 24\,\text{m}^3/\text{d}$

d. VSS in thickened sludge $= (1184)(0.80) = 947\,\text{kg/d}$

e. Filtrate flow $= (156 - 24)\,\text{m}^3/\text{d} = 132\,\text{m}^3/\text{d}$

f. TSS in filtrate $= (1246 - 1184)\,\text{kg/d} = 62\,\text{kg/d}$

g. Assuming that the BOD of the WAS solids is 50%,

BOD in filtrate $= (62\,\text{kg/d})(0.50) = 31\,\text{kg/d}$

h. BOD in thickened sludge = (1184 kg/d)(0.50) = 592 kg/d

7. Aerobic sludge digestion:

a. Operating parameters:

VSS destruction in digestion = 38%

BOD in supernatant = 500 g/m^3

TSS in suppernatant = 3000 g/m^3

Digested sludge draw-off concentration = 5%

b. TSS to digester = (2340 + 1184) kg/d (see 3.e and 6.b)
$= 3524$ kg/d

c. VSS to digester = (2340 + 1184) kg/d (see 3.h and 6.d)
$= 2468$ kg/d

d. Flow to digester = (39 + 24) m^3/d (see 3.j and 6.c)
$= 63$ m^3/d

e. Non-VSS to digester = (3524 − 2468) kg/d
$= 1056$ kg/d

f. VSS remaining after digestion = (2468 kg/d)(1 − 0.38)
$= 1530$ kg/d

g. TSS remaining after digestion = (1056 + 1350) kg/d
$= 2586$ kg/d

h. To determine the flow distribution between supernatant at 3000 g/m^3 (0.3%) concentration and digested sludge draw-off at 5% concentration, let Q_s be supernatant flow and Q_d be the digested sludge draw-off. Then

$$[(Q_s \text{ m}^3/\text{d})(0.003) + (Q_d \text{ m}^3/\text{d})(0.05)](10^3 \text{ kg/m}^3) = 2586 \text{ kg/d}$$

$$3Q_s + 50Q_d = 2586 \qquad \text{(i)}$$

$$Q_s + Q_d = 63 \qquad \text{(ii)}$$

Multiplying (ii) by 3 gives

$$3Q_s + 3Q_d = 189 \qquad \text{(iii)}$$

(i)−(iii) give

$$47Q_d = 2397$$

$$Q_d = \frac{2397}{47} = 51\,\text{m}^3/\text{d}$$

$$Q_s = 63 - 51 = 12\,\text{m}^3/\text{d}$$

i. BOD in supernatant $= \dfrac{(12\,\text{m}^3/\text{d})(500\,\text{g}/\text{m}^3)}{10^3\,\text{g}/\text{kg}} = 6\,\text{kg}/\text{d}$

j. TSS in supernatant $= \dfrac{(12\,\text{m}^3/\text{d})(3000\,\text{g}/\text{m}^3)}{10^3\,\text{g}/\text{kg}} = 36\,\text{kg}/\text{d}$

k. TSS in digested sludge draw-off $= (2586 - 36)\,\text{kg}/\text{d} = 2550\,\text{kg}/\text{d}$

Note: If there is no supernatant recycle, flow to and from the digester is the same ($63\,\text{m}^3/\text{d}$), and then

$$\text{TSS conc. in digested sludge} = \frac{2586\,\text{kg}/\text{d}}{(63\,\text{m}^3/\text{d})(10^3\,\text{kg}/\text{m}^3)} = 0.041\ (4.1\%)$$

8. Belt filter press (BFP) dewatering

a. Operating parameters:

Solids capture efficiency = 95%

Dewatered cake solids = 20%

Specific gravity of cake = 95%

Belt wash water flow and the weight of polymer are not considered in the mass balance calculations.

b. Sludge cake solids $= (2550\,\text{kg}/\text{d})(0.95) = 2423\,\text{kg}/\text{d}$

c. Cake volume $= \dfrac{2423\,\text{kg}/\text{d}}{1.05(0.20)(10^3\,\text{kg}/\text{m}^3)} = 12\,\text{m}^3/\text{d}$

d. Filtrate flow $= (52 - 12)\,\text{m}^3/\text{d} = 39\,\text{m}^3/\text{d}$

e. Filtrate TSS $= (2550 - 2423)\,\text{kg/d} = 127\,\text{kg/d}$

f. Assuming that the BOD of the filtrate is 50% of the filtrate TSS,

BOD in filtrate $= (127\,\text{kg}/\text{d})(0.50) = 64\,\text{kg}/\text{d}$

9. Total recycle quantity versus quantity assumed:

$$\text{Flow} = (132 + 12 + 39)\,\text{m}^3/\text{d} = 186\,\text{m}^3/\text{d vs. } 211\,\text{m}^3/\text{d}$$

$$\text{BOD} = (32 + 6 + 65)\,\text{kg/d} = 103\,\text{kg/d vs. } 101\,\text{kg/d}$$

$$\text{TSS} = (34 + 36 + 130)\,\text{kg/d} = 229\,\text{kg/d vs. } 225\,\text{kg/d}$$

10. These new recycle quantities should be used for the second iteration of mass balance, and if needed, additional iterations should be performed until the change in quantities of the recycle is less than 5%. If mass balance calculations are performed using a spreadsheet program, additional iterations are easier and can be repeated until the change in quantities is 1% or less.
11. The following quantities are total recycle quantities after a second iteration and are shown versus quantities in the first iteration:

$$\text{Flow} = (132 + 12 + 40)\,\text{m}^3/\text{d} = 186\,\text{m}^3/\text{d vs. } 211\,\text{m}^3/\text{d}$$

$$\text{BOD} = (32 + 6 + 65)\,\text{kg/d} = 103\,\text{kg/d vs. } 101\,\text{kg/d}$$

$$\text{TSS} = (63 + 36 + 130)\,\text{kg/d} = 229\,\text{kg/d vs. } 225\,\text{kg/d}$$

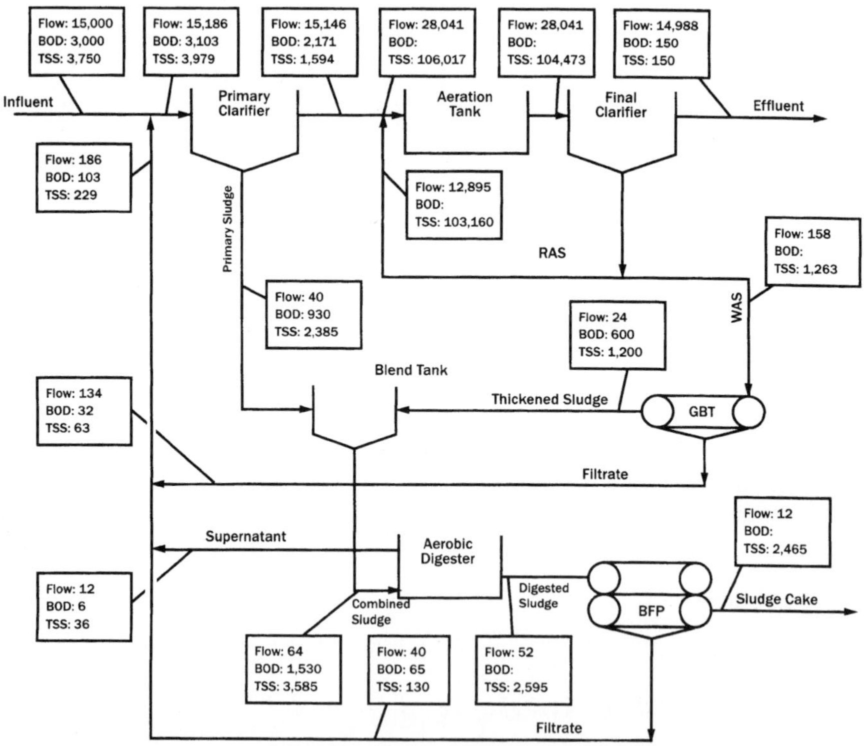

Figure 2.8 Mass balance of WWTP in Example 2.1. (Units: Flow in m^3/d, and BOD and TSS in kg/d.)

12. The quantities of flow (m^3/d), BOD (kg/d), and TSS (kg/d), after the second iteration, for all the unit processes of treatment are shown in Figure 2.8.

REFERENCES

Epstein, E. (1997), *The Science of Composting,* Technomic Publishing Co., Lancaster, PA.

Great Lakes–Upper Mississippi River Board of State Sanitary Engineering Health Education Services, Inc. (1997), *Recommended Standards for Wastewater Facilities,* Albany, NY.

Heukelekin, H., et al. (1951), Factors Affecting the Quantity of Sludge Production in the Activated Sludge Process, *Sewage and Industrial Waste,* Vol. 23, No. 8.

IEPA (1998), *Recommended Standards for Sewage Works,* Part 370 of Chapter II, Subtitle C: Water Pollution, Title 35, Illinois Environmental Protection Agency, Springfield, IL.

Koch, C., et al. (1990), Spreadsheets for Estimating Sludge Production, *Water Environment and Technology,* Vol. 2, p. 11.

Lue-Hing, C., Zeng, D. R., and Kuchenither, R. (Eds.) (1998), *Water Quality Management Library,* Vol. 4, *Municipal Sewage Sludge Management: A Reference Test on Processing, Utilization and Disposal*, Technomic Publishing Co., Lancaster, PA.

Metcalf & Eddy, Inc. (2003), *Wastewater Engineering: Treatment and Reuse,* 4th ed., Tchobanoglous, G., Burton, F. L., and Stensel, H. D. (Eds.), McGraw-Hill, New York.

U.S. EPA (1975), *Process Design Manual for Suspended Solids Removal,* EPA 625/1-75/003a.

——— (1979), *Process Design Manual for Sludge Treatment and Disposal,* EPA 625/1-79/011.

WEF (1998), *Design of Municipal Wastewater Treatment Plants*, 4th ed., Manual of Practice 8 (ASCE 76), Water Environment Federation, Alexandria, VA.

3

THICKENING AND DEWATERING

Wastewater Sludge Processing, By Izrail S. Turovskiy and P. K. Mathai

3.1 INTRODUCTION

Thickening is a process to increase the solids concentration of sludge and decrease its volume by removing a portion of the water. The thickened sludge remains in the fluid state and is capable of being pumped without difficulty. The purpose of reducing the volume by thickening is to increase the efficiency and decrease the costs of subsequent sludge-processing steps. Thickening of waste activated sludge is important because of its high volume and low solids concentration. Thickening from 1% solids concentration to 2%, for example, reduces the sludge volume by one-half. If it is concentrated to 5% solids, the volume is reduced by one-fifth of its original volume.

Dewatering is the removal of water from sludge to achieve volume reduction greater than that achieved by thickening. Dewatering the sludge results in a solid–semisolid material that is easier to handle. It can be shoveled, moved about with tractors fitted with buckets and blades, and transported by belt and screw conveyors. Dewatering of sludge is required before composting to improve airflow and texture, and before thermal drying or incineration to reduce fuel demand to evaporate excess moisture. Dewatering is also required prior to disposing of sludge in landfills to reduce leachate production at the landfill site.

Prior to dewatering, sludge requires conditioning by biological, chemical, and/or physical treatment to enhance water removal. Chemical conditioning to improve separation of solid and liquid phases is often required prior to thickening sludge using mechanical thickening equipment. Therefore, sludge conditioning is described below before thickening and dewatering are discussed.

3.2 CONDITIONING

Sludge conditioning refers to the process of improving solid–liquid separation. Conditioning is an important part of mechanical thickening and dewatering of sludge. Conditioning of sludge can be performed by inorganic or organic chemicals, power plant or sludge incinerator ash, or by physical processes such as heating, elutriation, and freezing and thawing. However, not all conditioning processes are equal. Although heating, elutriation, and freezing and thawing improve the dewaterability of sludge, it has to be augmented by chemical conditioning: at lower dosages, however. In addition to enhancing the separation of water from solids, some conditioning processes also disinfect sludge, affect sludge odors, alter the wastewater solids physically, provide limited solids destruction or addition, or improve solids recovery.

3.2.1 Factors Affecting Conditioning

Wastewater sludge consists of primary, secondary, and/or chemical solids with various organic and inorganic particles of mixed sizes. Depending on the sources, they have various internal water contents, degree of hydration, and surface chemistry. Sludge characteristics that affect thickening or dewatering and for which conditioning is employed include the following:

- Source
- Solids concentration
- Particle size and distribution
- pH and alkalinity
- Surface charge and degree of hydration
- Other physical factors

Source Sources such as primary sludge, waste activated sludge, chemical sludge, and digested biosolids are good indicators of the source of conditioner doses required for thickening or dewatering. Based on published data about chemical conditioning requirements, primary sludge, as a general rule, requires lower doses than those required by biological sludge. Among the varieties of secondary biological sludge, attached growth sludge requires lower doses than does suspended growth biological sludge. Conditioning requirements for aerobically and anaerobically digested sludge generally are the same as for secondary digested biological sludge. It should be noted that these are general rules and that the conditioning requirements for the same source of sludge may vary from plant to plant. Chemical sludge is difficult to classify as a single source for conditioning requirements because different types of chemical sludge require vastly different doses and types of conditioners.

Solids Concentration Municipal wastewater sludge contains a large number of colloidal and agglomerated particles, which have large specific surface areas. If the sludge has a low concentration of solids, these particles behave in a discrete manner with little interaction. In many applications, conditioning is the neutralization of the surface charge of sludge particles by the adsorption of oppositely charged organic polyelectrolyes or inorganic chemical complexes. With low interaction from the low concentration of solids, more coagulants are needed to overcome the surface charge. As the concentration of solids is increased, interaction increases. Therefore, on a mass of coagulant per mass of dry solids basis, the dosage can be reduced. For this reason, coagulant dose is usually expressed as a percentage of dry solids or as kilograms of coagulant per ton of dry solids. A higher suspended solids concentration produces effective conditioning over a wide range of dosage when organic polymers are used, which means that the higher the solids concentration, the less susceptible the process is to overdosing.

Particle Size and Distribution Particle size is considered the single most important factor influencing the dewaterability of sludge. For the same concentration of solids in sludge, the greater the number of small particles, the greater the surface area/volume ratio. Increased surface area means greater hydration, higher chemical demand, and increased resistance to dewatering. One of the objectives of conditioning is to increase particle size by combining the small particles into large aggregates.

pH and Alkalinity pH and alkalinity affect primarily the performance of inorganic conditioners. When added to water, inorganic conditioners reduce the water's pH. Therefore, the dosage of inorganic conditioners such as iron or aluminum, and the alkalinity or *buffering* of the sludge, determine the pH of the conditioning process. The resulting pH, in turn, determines the predominant coagulant species that will be present and the nature of the charged colloidal surface. The higher alkalinity of anaerobically digested biosolids is one of the reasons for the associated higher coagulant doses.

Surface Charge and Degree of Hydration For the most part, sludge solids repel rather than attract one another. This repulsion may be due to hydration or electrical effects. With hydration, a layer of water binds to the surface of the solid. This provides a buffer that prevents close approach between solids. In addition, sludge solids are negatively charged and thus tend to be mutually repulsive. Conditioning is used to overcome these effects of hydration and electrical repulsion.

Physical Factors Physical factors such as storage, pumping, mixing, and sludge treatment processes, including the types of thickening and dewatering devices to be used, also affect the thickening and dewatering characteristics of sludge. Sludge that has been stored for a long period of time requires more conditioning chemicals than fresh sludge does because of an increase in the degree of hydration and the fines content of the solids. Because of the fragile structure of sludge particles, some reduction in particle size typically results from the shear forces associated with the pumping process. Proper mixing and flocculation to evenly disperse the conditioning chemicals required also depend on the processing to which the sludge has been subjected and on the mechanics of thickening or dewatering process available.

3.2.2 Chemical Conditioning

Chemical conditioning is the most common conditioning process for sludge thickening and dewatering. Conditioning by adding chemicals can be viewed as coagulation or flocculation by neutralization of colloidal surface charge by oppositely charged organic polymers or inorganic chemicals. Particle size is the most important characteristic of the dewaterability of sludge. By adding chemicals, the particle size increases and the bound water decreases. Different sludges have different dewatering characteristics, and the same sludge source varies from plant to plant. Consequently, the type and dose of chemical addition must be determined on a case-by-case basis.

Tests for Conditioner Selection The most common and recommended procedures used to determine sludge conditioning effectiveness for thickening and dewatering are presented below.

The *Buchner funnel test* for determining the specific resistance of sludge is obtained by measuring the volume of filtrate collected from a sludge sample and the time it takes to filter. Using different conditioning chemicals and varying the doses of conditioning allow choosing the best reagents and the optimum dose for a particular sludge.

Capillary suction time (CST) is a rapid and simple method for screening dewatering aids. It relies on gravity and capillary suction of a piece of thick filter paper to draw out the water from a small sample of conditioned sludge. The sample is placed in a cylindrical cell on top of a chromatography-grade filter paper. The time in seconds it takes for the water in sludge to travel 10 mm in the paper between two fixed points is recorded electronically as the CST. This test is usually done to determine the optimum dose of polymer in a particular dewatering process. Because CST is a function of solids concentration and a dilute sludge usually has a low CST value, it should be compared for sludges of the same suspended solids concentration. CST for unconditioned sludge is about 200 seconds. A CST of 10 seconds or less is considered a good value for superior dewatering performance.

The *jar test* is the easiest and most common method used to evaluate chemical conditioning. In this test, 1-L samples of sludge with different conditioner concentrations are mixed rapidly. The samples are then flocculated by reducing the stir speed for a few minutes and then are allowed to settle. Liquid clarity and the densely settled sludge blanket indicate the ability of the chemical to condition sludge.

Among the foregoing tests, specific resistance using the Buchner funnel test is the best measure for comparing conditioning chemicals and dosages for different sludges. Studies conducted in Russia (Turovskiy, 2000) on the effect of digestion on the dewaterability of sludge have validated this. Other conclusions from the studies include: (1) the same type of sludge from different wastewater treatment plants has different dewatering characteristics; (2) raw sludge has less specific resistance than aerobically or anaerobically digested sludge; and (3) mesophylically digested sludge has less specific resistance than thermophylically digested sludge. The studies and their results are described below in detail.

Sludge Conditioning Studies The degree of sludge conditioning required depends on the sludge source (primary clarifiers, activated sludge treatment system, trickling filter, etc.), sludge quality, subsequent sludge treatment process, and in general, the degree of thickening or dewatering required. Much research has been done in the field of sludge conditioning, thickening, and dewatering. However, it is not well understood how the treatment of sludge affects its dewatering characteristics.

The studies were conducted using anaerobic digester simulators with sludge samples from various treatment plants in Russia, Lithuania, and Poland. Data obtained from treatment plants in Bulgaria, Finland, France, Germany, Hungary, and the United States were also analyzed. The digester simulators worked both in the mesophylic regime (35°C) and the thermophylic regime (55°C). The following parameters of raw and digested sludge samples were measured to determine the influence of the digestion process on the variations of sludge dewatering characteristics:

- Moisture
- Organic content
- Specific resistance
- Variations in the structure of solid particles
- Forms of bound water
- Variations in chemical composition

In addition to the above, samples of activated sludge were collected from clarifiers and thickeners in wastewater treatment plants and also from the thickening and dewatering simulators used in the studies. The following measurements were taken during the thickening process:

- Duration of thickening
- Variations in solids concentration

- Ash content
- Specific resistance

Specific resistance is usually expressed by the formula

$$r = \frac{2PF^2b}{\mu C} \tag{3.1}$$

where

r = specific resistance, m/kg
P = pressure of filtration, N/m^2
F = filtration area, m^2
b = t/V^2 (t = time of filtration in seconds, and V = volume of filtrate in m^3)
μ = dynamic viscosity of filtrate, $N{\cdot}s/m^2$
C = dry solids concentration, kg/m^3

In the studies, instead of r the modified specific resistance $R = r \times 10^{-11}$ was used.

To study the forms of bound water with the solid particles, the test method used was thermal drying of sludge. Freezing and thawing, refractometering, and viscosimetering were also investigated; however, the results were not included because of the difficulty in duplicating the test results.

Study Results The structure and sizes of sludge solid particles change due to digestion; a typical example is shown in Figure 3.1. The dewaterability of the same type of sludge varies from one plant to another. Specific resistances

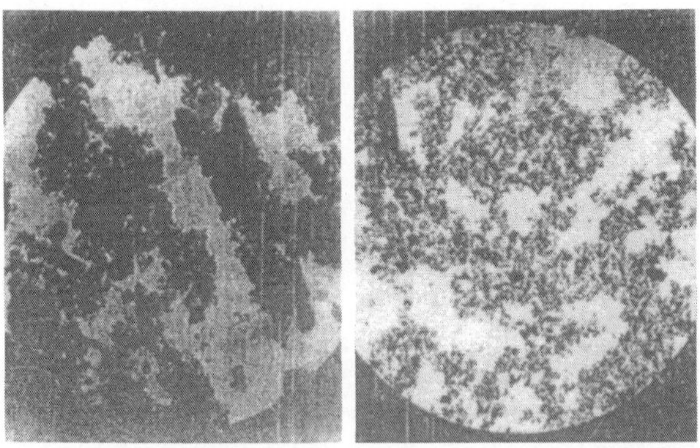

(1) Raw Sludge Solids (2) Digested Sludge Solids

Figure 3.1 Variation of sludge solids structure.

TABLE 3.1 Changes in Specific Resistance in Anerobic Digestion

Type of Sludge	Dry Solids (%)	Organics in Dry Solids (%)	Specific Resistance R (m/kg × 10^{-11})
Municipal primary sludge			
Raw	3.9–6.4	62.5–75.9	64–690
Digested	3.6–5.3	51.7–64.0	307–740
Sludge from municipal plus industrial wastewater treatment plant			
Raw	4.1–7.7	62–69	118–495
Digested	4.2–5.9	60–63	67–940
Sludge from municipal plus metallurgical wastewater treatment plant			
Raw	6.0–9.1	55–63	50–309
Digested	4.3–8.0	51–54	172–868
Mixture of municipal primary and thickened activated sludge			
Raw	3.7–4.6	70.0–75.1	2170–4035
Mesophilic digested (35°C)	2.0–4.1	62.2–70.0	3640–6750
Thermophilic digested (55°C)	2.3–3.2	61.2–67.0	8350–9500

TABLE 3.2 Changes in Specific Resistance (m/kg × 10^{-11}) During Aerobic Digestion

Type of Sludge	Dry Solids (%)	Number of Days of Digestion 0	5	10	15	40
Primary sludge	3.7–4.8	300–410	380–530	2100–4500	3700–6720	1070–1300
Thickened activated sludge	2.0–2.5	800–1130	1290–4500	5140–6250	4030–5700	970–1160
Mixture of primary and thickened activated sludge	3.0–4.5	602–775	2170–5170	2470–3760	3300–5220	830–1070

of anaerobically digested sludge and aerobically digested sludge are listed in Tables 3.1 and 3.2, respectively. The results in the tables indicate that:

- Raw primary sludge has less specific resistance than a mixture of raw primary sludge and thickened activated sludge.
- Raw sludge has less specific resistance than does digested sludge of the same type.
- A mixture of primary sludge and activated sludge has less specific resistance than that of anaerobically or aerobically digested sludge.

- Sludge digested in a mesophilic condition has less specific resistance than does the same sludge digested in a thermophilic condition.

The studies confirmed the findings of Gosh (1987), Lawler and Chung (1986), Parkin (1986), and Popel (1967) that the loading rate, periodicity of loading, digester mixing, and digestion duration and temperature are important factors that affect the digestion process. All of these factors influence the variation of dewaterability of sludge during and after the digestion process. Disintegration of organics from digestion brings the solids to a homogeneous grain structure; however, particle size decreases and the quantity of colloidal particles increases, both of which influence the dewaterability of digested sludge because of the increased degree of hydration. Experiments showed that the specific resistance increases in greater magnitude during continuous mixing of sludge with mechanical agitators than during periodic slow mixing or mixing by recirculation of gas. Experiments also indicated that aerobic digestion of primary sludge leads to a greater increase in specific resistance than does anaerobic digestion. Therefore, aerobic digestion of primary sludge should not be used if subsequent dewatering of digested sludge is required. An excessively long aerobic digestion time can result in a significant deterioration of sludge dewaterability.

The study of changes of particle size found that thickened activated sludge contains about 90% of particles that are less than 0.15 mm in size, while digested sludge has about 75% of these particles and primary sludge has only about 45% (see Figure 3.2). Specific resistance increases, as particle size dis-

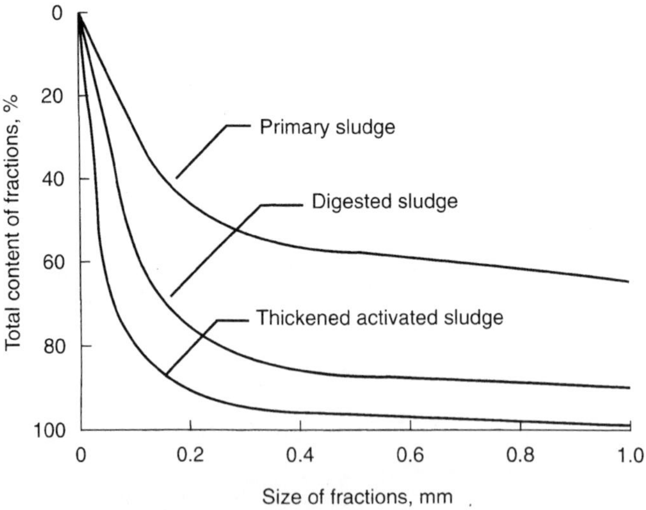

Figure 3.2 Solids size distribution in sludge.

tribution favors smaller particles. Reduction of specific resistance can be achieved by removing small particles. Small particles can be removed by *elutriation* of digested sludge: that is, washing the sludge with a stream of water. A reduction in specific resistance by elutriation can be expressed by the formula

$$\log R_n = \log R_0 e^{-an} \tag{3.2}$$

where

R_n = specific resistance of washed digested sludge, m/kg
R_0 = specific resistance of unwashed digested sludge, m/kg
a = coefficient that relates to solids washed out and size fraction removed, usually 0.04 to 0.14
n = quantity of water in sludge, m^3/m^3

The specific resistance can also be reduced by the coagulation of small particles in sludge. The effects of coagulation by ferric chloride on specific resistance for various types of sludge are shown in Figures 3.3 and 3.4.

Water in sludge exists as either free water or bound water. Bound water is water bound to the sludge particles physically or chemically. The more the bound water exists in sludge, the more the energy that must be spent to remove it. Figure 3.5, developed from the study, presents the effect of bound water on thermal drying of sludge. Straight lines a–b in the figure show the consumption of power for warming up the sludge, lines b–1cr represent the free water that exudes from the sludge, and lines 1cr–2cr represent the bound water that exudes from the sludge. Curved lines 2cr–c show that as the consumption of power is rising, a part of the power is spent to overcome the internal forces that bind the water to the solid particles in sludge.

As can be seen from lines b–1cr, the moisture from the thickening of activated sludge goes down from 98% to 87.5%; for the mixture of digested primary and activated sludge, it goes from 97.5% to 84.6%; and for the primary sludge, it goes from 94.6% to 73%. Despite these differences, 85.6% of the water is removed from activated sludge and 85% is removed from primary sludge. However, the ratio of water to dry solids in the activated sludge is 7:1, and in the primary sludge, it is 2.7:1. Therefore, the thickened activated sludge contains more bound water than do digested sludge and primary sludge (1cr in Figure 3.5). Experiments showed that there is a close relationship between specific resistance and bound water: The less the moisture in the sludge, the less the specific resistance, which is characterized by the first critical points 1cr.

In the study, during the process of coagulation of sludge with inorganic chemicals or polymers, bound water was separated and the structure of the sludge was altered. Part of the water that was physically and chemically bound with solids exuded from the solids, thereby decreasing the quantity of absorbed water. The position of the first critical point changed accordingly. This change

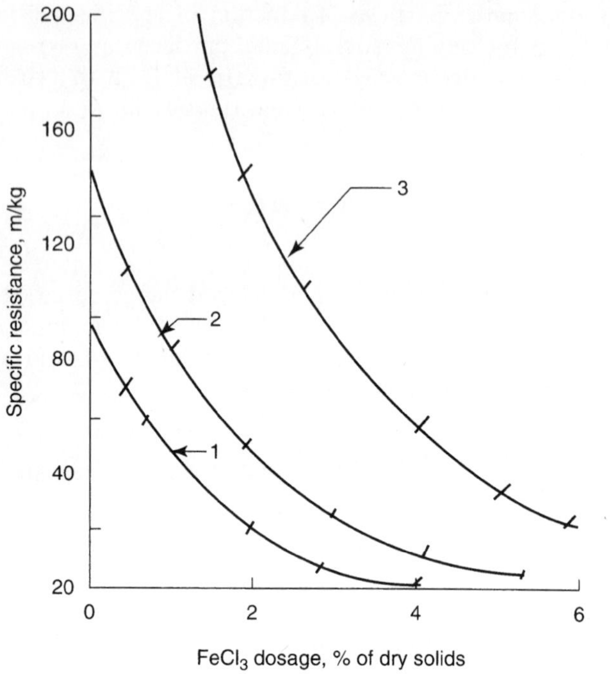

1. Raw sludge (W = 92.1 %; R = 88 m/kg)
2. Raw sludge (W = 94.8 %; R = 153 m/kg)
3. Raw sludge (W = 93.2 %; R = 350 m/kg)
W = Moisture of sludges
R = Specific resistance

Figure 3.3 Effects of coagulation on raw sludge.

is presented in Figure 3.6. This condition allows removing more water by mechanical methods after the coagulation of sludge.

There are certain differences between the specific resistance and critical points. Specific resistance shows the speed of separation of water from dry solids, whereas the first critical point shows the limit of sludge dewatering by mechanical methods. Knowledge of critical point for a sludge allows the most effective conditioning process for the sludge.

Inorganic Chemical Conditioning Until the 1970s the most useful chemical conditioners for sludge dewatering were inorganic compounds such as ferric chloride, ferrous sulfate, and aluminum chloride, one of which is added to sludge first and then usually followed with lime. Alkalinity is an important sludge characteristic that affects inorganic conditioners. Ferric chloride works

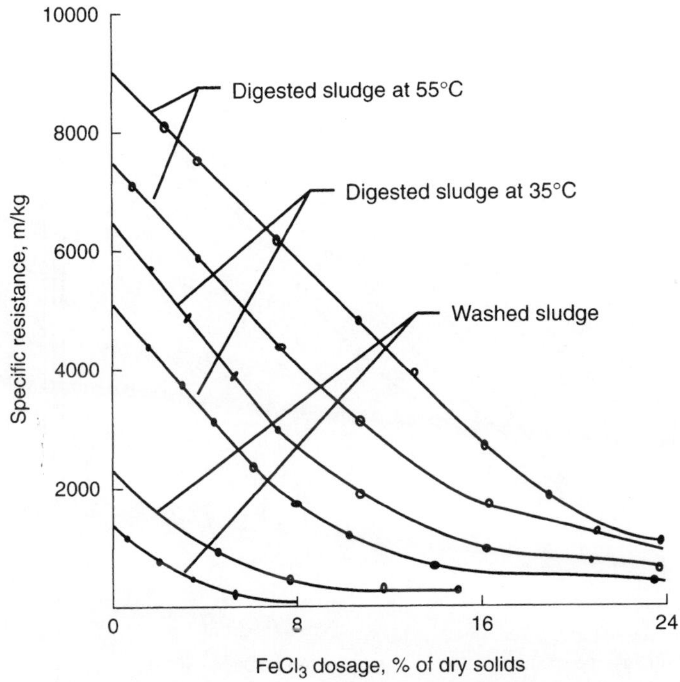

Figure 3.4 Effects of coagulation on digested sludge.

better with the pH of sludge at 6.0 to 6.5 and it reduces the pH to 4.5 to 6.0. Ferrous sulfate or aluminum chloride usually requires a higher dosage than ferric chloride. Lime following iron or aluminum salts increases the pH to 10.5 to 11.5. Generally, the required dosage is an approximate 1:3 ratio of ferric chloride ($FeCl_3$) to lime (CaO). The dosage of reagents for dewatering sludge on vacuum filters or on pressure filters depends on the specific resistance of sludge. The higher the specific resistance, the higher the dosage of reagents required. The dosage of chemicals in each case is established experimentally by measuring the specific resistance of the sludge.

The dosage of lime for conditioning sludge for dewatering can be determined by the equation

$$D = 0.3\left[R^{1/2} + \left(\frac{B}{C} + 0.001A\right)^{1/2}\right] \tag{3.3}$$

where

D = dosage of lime as CaO, % of dry solids
R = adjusted value of the specific resistance ($R = r \times 10^{-11}$, where r is the specific resistance of the sludge, m/kg)
B = sludge moisture, %

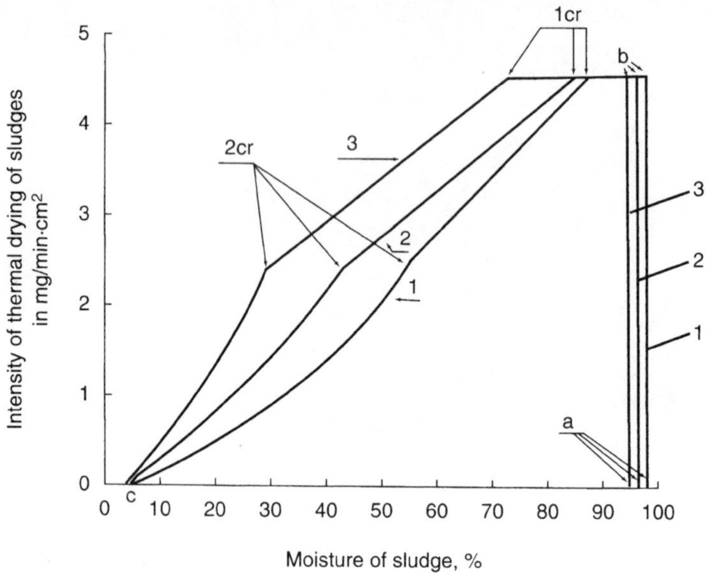

1 - Activated sludge - 98%
2 - Digested mixture of primary and activated sludge - 97.5%
3 - Primary sludge - 94.6%
1cr - First critical point of moisture in sludges.
2cr - The second critical point of moisture in sludges.
Line "b - 1cr." - Removal of free water from sludges.
Line "1cr - 2cr - c" - Removal of bound water.
"c" - The end of the drying process.

Figure 3.5 Effects of bound water on thermal drying of sludge.

C = concentration of dry solids in sludge, %
A = alkalinity of sludge before coagulation, mg/L as $CaCO_3$

The dosage of ferric chloride is usually 30 to 40% of the dosage of CaO computed by equation (3.3).

Typical values of ferric chloride and lime for various types of sludge for vacuum filter and recessed plate pressure filters are shown in Table 3.3. The concentration of ferric chloride in commercial ferric chloride liquid is between 30 and 35% by weight in water. A 30% ferric chloride solution has a specific gravity of 1.39 at 30°C and contains 1.46 kg (3.24 lb) of ferric chloride.

The costs of chemical reagents for conditioning of sludge comprise the principal fraction of the operating costs for dewatering sludge on vacuum or pressure filters. Therefore, the chemical dosage should be minimized but still provide adequate conditioning with satisfactory dewatering results.

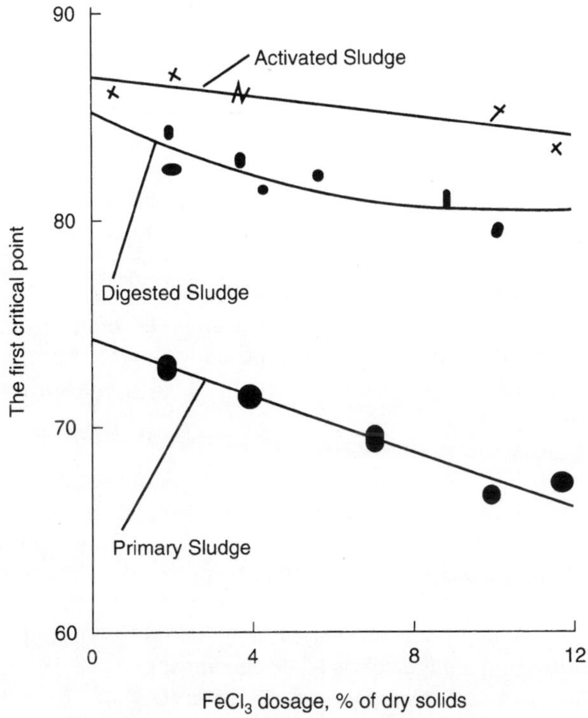

Figure 3.6 Changes in critical points.

TABLE 3.3 Typical Dosages of Ferric Chloride and Lime for Vacuum Filter and Recessed Plate Filter Press Dewatering

Method	Sludge Type	Ferric Chloride ($FeCl_3$) (% dry solids)	Lime (CaO) (% dry solids)
Vacuum filter dewatering	Raw primary	1.6–3.5	4.8–9.4
	Raw WAS	5–9	16–25
	Raw (primary + WAS)	1.6–3.5	4.8–9.4
	Anaerobically digested primary	2–5	6–15
	Anaerobically digested (primary + WAS)	3.7–7.0	8–19
Recessed plate filter press dewatering	Raw primary	1.8–4.3	5.2–12.0
	Raw WAS	6–10	20–30
	Raw (primary + WAS)	2–9	8–25
	Anaerobically digested primary	2.8–4.8	9–13
	Anaerobically digested (primary + WAS)	3.5–8.2	10.2–23.6

Source: Adapted in part from U.S. EPA, 1979.

Lime is available in two dry forms: quick lime (CaO), pebble or granular; and powdered hydrated lime [$Ca(OH)_2$]. Lime raises the pH reduced by ferric chloride or aluminum chloride addition and increases sludge porosity. In addition, lime provides a certain degree of odor reduction because sulfides in solution are converted from hydrogen sulfide to the sulfide and bisulfate ions, which are nonvolatile at alkaline pHs. Lime can also be beneficial because of its sludge stabilization effect. Although inorganic chemicals are effective conditioners for control of pH, odor, and specific resistance, they have many disadvantages. Ferric chloride is a very corrosive material. Therefore, special care should be taken in selecting the material for storage tank, piping, and metering pumps. Lime also needs special equipment for storage and feeding. Inorganic chemical conditioning increases sludge mass. Approximately one part of additional solids can be expected for each pound of ferric chloride and lime added.

Design Example 3.1 A wastewater treatment plant produces anaerobically digested sludge that needs to be dewatered using recessed plate pressure filters. At a solids concentration of 3%, the sludge contains 12,000 lb (5443 kg) of solids. The sludge is a mixture of 40% primary and 60% waste activated sludge. The press will operate one shift a day (7 hours effective) five days a week. Determine the amount of ferric chloride and lime required for conditioning, and the total sludge solids after dewatering.

1. Maximum amount of sludge to be dewatered

$$= \frac{(12{,}000\,\text{lb/d})(7\,\text{d/wk})}{(5\,\text{d/wk dewatering})(7\,\text{h/d})}$$
$$= 2400\,\text{lb/hr}\ (1089\,\text{kg/h})$$

2. Based on the data in Table 3.3, use a ferric chloride ($FeCl_3$) requirement of 5% of dry solids and a lime (CaO) requirement of 20% of dry solids. These chemical dosages convert to 100 lb/ton (50 kg/ton) of ferric chloride and 400 lb/ton (200 kg/ton) of lime.

3. Ferric chloride required $= \dfrac{(2400\,\text{lb/hr})(100\,\text{lb/ton})}{2000\,\text{lb/ton}}$

$$= 120\,\text{lb/hr}\ (54\,\text{kg/h})$$

Assuming that ferric chloride is available as 35% solution [4.13 lb $FeCl_3$ per gallon (0.5 kg/L) of solution],

$$\text{ferric chloride solution required} = \frac{120\,\text{lb/hr}}{4.13\,\text{lb/gal}}$$
$$= 29.1\,\text{gal/hr}\ (110\,\text{L/h})$$

4. $$\text{CaO required} = \frac{(2400\,\text{lb/hr})(400\,\text{lb/ton})}{2000\,\text{lb/ton}}$$
$$= 480\,\text{lb/hr}\ (217.8\,\text{kg/h})$$

Quicklime is available at 90% CaO; therefore,

$$\text{quicklime required} = \frac{480\,\text{lb/hr}}{0.9\,\text{lb/lb}}$$
$$= 533\,\text{lb/hr}\ (242\,\text{kg/h})$$

5. The amount of extra sludge produced from chemicals is 1 lb (0.45 kg) for each pound of $FeCl_3$ and quicklime added; therefore,

$$\text{total daily solids for disposal} = (2400\,\text{lb sludge} + 120\,\text{lb ferric chloride} + 533\,\text{lb quicklime})\left(\frac{7\,\text{hr/d}}{2000\,\text{lb/ton}}\right)$$
$$= 10.7\,\text{tons/d}\ (9.7\,\text{Mg/d})$$

Organic Polyelectrolytes Since the 1960s, organic polyelectrolytes (polymers) as sludge-conditioning agents have become popular. They have many advantages over inorganic chemicals. They are easy to handle, and the feed system takes up less space. To obtain the same degree of reduction in specific resistance, doses of polyelectrolytes are several times less than those of inorganic reagents. All these reduce conditioning costs.

Organic polyelectrolytes are water-soluble large organic molecules repeated in a long chain. Common sludge-conditioning polyelectrolytes can be classified by the polymer compound's charge as anionic, nonionic, or cationic; by molecular weight; and by polymer form as dry, liquid, emulsion, or gel. A combination of molecular weight and polymer molecule's charge is the most useful feature in comparing performance.

Sludge conditioning with polymers is a process of destabilizing small particles and converting them to bigger particles by flocculation. Figure 3.7 illustrates how dry polymer is prepared for use in sludge thickening or dewatering processes. Equipment required for preparing a solution of dry polymer includes dry product metering, flocculent dispenser, polymer dissolving tank, storage or day tank, low-speed mixer, and solution metering pumps. Most automatic dry polymer feed systems rely on air to convey the dry polymer to the point where polymer is first wetted with water. Depending on the type and dewaterability of sludge, dosages of polymer can vary from 1 to 10 g/kg (2 to 20 lb/ton) of dry solids. Polymers are diluted by water to a concentration of 0.1 to 0.2%. Polymer conditioning is the common practice for sludge thickening or dewatering with centrifuges and for dewatering with belt filter

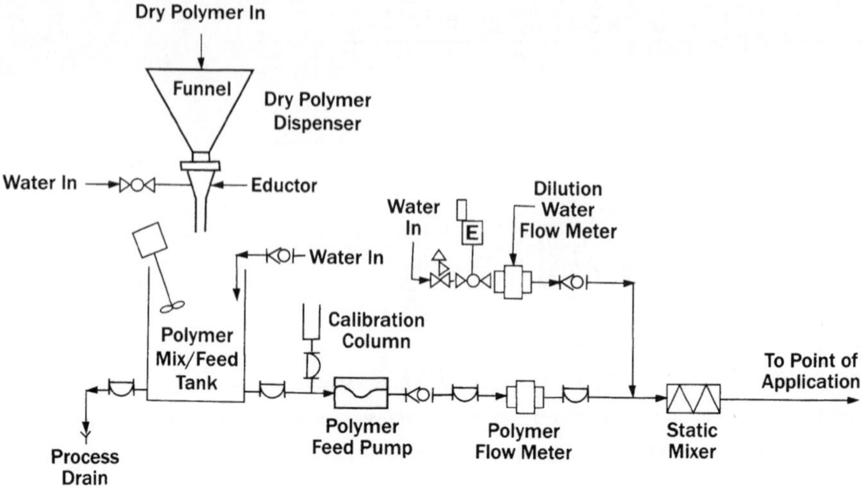

Figure 3.7 Typical polymer solution makeup and feed system. (Reprinted with permission from WEF, 1998.)

TABLE 3.4 Typical Dosages of Polymer for Thickening Sludge

Method	Sludge Type	Polymer Dosage	
		g/kg Dry Solids	lb/ton Dry Solids
Gravity belt thickening	Raw primary	1–2	2–4
	Raw WAS	2–4	4–8
	Raw (primary + WAS)	1–3	2–6
	Digested (primary + WAS)	1.5–3.0	3–6
Rotary drum thickening	Raw WAS	1.0–2.5	2–5
	Digested (primary + WAS)	1.5–3.0	3–6
Dissolved air flotation thickening	Raw WAS	0–2	0–4
	Raw (primary + WAS)	0–3	0–6
	Raw WAS	1–3	2–6
Solid bowl centrifuge thickening	Raw (primary + WAS)	1–3	2–6
	Aeobically digested WAS	1–4	2–8
	Anaerobically digested (primary + WAS)	2.5–3.5	5–7

Source: Adapted in part from U.S. EPA, 1979.

presses. Table 3.4 shows the dosages of polymer for various thickening processes, and Table 3.5 shows the dosages for sludge dewatering. It is important to note that increasing the dosages beyond the optimum values worsens the dewaterability of sludge.

TABLE 3.5 Typical Dosages of Polymer for Dewatering Sludge

Method	Sludge Type	Polymer Dosage	
		g/kg Dry Solids	lb/ton Dry Solids
Belt filter press dewatering	Raw primary	1.2–3.3	2.4–6.6
	Raw WAS	2.6–6.5	5.2–13.0
	Raw (primary + WAS)	3–6	6–12
	Raw (primary + trickling filter)	3–6	6–12
	Anaerobically digested primary	2–4	4–8
	Anaerobically digested (primary + WAS)	3–8	6–16
Solid bowl centrifuge dewatering	Raw primary	0.5–2.3	1.0–4.6
	Raw WAS	3–8	6–16
	Raw (primary + WAS)	2–6	4–12
	Anaerobically digested primary	3.6–9.0	7.2–18.0
	Anaerobically digested WAS	5.0–9.7	10.0–19.4
	Anaerobically digested (primary + WAS)	2.5–8.0	5–16
Vacuum filter dewatering	Raw primary	2–5	4–10
	Raw (primary + WAS)	4–7	8–14
	Anaerobically digested primary	3.6–9.0	7.2–18.0
	Anaerobically digested (primary + WAS)	4.4–8.7	8.8–17.4
Recessed plate filter press dewatering	Raw (primary + WAS)	2–7	4–14

Source: Adapted in part from U.S. EPA, 1979.

3.2.3 Other Conditioning Methods

Other sludge conditioning methods that have been used include adding nonchemical conditioning aids, thermal conditioning, freeze–thaw conditioning, and elutriation.

Nonchemical Conditioning Aids Power plant or sludge incinerator ash has been used successfully to improve the dewatering performance of vacuum filters and pressure filter presses. The properties of ash that improve dewatering of sludge include solubilization of its metallic constituents and its ability to attach to small solid particles of sludge and change their structure. The advantages of ash addition for sludge dewatering include elimination or substantial reduction in other chemical conditioning agents, increase in cake dryness, and significant improvements in filtrate quality. Disadvantages include the addition of a sizable quantity of inerts to the sludge cake and additional material handling. For installations where landfilling of sludge cake follow dewatering, the use of ash to improve the total solids content should be evaluated. The addition of ash to the sludge produces a drier cake, but it does nothing for the fuel value of the cake to be incinerated. Ash has no heating value, and it lowers the percent volatile solids in the cake; therefore, fuel use will increase. Pulverized coal is also a good sludge conditioning aid, especially if incineration is to follow the dewatering.

Other conditioning aids include diatomaceous earth and cement kiln dust. These agents are used primarily as a precoat in pressure filter presses. Sawdust is sometimes used as a conditioning agent, especially for dewatering sludge before composting.

Thermal Conditioning Thermal conditioning of sludge involves heating the sludge in the temperature range 170 to 220°C at a pressure of 1.2 to 2.5 MPa for 15 to 30 minutes. A thermal sludge conditioning system is illustrated in Figure 3.8. The influent sludge is ground before treatment to obtain particle sizes no greater than 4 to 5 mm. Plunger or screw (augur) pumps with working pressure up to 2.5 MPa are used for conveying the sludge to the thermal treatment system. The sludge and air mixture is heated in two stages: first in the heat exchanger by the heat of treated sludge from the reactor, and then by an external heat source in the reactor. The final heating of the sludge in the

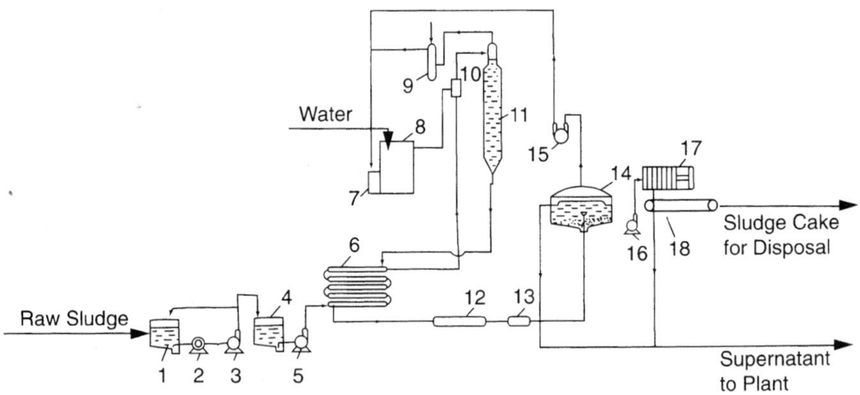

1 - Sludge storage, 2 - Grinder, 3 - Transfer pump, 4 - Day tank,
5 - High pressure feed pump, 6 - Heat exchanger, 7 - Furnace,
8 - Steam boiler, 9 - Separator, 10 - Eductor, 11 - Reactor, 12 - Reducer,
13 - Cooler, 14 - Thickener, 15 - Ventilator, 16 - Thickened sludge pump,
17 - Filter press, 18 - Conveyor.

Figure 3.8 Schematic of a thermal conditioning system.

reactor can be conducted by several methods. The simplest and most effective method is heating with steam, which enters the sludge pipe through an eductor before the reactor. The advantage of this method is in the use of comparatively low-pressure steam at a temperature approaching the sludge treatment temperature. The conditioned sludge is then discharged back through the heat exchanger where it cools to 45 to 55°C. The sludge is then thickened in a gravity thickener before dewatering. Due to the evaporation of water from the surface of the thickener, unpleasant odors are generated. To decrease the degree of evaporation, sludge is additionally cooled in a cooler to a temperature of 30 to 35°C. The thickener may be provided with a cover with forced ventilation to contain the evaporated air. The mechanical dewatering of thermally treated sludge is conducted predominantly in pressure filters. The use of pressure filters permits dewatering sludge to about 50 to 60% solids. The values of the thermal treatment parameters vary depending on the system used and have to be determined experimentally and on the basis of the decrease in the specific resistance of sludge.

In the thermal treatment process, some decomposition of sludge volatile solids occurs. The degree of decomposition depends on the initial properties of sludge. Approximately 75 to 80% of decomposing organics dissolve in the liquid, and 20 to 25% volatilize. Decanted water from the thickener and the filtrate from the filter press contain a high 2000 to 6000 mg/L solids, which can increase the loading to the treatment plant up to 10 to 25%. Some of these solids in sidestreams are difficult to oxidize. Therefore, before discharging to the plant influent, the sidestream may have to be treated by adding chemicals to reduce the organic loads.

Thermal conditioning of wastewater sludge has the following advantages:

- Except for straight waste activated sludge, the process will produce sludge with good dewatering characteristics. Cake solids concentrations of 50 to 60% are typically obtained with mechanical dewatering equipment following thermal conditioning.
- Additional chemical conditioning is normally not required.
- The process sterilizes the sludge, rendering it free of pathogens.
- The process is suitable for many types of sludge that cannot be stabilized biologically because of the presence of toxic materials.

Disadvantages of thermal conditioning include the following:

- The process has a high capital cost, due to the use of corrosion-resistant materials such as stainless steel in heat exchangers. Other support equipment is required for odor control and high-pressure transport.
- The process requires supervision, skilled operators, and a strong preventive maintenance program.
- The process produces an odorous gas stream that must be collected and treated before release.

- The process produces sidestreams with high concentrations of organics, ammonia nitrogen, and color.
- Scale formation in heat exchangers, pipes, and reactor requires acid washing.

Thermal conditioning is being practiced with positive results in the Asher Wastewater Treatment Plant in Paris, France, and the Luberetzkay Wastewater Treatment Plant in Moscow, Russia. At both plants, sludge is dewatered without further chemical conditioning, dewatered cake has a high solids content, biosolids are disinfected effectively, and the stability is retained even after long storage. Several wastewater treatment plants in England, Germany, and the United States have discontinued their thermal conditioning practice because of the disadvantages discussed above.

Freeze–Thaw Conditioning Freezing and subsequent thawing of sludge result in a change in its structure and the conversion of bound water to free water. This has been observed in the natural freezing and thawing of sludge in drying beds in colder climates. This increases the sludge dewaterability significantly. Freezing and thawing of sludge decrease its specific resistance and permit mechanical dewatering of sludge without coagulation or with significant reduction in the quantity of reagents required for coagulation.

In the artificial freezing of sludge, the optimal value of specific heat flux is 230 to 1000 W/m^2·h. At higher heat flow values, the specific resistance of sludge decreases insufficiently due to rapid freezing, and at low values it is necessary to increase the surface area of the heat-exchange equipment, with a consequent increase in the capital cost. Effects of artificial freezing depend on temperature and duration of freezing. Slow freezing decreases specific resistance more rapidly.

Artificial freezing of sludge can be conducted in direct-contact freezers using ice generators of the drum or panel type. To decrease power use in the sludge freezing and thawing process, phase-conversion heat recovery should be used in thawing the sludge, that is, using the heat given off during freezing. The electrical energy consumption for artificial freezing of 1 m^3 of sludge is about 50 kWh. After thawing, the sludge can be dewatered on pressure filter presses or on sludge drying beds. Filter presses can produce sludge cake as high as 50 to 60% solids. The loading rate on sludge drying beds can be as much as 5 m^3/m^2·yr.

Elutriation *Elutriation* is the term commonly used to refer to the washing of anaerobically digested sludge before dewatering. Washing causes a dilution of the bicarbonate alkalinity in the sludge and therefore reduces the demand for acidic metal salt by as much as 50%. Two to four volumes of washwater, typically plant effluent, flow countercurrent to one volume of anaerobically digested sludge. Elutriation tanks are designed to act as gravity thickeners, with a mass solids loading of 8 to 10 lb/ft^2-d (39 to 48.8 kg/m^2·d).

Presently, the process is not used as extensively as it had been because in addition to reducing alkalinity, it washes out 10 to 15% of the solids from the sludge stream. These solids, when recycled back to the plant influent, can degrade the plant effluent through additional solids and organic loading on the plant if this load has not been accounted for in the design.

3.3 THICKENING

Thickening of sludge is a process to increase its solids concentration and to decrease its volume by removing some of the free water. The resulting material is still fluid. Thickening is employed prior to subsequent sludge-processing steps, such as digestion and dewatering, to reduce the volumetric loading and increase the efficiency of subsequent processes.

The most commonly used thickening processes are gravity thickening, dissolved air floatation thickening, gravity belt thickening, and rotary drum thickening. Table 3.6 presents a comparison of these thickening processes. Selection of a particular thickening process sometimes depends on the size of the wastewater treatment plant and the downstream train chosen. The main design variables of any thickening process are:

- Solids concentration and flow rate of the feed stream
- Chemical demand and cost if chemicals are used for conditioning
- Suspended and dissolved solids concentrations and flow rate of the clarified stream
- Solids concentration and flow rate of the thickened sludge

3.3.1 Gravity Thickening

Gravity thickening is the simplest and most commonly used method for sludge thickening in wastewater treatment plants. Circular concrete tanks are the most common configuration for gravity thickeners, although rectangular concrete tanks have also been used. Figure 3.9 shows a cross-sectional view of a typical circular gravity thickener.

Design Considerations A gravity thickener is similar in design to a conventional sedimentation tank but has a more steeply sloping floor. Tanks usually range from 10 to 24 m (33 to 80 ft) in diameter. Side-water depths vary from 3 to 4 m (10 to 13 ft) and floor slopes vary from 1:4 to 1:6, both depending on the period of time required to thicken the sludge to the concentration and storage volume required to compensate for fluctuations in the solids loading rate period. The steeper slope also reduces sludge raking problems by allowing gravity to do a greater portion of the work of moving the settled solids to the center of the thickener.

TABLE 3.6 Comparison of Thickening Methods

Method	Advantages	Disadvantages
Gravity thickening	Least operation skill required Low operating costs Minimum power consumption Ideal for small treatment plants Good for rapidly settling sludge such as WAS and chemical Conditioning chemicals typically not required	Large space required Odor potential Erratic and poor solids concentration (2 to 3%) for WAS Floating solids
Dissolved air flotation thickening	Provides better solids concentration (3.5 to 5%) for WAS than that of gravity thickening Require less space than a gravity thickener Will work without chemicals or with low dosages of chemicals Relatively simple equipment components	Operating costs higher than for a gravity thickener Relatively high power consumption Moderate operator attention required Odor potential Larger space requirements compared to other mechanical methods Has very little storage capacity compared to a gravity thickener Not very efficient for primary sludge Requires polymer conditioning for higher solids capture or increased loading
Centrufugal thickening	Effective for thickening WAS to 4 to 6% solids concentration Control capability for process performance Least odor potential and housekeeping requirements because of contained process Less space required	High capital cost High power consumption Requires moderate operator attention Sophisticated maintenance requirements Requires polymer conditioning for higher solids capture
Gravity belt thickening	Effective for WAS with 0.4 to 6% solids concentration Control capability for process performance High solids capture efficiency Relatively low capital cost Relatively low power consumption	Polymer dependent Housekeeping requirements Odor potential Moderate operator attention required Building commonly required
Rotary drum thickening	Effective for WAS with 0.4 to 6% solids concentration Less space required Low power consumption	Polymer dependent and sensitive to polymer type Housekeeping requirements Odor potential Moderate operator attention required Building commonly required

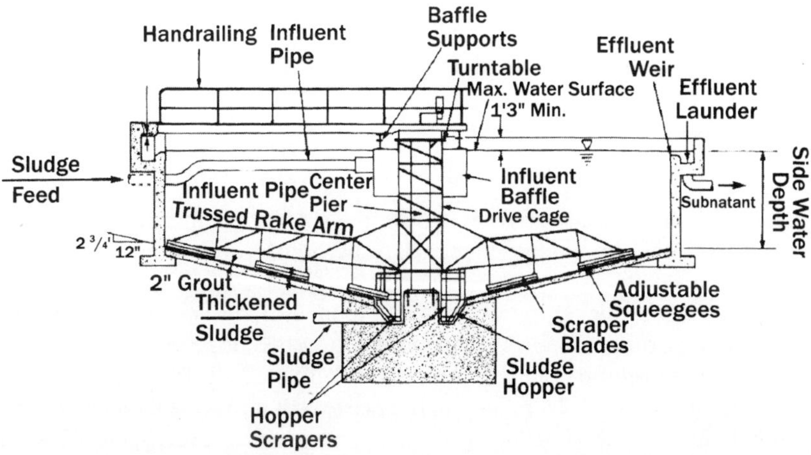

Figure 3.9 Cross-sectional view of a typical circular gravity thickener.

Gravity thickener mechanisms are similar in design to primary clarifiers. The inlet to the thickener is to a center-feed well through a bottom-feed, side-entry (Figure 3.9), or overhead inlet. The rake support truss is often provided with pickets that are thought to help release water from the solids. However, the rake support truss can provide sufficient sludge mixing to make pickets unnecessary. Drive mechanisms are heavier than those required for primary settling tanks, to overcome the problem of island formation of highly viscous solids. Thickeners are equipped with a skimming mechanism and baffling because of the inherent floating of the scum layer associated with sludge.

Design Criteria The most important aspect of thickener design is the establishment of the area required to achieve a desired degree of thickening. If sludge from a particular facility is available for testing, the required surface area can be found using a batch settling column and developing a solids flux for the particular sludge. Solids flux is the mass of solids passing through a unit area per unit time. The required thickener area is calculated using the equation

$$A = \frac{C_0 Q_0}{G_T} \tag{3.4}$$

where

A = thickener area, m^2 (ft^2)
C_0 = influent dry solids concentration, kg/m^3 (lb/ft^3)
Q_0 = influent flow, m^3/d (ft^3/hr)
G_T = solids flux, kg/m^2·d (lb/ft^2-hr)

The change in sludge volume in a gravity thickener can be determined from the formula

$$V_2 = V_1 \frac{C_1}{C_2} \tag{3.5}$$

where V_1 and V_2 are initial and final (thickened sludge) volumes, respectively; and C_1 and C_2 are the sludge concentrations before and after thickening, respectively.

In the prolonged thickening of activated sludge, there is a sharp increase in its specific resistance due to the decomposition of sludge and the increase in the quantity of bound water. Examples of activated sludge specific resistance are shown in Figure 3.10. During the process of gravity thickening waste activated sludge, as the concentration of dry solids increases from 0.2% to 2%, specific resistance increases from 35 m/kg to 1480 m/kg, but the volume of sludge decreases tenfold. However, when concentration increases from 2% to 3.2%, volume decreases only 1.2 times, while specific resistance increases from 1480 m/kg to 7860 m/kg. From Figure 3.10 it can be seen that the specific resistance of activated sludge is closely related to the concentration of dry solids in sludge.

Long-term gravity thickening of activated sludge results in a sharp increase in specific resistance, which significantly worsens the dewaterability of sludge. However, dewatering of nonthickened activated sludge does not make sense

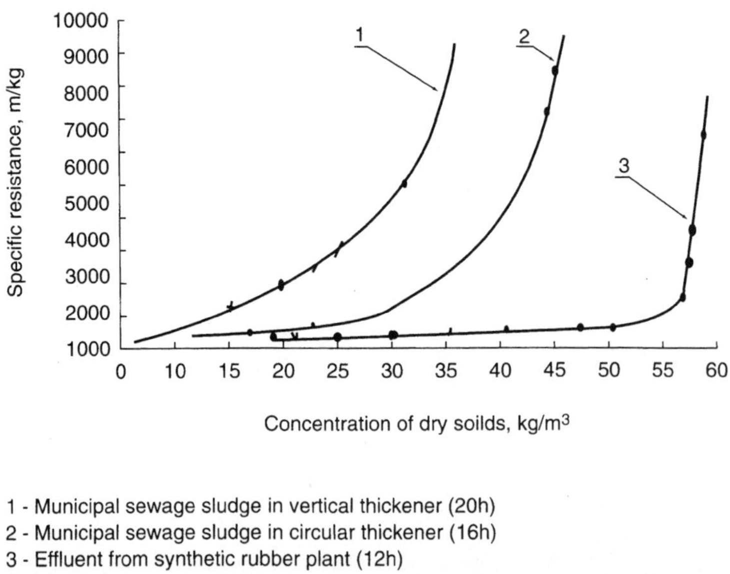

Figure 3.10 Effects of thickening of activated sludge on specific resistance.

because of the initial high volume and low solids concentration. The kinetics of activated sludge allows determining the optimum concentration of dry solids that corresponds to the best efficiency of the dewatering equipment. The optimum thickened sludge concentration from vertical thickeners is 2.0 to 2.5% and from radial thickeners is 2.9 to 3.4%. The corresponding optimum time for thickening is 10 to 14 hours for vertical thickeners and 9 to 11 hours for radial thickeners equipped with sludge rakes. Mixed liquor from aeration tanks thickens faster than activated sludge from secondary settling tanks.

In most cases the sludge to be thickened is not available for settling testing. In such instances, thickeners are designed based on established solids loading and thickener overflow rates. Table 3.7 provides the typical rates for calculating the required surface area. Based on the type and solids concentration of sludge to be thickened, the thickener can be designed for the thickened sludge underflow concentration required for downstream processing. Recommended hydraulic overflow rates range from 15.5 to 31 $m^3/m^2 \cdot d$ (382 to 760 gpd/ft^2-d) for primary sludge, 4 to 8 $m^3/m^2 \cdot d$ (100 to 200 gpd/ft^2-d) for waste activated sludge, and 6 to 12 $m^3/m^2 \cdot d$ (150 to 300 gpd/ft^2-d) for combined primary and waste activated sludges. A high overflow rate can result in excessive solids carryover. Conversely, a low overflow rate means high thickener detention time, which can produce floating sludge (when methane produced by the anaerobic breakdown of solids buoys the sludge blanket), and odors from septic conditions.

Operational Considerations If thickener is used on a continuous basis without undo storage of sludge, septic or odor problems should be avoided.

TABLE 3.7 Gravity Thickener Design Criteria

Type of Sludge	Influent Solids Conc. (%)	Thickening Time (h)	Thickened Solids Conc. (%)	Dry Solids Loading	
				kg/m^2·d	lb/ft^2-d
Primary (PRI)	3–6	5–8	4–8	100–200	20–40
Trickling filter (TF)	1–4	8–16	3–6	40–50	8–10
Rotating biological contactor (RBC)	1.0–3.5	8–16	2–5	35–50	7–10
WAS	0.4–1.0	5–15	2.0–3.5	25–80	5–16
PRI + TF	2–6	5–10	5–9	60–100	12–20
PRI + RBC	26	5–12	5–9	60–100	12–20
PRI + WAS	0.6–4.0	5–15	3–7	25–200	5–40
Aerobically digested WAS	0.5–1.0	1.5–12.0	2–5	50–200	10–40
Anaerobically digested PRI	4–7	20–1440	6–13	—	—
Anaerobically digested PRI + WAS	2–4	20–1440	8–11	—	—

Source: Adapted from U.S. EPA, 1979.

Depending on temperature, primary sludge can be retained in the thickener for two to four days before upset conditions develop. Best practice is to maintain a sludge detention time of one to two days. The activated sludge thickening detention time has to be limited to a maximum of 15 hours.

The development of septic conditions during thickening results in floating solids that passes over into the thickener overflow, foul odors, and reduced underflow concentration. Waste activated sludge with a sludge volume index (SVI) of less than 100 indicates an older, denser, fast-settling sludge. SVI of more than 150 indicates a young, low-density, slow-settling sludge. Typically, a sludge with an SVI greater than 200 is considered to be a *bulking sludge*. Bulking sludge is characterized by a rapid and obvious rise in the sludge blanket and production of dilute underflow concentration during otherwise normal operation. Generally, sludge bulking is not a function of thickener operations, and the problem must be cured by correcting the basic cause within the plant. Thickener performance problems can also be corrected by the addition of chlorine to the thickener influent.

Provisions for dilution water should be included, especially for primary and secondary sludge mixes, to improve process performance by maintaining proper hydraulic loading. If the sludge temperature never exceeds 15 to 20°C, a dilution-to-sludge volume ratio of up to 4:1 is satisfactory. Higher temperature requires more dilution. As an alternative to using a large volume of dilution liquid and recycling it through the plan treatment process, the thickener overflow liquid can be aerated for 15 to 20 minutes and then used as dilution liquid. An added benefit to this is the retention of about 99% of the solids in the thickener because net overflow volume is negligible.

The strength of the thickener overflow or supernatant, as measured by suspended solids, can vary significantly. Reported values vary from 200 to 2500 mg/L. Addition of polymer to gravity thickener feed has been practiced at several plants. Results indicate that the addition of polymers increases solid capture but has very little or no effect in increasing solids concentration.

Design Example 3.2 Design a gravity thickener for the combined primary and waste activated sludge for a plant with average primary sludge flow of 8000 gal/d (30.3 m^3/d) at 4% solids, and average waste activated sludge flow of 50,000 gal/d (189.2 m^3/d) at 0.8% solids.

1. At average design conditions (neglect specific gravity):

$$\begin{aligned}\text{primary sludge solids} &= (8000\,\text{gal/d})(8.34\,\text{lb/gal})(0.04\,\text{lb/lb})\\ &= 2669\,\text{lb/d}\,(1211\,\text{kg/d})\end{aligned}$$

$$\begin{aligned}\text{WAS solids} &= (50{,}000\,\text{gal/d})(8.34\,\text{lb/gal})(0.008\,\text{lb/lb})\\ &= 3336\,\text{lb/d}\,(1513\,\text{kg/d})\end{aligned}$$

$$\text{combined solids} = (2669+3336)\,\text{lb/d}$$
$$= 6005\,\text{lb/d}\,(2724\,\text{kg/d})$$

$$\text{combined sludge flow} = (8000+50{,}000)\,\text{gal/d}$$
$$= 58{,}000\,\text{gal/d}\,(219.5\,\text{m}^3/\text{d})$$

2. Combined solids concentration:

$$\text{sludge} = \frac{6005\,\text{lb/d}}{(8.34\,\text{lb/gal})(58{,}000\,\text{gal/d})} \times 100\%$$
$$= 1.44\%$$

3. From Table 3.7, select the lowest solids loading rate of $5\,\text{lb/ft}^2\text{-d}$ ($25\,\text{kg/m}^2\cdot\text{d}$) to compensate for the peak sludge flow conditions:

$$\text{area required} = \frac{6005\,\text{lb/d}}{5\,\text{lb/ft}^2\text{-d}} = 1201\,\text{ft}^2\ (112\,\text{m}^2)$$

4. Diameter of thickener $= \sqrt{(1201\,\text{ft}^2)(4/\pi)} = 40\,\text{ft}\ (12.2\,\text{m}^2)$

5. From Table 3.7, the expected underflow (thickened sludge) concentration is about 5%. Assume a solids-capture efficiency of 90%.

$$\text{thickened sludge solids} = (6005\,\text{lb/d})(0.90)$$
$$= 5404.5\,\text{lb/d}\ (2452\,\text{kg/d})$$

$$\text{thickened sludge flow} = \frac{5404.5\,\text{lb/d}}{(8.34\,\text{lb/gal})(0.05\,\text{lb/lb})}$$
$$= 12{,}960\,\text{gal/d}\ (49\,\text{m}^3)$$

$$\text{solids in supernatant} = (6000 - 5404.5)\,\text{lb/d}$$
$$= 600.5\,\text{lb/d}\ (272\,\text{kg/d})$$

$$\text{solids concentration in supernatant} = \frac{600.5\,\text{lb/d}}{(8.34\,\text{lb/d})(45{,}040\,\text{gal/d})} \times 100\%$$
$$= 0.16\% = 1600\,\text{mg/L}$$

6. $\text{Hydraulic rate} = \dfrac{58{,}000\,\text{gal/d}}{1201\,\text{ft}^2} = 48.3\,\text{gal/ft}^2\text{-d}\ (2.0\,\text{m}^3/\text{m}^2\cdot\text{d})$

The recommended minimum hydraulic rate is 380 gal/ft^2-d (15.5 m^3/m^2·d). Therefore, about 330 gal/d (1.25 m^3/d) of dilution water should be provided.

3.3.2 Dissolved Air Floatation Thickening

In the dissolved air floatation (DAF) thickening process, air is introduced to the sludge at a pressure in excess of atmospheric pressure. When the pressure is reduced to atmospheric pressure and turbulence is created, air in excess of that required for saturation leaves the solution as fine bubbles 50 to 100 μm in diameter. These bubbles attach to the suspended solids or become enmeshed in the solids matrix. Since the average density of solids–air aggregates is less than that of water (0.6 to 0.7), they rise to the surface. Good solids floatation occurs with a solids–air aggregate specific gravity of 0.6 to 0.7. The floating solids are collected by a skimming mechanism similar to a scum skimming system.

DAF thickening is used most efficiently for waste activated sludge. Although other types of sludge, such as primary sludge and trickling filter sludge, have been floatation thickened, gravity thickening of the sludge is more economical than DAF thickening.

The schematic of a typical DAF thickening system is presented in Figure 3.11. The major components of a DAF system are the pressurization system with an air saturation tank, a recycle pump, an air compressor and pressure release valve, and a DAF tank equipped with surface skimmer and bottom solids removal mechanism. Figure 3.12 illustrates two models of flotation thickeners that are in use in Russia and Ukraine.

There are three ways in which the pressurization system can be operated. In the method called *total pressurization*, the entire sludge flow is pumped through the pressurization tank and the air-saturated sludge is then passed through a reduction valve before entering the floatation tank. In the second method, called *partial pressurization*, only a part of the sludge flow is pumped through the pressurization tank. After pressurization, the pressurized and unpressurized streams are combined and mixed before they enter the floatation tank. In the third method, called the *recycle pressurization* (Figure 3.11), a portion of the subnatant is saturated with air in the pressurization tank and then combined and mixed with the sludge feed before it is discharged into the floatation tank.

The major advantage of the recycle pressurization system is that it minimizes high-shear conditions, an important parameter when dealing with flocculant-type sludge. The recycle pressurization system also eliminates clogging problems with the pressurization pump, air saturation tank, and pressure release valve from the stringy material in the feed sludge. For these reasons, recycle pressurization is the most commonly used. The recycle flow can also be obtained from the secondary effluent, which has the advantage of lower suspended solids and a lower grease content than the subnatant from the DAF tank.

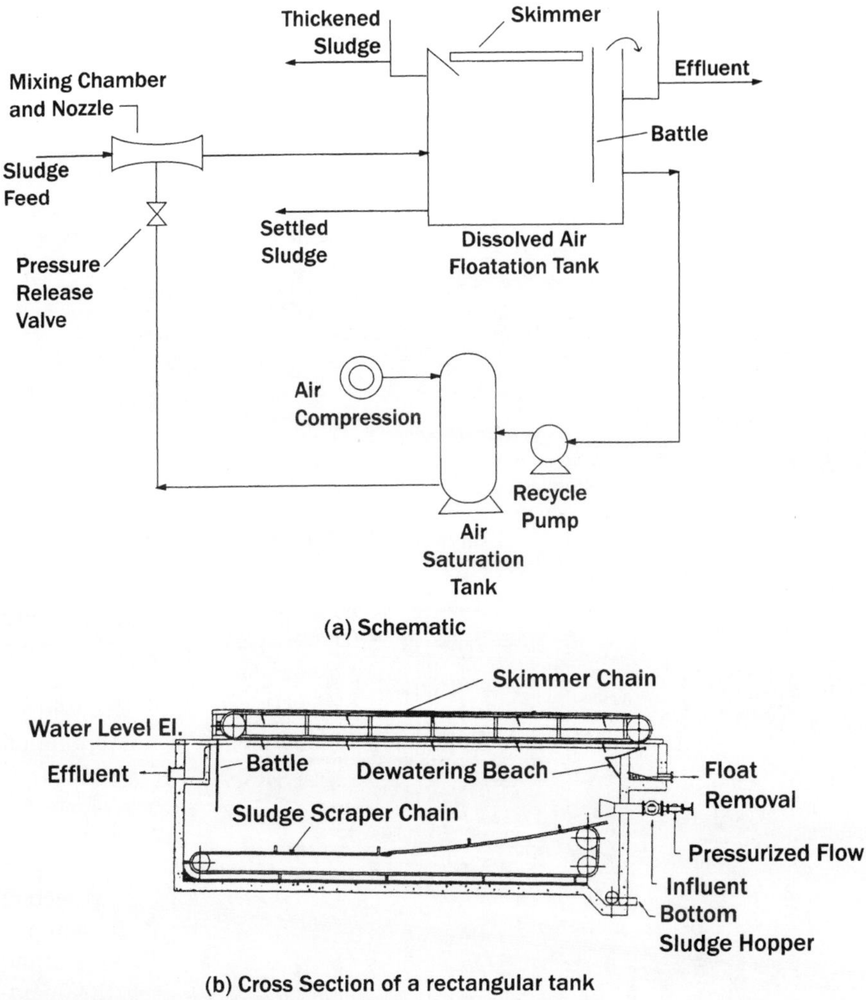

Figure 3.11 Dissolved air flotation system.

The floatation tank can be circular or rectangular and made of steel or concrete. Smaller tanks are usually steel and come completely assembled. For large installations requiring multiple tanks or large tanks, concrete tanks are more economical. Rectangular tanks have several advantages over circular units. In rectangular tanks, skimmers skim the entire surface and the flights can be closely spaced, allowing more efficient skimming. In a rectangular tank, bottom sludge flights are usually driven by a separate unit and hence can be operated independent of the skimmer flights. The main advantage of circular units is their lower cost in terms of structural concrete and mechani-

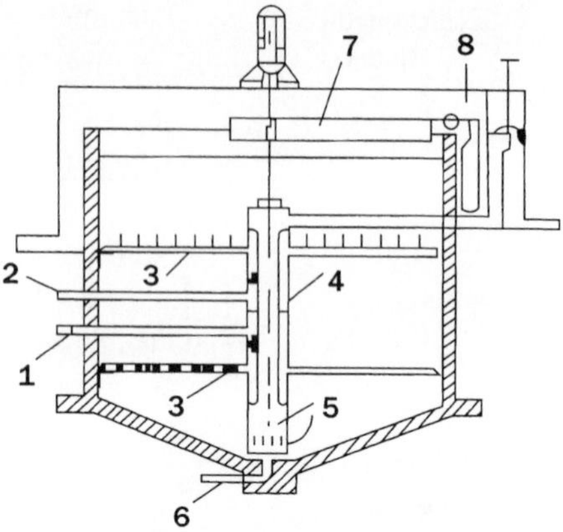

a. Flotation thickener designed by UIIVKh:
1. Water-air mixture inflow; 2. Sludge inflow;
3. Perforated pipes; 4. Distribution device;
5. Clarified water discharge; 6. Thickener evacuation; 7. Rake; 8. Peripheral channel.

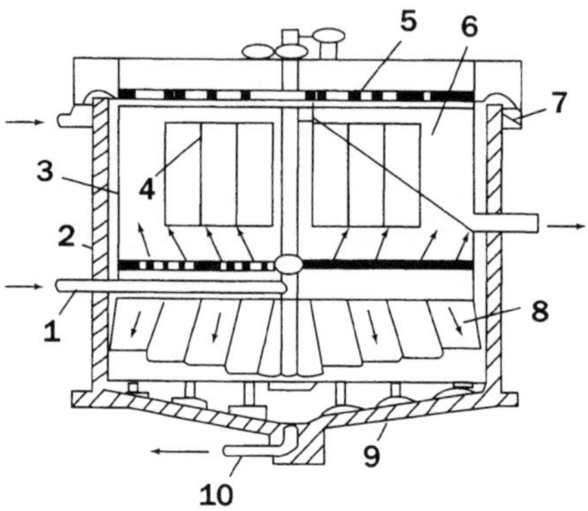

b. Flotation thickener: 1. Inflow of sludge-air mixture; 2. Rotation perforated distributor;
3. Peripheral partition; 4. Cylindrical partitions;
5. Sludge rake; 6. Sludge collecting channel;
7. Circular spillway; 8. Conical partitions;
9. Rake device; 10. Discharge of floating sludge and evacuation of thickener.

Figure 3.12 Circular flotation thickeners.

cal equipment. However, shipping problems limit a completely assembled steel circular unit to about 9 m^2 (100 ft^2) or less.

Process Design Considerations Numerous factors affect DAF process performance, including the following:

- Type and characteristics of feed sludge
- Solids loading rate
- Hydraulic loading rate
- Air-to-solids ratio
- Polymer addition

Although sufficient data are available from operating units in more than 40 years to size DAF systems, bench- and pilot-scale testing can provide valuable information. Most manufacturers of DAF systems have designed and built bench-scale units for evaluations. These manufacturers have scale-up criteria for their equipment to predict full-scale operational parameters. Consideration should also be given to renting a pilot unit available from most manufacturers to test and evaluate the effect of such parameters as solids and hydraulic loadings, recycle ratio, air-to-solids ratio, and polymer type and dosage. If sludge is not available for testing, a detailed review must be made of experience at installations where a similar type of sludge is being thickened by DAF thickeners.

Type and Characteristics of Feed Sludge A variety of sludges can be thickened effectively by flotation. These include conventional WAS, extended aeration sludge, pure-oxygen activated sludge, and aerobically digested sludge. The first step in designing a DAF system is to evaluate the characteristic of the feed sludge. Information is needed about the range of solids concentration that can be expected. If WAS is to be thickened, the mixed liquor sludge volume index (SVI) must be determined because SVI can significantly affect the DAF thickening performance. The SVI should be less than 200 if a float concentration of 4% is required with nominal polymer dosage.

Solids Loading Rate The solids loading rate is expressed as the weight of solids per hour per unit effective floatation area. Typical solids loading rate are given in Table 3.8. The loading rates shown will normally result in a minimum of 4% concentration in the float. As can be seen from the table, the solids loading rate can generally be increased up to 100% with polymer addition.

Hydraulic Loading Rate The hydraulic loading rate is expressed as combined flow rates of feed sludge and recycle per unit effective floatation area ($m^3/m^2{\cdot}d$ or gpm/ft^2). When like units are canceled, it becomes a velocity gradient to the average downward velocity of water as it flows through the floatation tank. The maximum hydraulic loading rate must always be less than the minimum rise rate of the sludge–air particles to ensure that all the particles will reach the surface before they reach the effluent end of the tank. Suggested hydraulic loading rates range from 30 to $120\,m^3/m^2{\cdot}d$ (0.5 to $2\,gpm/ft^2$).

TABLE 3.8 Typical Solids Loading Rates for DAF Thickening

	Loading Rate ($kg/m^2 \cdot h$)		Loading Rate (lb/ft^2-hr)	
Type of Sludge	Without Polymer	With Polymer	Without Polymer	With Polymer
Primary	4–6	Up to 12.5	0.83–1.25	Up to 2.5
WAS (air activated)	2–4	Up to 10	0.4–0.8	Up to 2.0
WAS (oxygen activated)	3–4	Up to 10	0.6–0.8	Up to 2.2
Trickling filter	3–4	Up to 10	0.6–0.8	Up to 2.0
Primary + WAS (air)	3–6	Up to 10	0.6–1.25	Up to 2.0
Primary + trickling filter	4–6	Up to 10	0.83–1.25	Up to 2.5

Source: Adapted from U.S. EPA, 1979.

Air-to-Solids Ratio The air-to-solids ratio is perhaps the single most important factor affecting DAF performance. It is defined as the weight ratio of air to the solids in the feed stream. The ratio for a particular application is a function of the characteristics of the sludge, principally the SVI, the air-dissolving efficiency of the pressurization system, and distribution of the air–solids mixture into the floatation tank. For domestic wastewater sludge, reported values of air-to-solids ratios range from 0.01 to 0.4, with most systems operating at a value under 0.06.

Polymer Addition Chemical conditioning with polymer has a marked effect on DAF thickener performance. The particles in a given sludge may not be amenable to the floatation process because their small size will not allow proper air bubble attachment. The surface properties of the particles may also have to be altered before effective floatation can occur. Sludge particles can be surrounded by electrically charged layers that disperse the particles in the liquid phase. Polymers can neutralize the charge, causing the particles to coagulate so that the air bubbles can attach to them for effective floatation.

Bench- or pilot-scale testing is the most effective method to determine the optimal amount of polymer required and the point of addition (in the feed stream or the recycle stream) for a particular installation. Typical polymer dosages range from 2 to 5 g polymer per kilogram of dry solids (4 to 10 lb/ton).

In the lower ranges of solids and hydraulic loading rates, polymer addition typically is not necessary. Polymer conditioning usually affects solids capture to a greater extent than float solids concentration. With polymer addition, float solids can be increased by about 0.5%; however, the solids capture efficiency can be increased from about 90% to better than 95%.

Design Example 3.3 Design a DAF thickener to thicken sludge with a maximum of 2500 lb (1134 kg) per day of WAS solids. The solids concentration of the sludge is 0.6%. The thickening operation will take place 7 hours per day, 5 days per week.

1. Surface area:

$$\text{net hourly load to DAF} = \frac{(2500\,\text{lb/d})(7\,\text{d/wk})}{(5\,\text{d/wk})(7\,\text{h/d})}$$
$$= 500\,\text{lb/hr}\ (227\,\text{kg/h})$$

From Table 3.8, the solids loading rate for WAS thickening is 0.4 to 0.8 lb/ft²-hr without the use of polymer. Use a value of 0.5 lb/ft²-hr:

$$\text{effective surface area} = \frac{500\,\text{lb/hr}}{0.5\,\text{lb/ft}^2\text{-hr}}$$
$$= 1000\,\text{ft}^2\ (93\,\text{m}^2)$$

2. Flow streams:

$$\text{sludge feed} = \frac{500\,\text{lb/hr}}{(8.34\,\text{lb/gal})(0.006\,\text{lb/lb})(60\,\text{min/hr})}$$
$$= 166.5\,\text{gpm}\ (38\,\text{m}^3/\text{h})$$

Based on information from DAF manufacturer, a 50% recycle flow (83.3 gpm) is selected:

$$\text{total flow to DAF} = (166.5 + 83.3)\,\text{gpm}$$
$$= 249.8\,\text{gpm}\ (67\,\text{m}^3/\text{d})$$

A float solids concentration of 4% and a capture efficiency of 90% are expected.

$$\text{thickened solids} = (500\,\text{lb/hr})(0.9)$$
$$= 450\,\text{lb/hr}\ (204\,\text{kg/h})$$

$$\text{thickened sludge flow} = \frac{450\,\text{lb/hr}}{(8.34\,\text{lb/gal})(0.04)(60\,\text{min/hr})}$$
$$= 22.5\,\text{gpm}\ (5\,\text{m}^3/\text{h})$$

Note: The solids in recycle flow are neglected.

3. $$\text{Hydraulic loading rate} = \frac{249.8\,\text{gpm}}{1000\,\text{ft}^2}$$
$$= 0.25\,\text{gpm/ft}^2\ (0.6\,\text{m}^3/\text{m}^2\cdot\text{d})$$

This rate is well below the minimum acceptable rate of 0.5 gpm/ft^2. Therefore, the DAF system might be able to handle a higher solids loading rate, or fewer hours of operation, provided that downstream processing requirements are not affected adversely.

4. Float tank size: Provide two units.

$$\text{effective surface area} = \frac{1000\,\text{ft}^2}{2} = 500\,\text{ft}^2/\text{unit}$$

   Use two circular DAF tanks 26 ft (8 m) in diameter each, or two rectangular tanks 34 ft long × 16 ft wide (10 m × 5 m) each.
   Note: Factory-built DAF tanks are usually about 10 ft wide. Therefore, four tanks of 25 ft long × 10 ft wide will also be appropriate.

5. Air requirement: Use an air-to-solids ratio of 0.06 by weight:

$$\text{air required} = \frac{(500\,\text{lb/hr})(0.06\,\text{lb/lb})}{(0.075\,\text{lb/scfm})(60\,\text{min/hr})}$$
$$= 6.7\,\text{scfm}\ (0.2\,\text{m}^3/\text{min})$$

3.3.3 Centrifugal Thickening

Centrifugal thickening is the acceleration of sedimentation through the use of centrifugal force. In a gravity thickener, solids settle by gravity. In a centrifuge, force 500 to 3000 times of gravity is applied; therefore, a centrifuge acts as a highly effective gravity thickener.

Centrifuges are commonly used for thickening WAS. Primary sludge is seldom thickened by centrifuges because it commonly contains abrasive material that is detrimental to a centrifuge. In addition to being very effective for thickening WAS, centrifuges have the additional advantages of less space requirement and the least odor potential and housekeeping requirements because of the contained process. However, capital cost and maintenance and power costs can be substantial. Therefore, the process is usually limited to large treatment plants.

There are basically three types of centrifuges: disk nozzle, imperforate basket, and solid bowl. Figure 3.13 shows schematics of all three types of centrifuges. Disk nozzle centrifuges require extensive prescreening and degritting of feed sludge. They can be used on sludges with particle sizes of 400 μm or less. Imperforate basket centrifuges can be used for batch operation only and not continuous feed and discharge. They suffer from high bearing wear and require significant maintenance. For these reasons, disk nozzles and imperforate basket centrifuges are being replaced by solid bowl centrifuges.

Solid bowl centrifuges (often referred to as *continuous decanter scroll* or *helical screw conveyor centrifuges*) are made in two basic configurations: countercurrent and concurrent. The primary differences between the two are

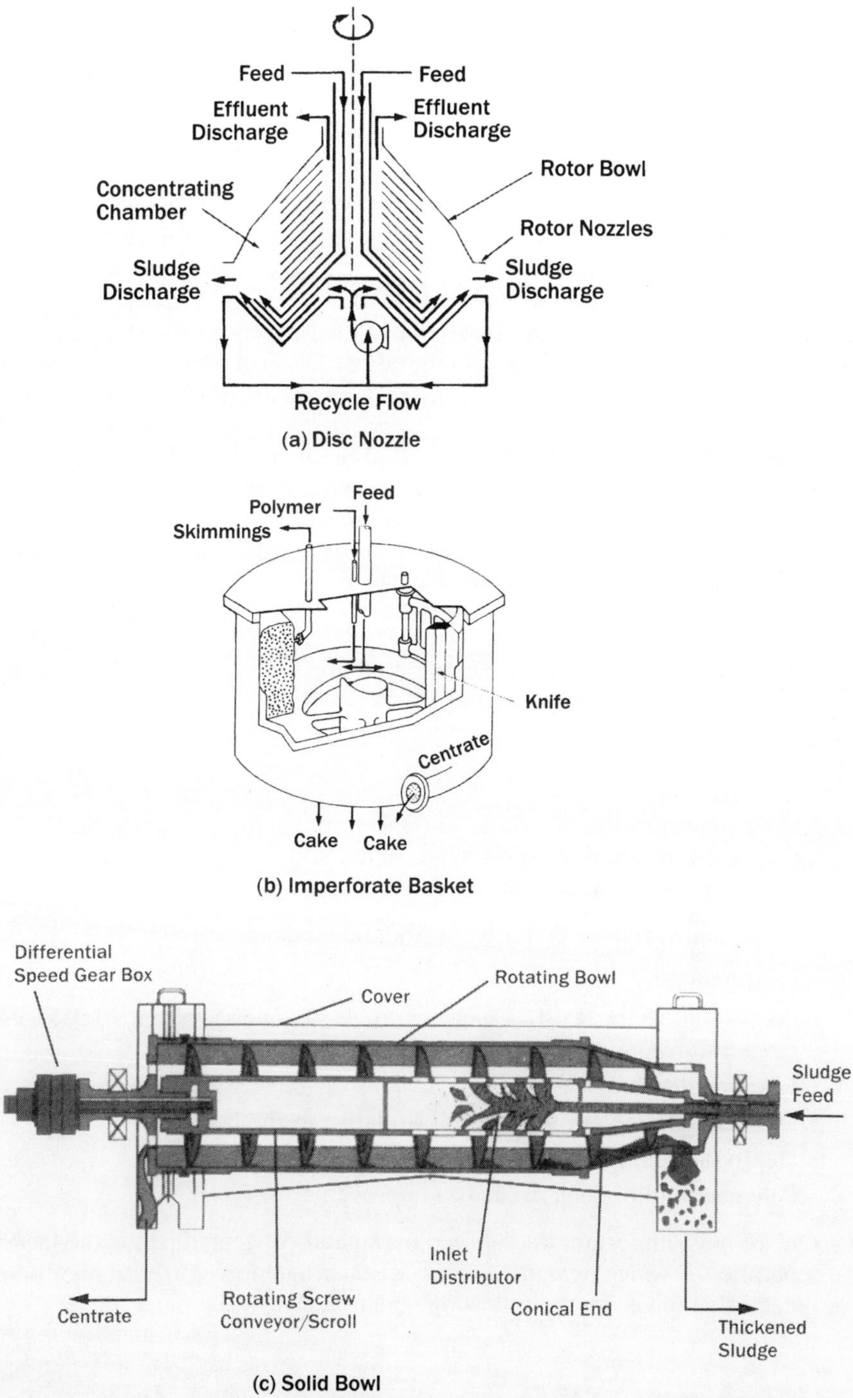

Figure 3.13 Schematics of centrifuges for thickening. [Part (c) from Centrisys, Kenosha, WI.]

the configuration of the conveyor (scroll) toward the liquid discharge end of the machine, and the location and configuration of the solids discharge port. Sludge feed enters the bowl through a concentric tube at one end of the centrifuge. The liquid depth in the centrifuge is determined by the discharge weir elevation relative to the bowl wall. The weir is typically adjustable. As the sludge particles are exposed to the gravitational field, they start to settle out on the inner surface of the rotating bowl. The lighter liquid (centrate) pools above the sludge layer and flows toward the centrate outlet ports located at the larger end of the machine. The settled sludge particles on the inner surface of the bowl are transported by the rotating conveyor (scroll) toward the opposite end (conical section) of the bowl. The main difference between a thickening and a dewatering centrifuge is in the construction of the conveyor and the conical part of the bowl. The slope of the conical part is less in a thickening centrifuge.

Performance of a centrifuge is measured by the thickened sludge concentration (sludge cake concentration for a dewatering centrifuge) and the solids recovery (often called *solids capture*). The recovery is calculated as the thickened dry solids as a percentage of the feed dry solids. Using the commonly measured solids concentrations, the recovery (capture efficiency) is calculated using the following equation (WEF, 1998):

$$R = \frac{C_k (C_s - T_c)}{C_s (C_k - T_c)} \times 100 \tag{3.6}$$

where

R = recovery, %
C_k = concentration of thickened (dewatered) sludge, % dry solids
C_s = concentration of feed sludge, % dry solids
T_c = concentration of centrate, % dry solids

Operational variables that affect thickening include:

- Feed flow rate
- Feed sludge characteristics, such as particle size and shape, particle density, temperature, and sludge volume index
- Rotational speed of the bowl
- Differential speed of the conveyor relative to the bowl
- Depth of the liquid pool in the bowl
- Polymer conditioning, needed to improve performance

One of most important operational parameters of centrifuges is the factor of separation F, which demonstrates how centrifugal forces are stronger than sedimentation forces by the following equation:

$$F = \frac{a}{g}, \qquad a = wr, \qquad \text{or} \qquad F = r\frac{n}{g} \tag{3.7}$$

where

F = separation factor
a = speed of centrifugal force, m/s^2
g = speed of sedimentation force, m/s^2
w = angle speed of bowl (rotor), min^{-1}
r = inside radius of bowl, m
n = speed of bowl (rotor) rotation, min^{-1}

Increasing the speed of bowl (rotor) rotation allows an increase in the factor of separation. However, the high bowl rotation speed can decrease the sludge particle sizes, increase the polymer demand, and decrease the effectiveness of flocculation. Therefore, centrifuges typically operate at speeds between 1500 and 2500 rpm, with the factor of separation between 600 and 1600. At the lower values of factor of separation, thickened sludge concentration and solids recovery values are lower.

The design features of centrifuges differ substantially among centrifuge manufacturers. Therefore, interrelationships of these variables will be different in each location, and specific design recommendations are not available. The most common approaches for estimating centrifugal thickening performance include bench-scale testing by centrifuge manufacturers and field pilot tests. Table 3.9 presents the performance of thickening centrifuges in various places in the world. Table 3.10 indicates performance at selected locations in North America.

TABLE 3.9 Performance of Solid Bowl Centrifuge for Thickening Activated Sludge

Country or Region	Dry Solids Conc. (%)	SVI	Polymer Use (g/kg dry solids)	Thickened Sludge Dry Solids Conc. (%)	Solids Recovery (%)
United States	0.2–0.6	110–190	None	1.9–7.9	47–91
	1.0–1.5	70–80	None	5–6	90–92
	0.3–0.8	100–190	0.20–0.25	3.6–10.0	57–97
	0.6–0.8	100–300	3.3–3.5	6–9	88–95
Canada	0.70–0.75	80–100	None	4.7–6.1	65–71
Europe	0.4–1.0	100–150	None	3.5–8.0	60–94
	0.4–1.5	100–150	0.5–1.0	4–9	80–85
Japan	0.3–0.6	—	None	4–6	80–86
Russia	0.5–0.6	80–120	None	4–7	63–90
	0.4–0.6	80–120	0.8–1.0	5.0–6.4	86–95
	0.7–1.2	100–140	0.8–1.0	6.0–7.5	81–86

TABLE 3.10 Reported Operating Results of Activated Sludge Thickening with Solid Bowl Centrifuges

Location	Activated Sludge Type	Feed Solids Concentration (mg/L)	SVI	Feed Flow Rate (L/m)	Thickened Solids Concentration (%)
Atlantic City, NJ	Air	3,000	100	1,230	10
Hyperion, Los Angeles, CA	Air	4,800–6,000	110–190	2,300–3,000	3.7–5.7
					3.6–6.0
	Air	4,800–6,000	110–190	2,300–3,000	1.9–7.9
					1.7–8.2
East Bay MUD, Oakland, CA	Oxygen	5,000	250–400	4,200	7
Naples, FL	Air	10,000–15,000	70–80	380	6
Jones Island, Milwaukee, WI	Air	6,000–8,000	80–150	1,100–1,900	3.0–5.5
Littleton, CO	Air	6,000–8,000	100–300	570–1,100	6–9
Lakeview, Ontario,	Air	7,560	80–120	840	4.7
Canada	Air	7,120	80–120	1,350	6.1

Source: WEF, 1998

Solid bowl centrifuges are quite versatile in their application. Major applications have been for thickening air or oxygen waste activated sludge. Aerobically and anaerobically digested sludge have also been thickened successfully by centrifuges. Particle size and particle distribution within the feed sludge have a significant effect on thickening performance. Naturally, well-flocculated solids in sludge often do not have sufficient strength to stay together under the high shearing forces encountered within a centrifuge. Therefore, polymer flocculation may be necessary to formulate tougher floc aggregate. Polymer use can improve solids capture efficiency typically to a range of 90 to more than 95%.

In general, significant design considerations for centrifuge thickening facilities include the following (WEF, 1998):

- Provide effective wastewater degritting and screening, or grinding. In case where wastewater screening or grinding is inadequate, sludge should be ground before feeding the centrifuge to avoid plugging.
- Use an adjustable-rate feed pumping with positive flow rate control from a feed source that is relatively uniform in consistency; a mixed storage or blend tank is appropriate.
- Thickened sludge should be handled by direct discharge to a well for subsequent pumping using a positive-displacement pump, direct discharge

TABLE 3.10 *Continued*

Solids Capture (%)	Polymer Use (g/kg)	Machine Size: Bowl Diameter × Length (mm)	Bowl Speed (rpm)	Centrifuge Configuration
95	2.5	740 × 2,340	2,600	Countercurrent
88–91	None	1,100 × 4,190	1,600	Cocurrent
77–96	0.2–2.2	1,100 × 4,190	1,600	Cocurrent
47–89	None	1,000 × 3,600	1,995	Countercurrent
57–97	0.4–1.4	1,000 × 3,600	1,995	Countercurrent
66	6	1,000 × 3,600	—	Countercurrent
90–92	None	740 × 3,050	2,000	Countercurrent
92–93	—	—	1,000	Cocurrent
88–95	3.3–3.5	740 × 2,340	2,300	Countercurrent
77	None	740 × 2,340	2,300	Countercurrent
65	None	740 × 3,050	2,600	Countercurrent

to an open-throat progressing cavity pump, or direct discharge to a screw conveyor.

- Discharge centrate to either primary or secondary treatment process; centrate handling may require a foam suppression system.
- Consider structural aspects such as static and dynamic loadings from the centrifuge, vibration isolation, and provision of an overhead hoist for equipment maintenance.
- Provide water for centrifuge flushing when equipment shutdown occurs.
- Consider the need for a heated water supply to flush grease buildup periodically.
- Provide proper centrifuge venting and consider the need for odor control.
- Consider struvite formation potential in anaerobic digesters when thickening anaerobically digested sludge.

3.3.4 Gravity Belt Thickening

Gravity belt thickening utilizes the principle of solid–liquid separation by coagulation and flocculation of solids and drainage of free water from the slurry through a moving fabric-mesh belt. It is dependent on conditioning of the sludge, typically with a cationic polymer to neutralize the negative charge of the biological solids.

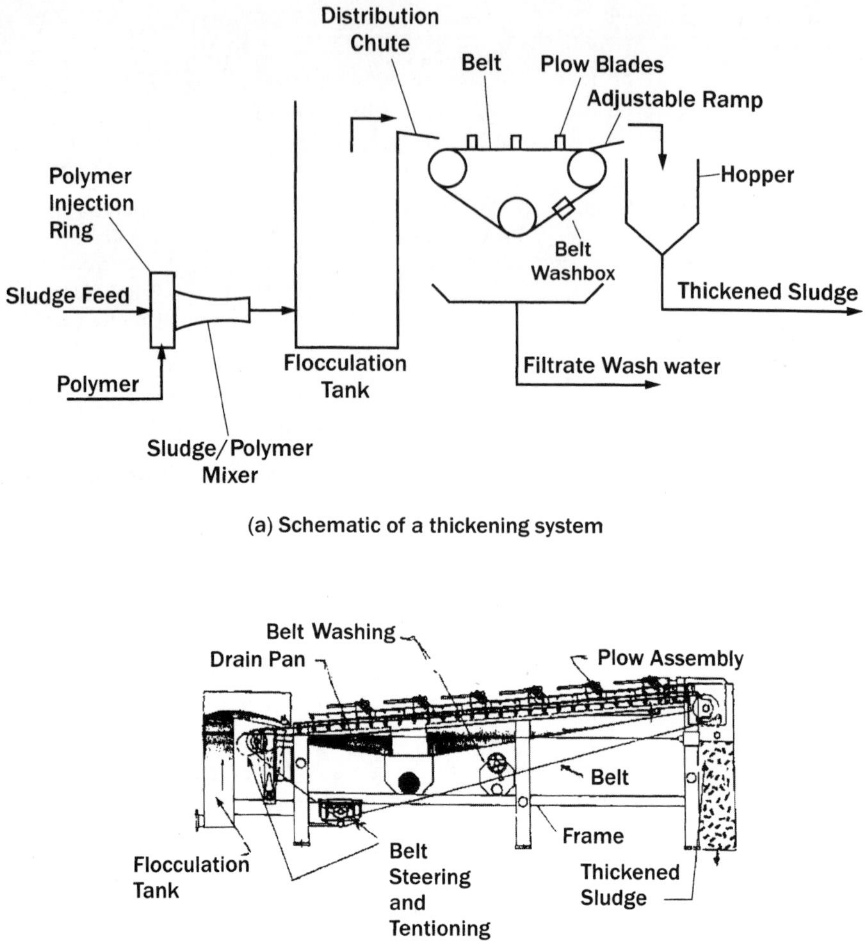

(a) Schematic of a thickening system

(b) Schematic cross section of a thickener

Figure 3.14 Gravity belt thickening.

The gravity belt thickener (GBT), introduced in 1980, is a modification of the upper gravity drainage section of the belt filter press. Figure 3.14 is a schematic of a GBT. Polymer is added and mixed with the feed sludge, normally with a polymer injection ring and variable orifice mixer installed in series in the inlet line, before it enters a flocculation tank. For heavy floc formation, polymer addition is essential to successful thickening. The amount of polymer used depends on the type of sludge to be thickened and the solids characteristics. The amount of mixing energy applied to the slurry to mix the polymer thoroughly also affects the formation of the floc.

From the flocculation tank, the conditioned slurry is distributed evenly across the width of the belt. Here the solids and the water begin to be physically separated. The solids–water mixture sits on the moving belt, allowing the water to drain through it. The water is collected in drain pans and routed to a sump. As the slurry moves on the belt, it is turned over by plough blades placed on top of the belt. The plough blades greatly increase the gravity thickening process by clearing places for the water to drain and by turning the solids mass on the porous belt to release additional water. The slurry is stopped from running off the sides of the belt by restrainers and rubber seals. An adjustable ramp located at the discharge end of the thickener provides an increase in residence time plus a shearing action to the solid material enhancing the thickening. The thickened sludge is removed from the belt by a scraper blade. The belt moves past the scraper to a washbox, where the belt is washed to remove the solids that have been forced into the porous of the weave.

The GBT can be used to thicken sludge of initial concentration as low as 0.4%. Solids capture efficiency of greater than 95% can be achieved, and as high as 99% has been reported. Other advantages of GBT include relatively moderate capital cost and relatively low power consumption. However, gravity belt thickening is polymer dependent. Polymer addition of 1.5 to 6 g/kg (3 to 12 lb/ton) on a dry weight basis is required. Other disadvantages include odor potential and the requirement of moderate operator attention to optimize polymer feed and belt speeds. A building is commonly required for the GBT to protect the thickening operation from inclement weather.

The GBT has been used for thickening waste activated sludge, aerobically and anaerobically digested sludge, and some industrial sludge. Table 3.11

TABLE 3.11 Gravity Belt Thickener Design Criteria and Performance

Type of Sludge	Dry Solids (%)	Dry Solids Loading		Polymer Dosage		Thickened Dry Solids (%)
		kg/m·h	lb/ft-hr	g/kg dry solids	lb/ton dry solids	
Primary (P)	2–5	900–1400	1980–3190	1.5–3	3–6	8–12
Secondary (S)	0.4–1.5	300–540	660–1200	3–5	6–10	4–6
Combined: 50% P & 50% S	1.0–2.5	700–1100	1540–2420	2–4	4–8	6–8
Anaerobically digested: 50% P & 50% S	2–5	600–790	1320–1740	3–5	6–10	5–7
Anaerobically digested: 100% S	1.5–3.5	500–700	1100–1540	4–6	8–12	5–7
Aerobically digested: 100% S	1.0–2.5	500–700	1100–1540	3–5	6–10	5–6

Source: Adapted from WEF, 1998.

presents design criteria and performance for various types of municipal wastewater sludge. Typical hydraulic loading rates are 380 to 900 L/min (100 to 250 gpm) per meter of effective belt width. GBTs for sludge thickening are available in 0.5-, 1.0-, 1.5-, 2.0-, and 3.0-m effective belt widths. Testing is recommended to verify that the sludge can be thickened at typical polymer dosages.

3.3.5 Rotary Drum Thickening

A rotary drum thickener (RDT), similar to a gravity belt thickener, achieves solid–liquid separation by coagulation and flocculation of solids and drainage of free water through a rotating porous media. The porous media can be a drum with wedge wires, perforations, stainless steel fabric, polyester fabric, or a combination of stainless steel and polyester fabric. Thickening is dependent on conditioning of the sludge, typically with a cationic polymer.

The thickener consists of an internally fed rotary drum with an integral internal screw, which transports the thickened sludge out of the drum. The drum rotates on trunnion wheels and is driven by a variable-speed drive. The polymer-conditioned sludge enters through an inlet pipe, where it is fed to the inside of the drum. The free water passes through the drum perforations into a collection trough, leaving the thickened sludge inside the drum. A spray bar extends the entire length of the drum to clean the drum to prevent blinding of the perforations. A cover, usually stainless steel, is provided for housekeeping and odor containment. Figure 3.15 shows a rotary drum thickener.

The RDT can be used to thicken sludge of initial concentration as low as 0.5%, with high solids capture efficiency. Advantages of RDT include fewer space requirements and relatively low capital cost and power consumption. In addition to being polymer dependent, it is sensitive to polymer type because of the shear potential of flocs in the rotating drum. Although there is odor potential, the unit is regularly enclosed to control odors. A building is commonly required for the RDT.

RDT can be used for thickening waste activated sludge, anaerobically and aerobically digested sludge, and some industrial sludge. It is typically used in small to medium-sized wastewater treatment plants. Units are available up to 1420 L/min (400 gpm) capacity. Table 3.12 presents typical performance data for rotary drum thickeners.

3.3.6 Miscellaneous Thickening Methods

Combined Thickening in Primary Clarifiers Primary clarifiers are often used for combined thickening of raw primary sludge and secondary sludge. This allows the blending of the two types of sludge on a continuous basis. Although combined sludge thickening has been used for WAS, it is more

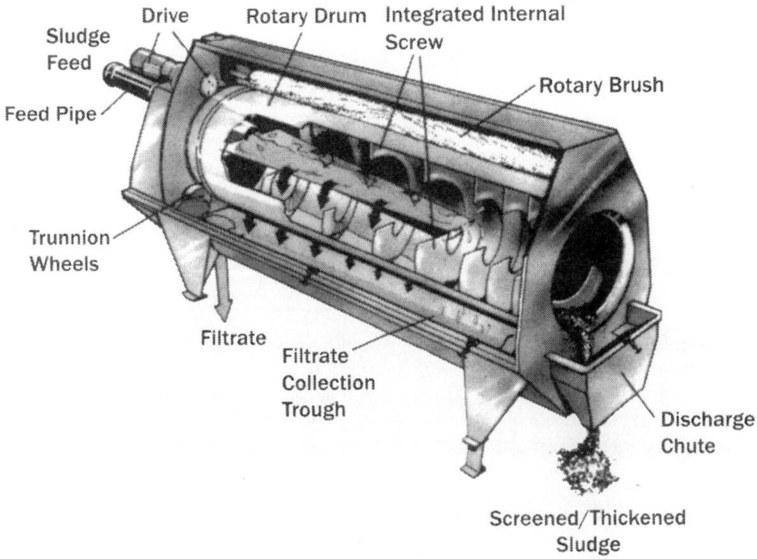

Figure 3.15 Rotary drum thickener. (From Parkson Corporation, Fort Lauderdale, FL.)

TABLE 3.12 Typical Performance Data for Rotary Drum Thickeners

Sludge Type	Feed Solids (%)	Water Removed (%)	Thickened Solids (%)	Solids Recovery (%)
Primary	3.0–6.0	40–75	7–9	93–98
WAS	0.5–1.0	70–90	4–9	93–99
Primary and WAS	2.0–4.0	50	5–9	93–98
Aerobically digested	0.8–2.0	70–80	4–6	90–98
Anaerobically digested	2.5–5.0	50	5–9	90–98
Paper fibers	4.0–8.0	50–60	9–15	87–99

Source: WEF, 1998.

feasible for sludge from an attached growth biological process such as trickling filter sludge. Figure 3.16 is a schematic of combined thickening. Clarifiers for combined thickening are often designed with steeper floor slope (as high as 2.75:12, which is standard for gravity thickeners) to reduce the sludge blanket depth over the sludge withdrawal point. However, a thicker sludge blanket means a longer retention time, which can cause septic conditions and gasification from biological action.

Lagoons Although out of favor with designers, facultative sludge lagoons can provide an effective means for further concentrating anaerobically digested sludge. They cannot function properly without major environmental

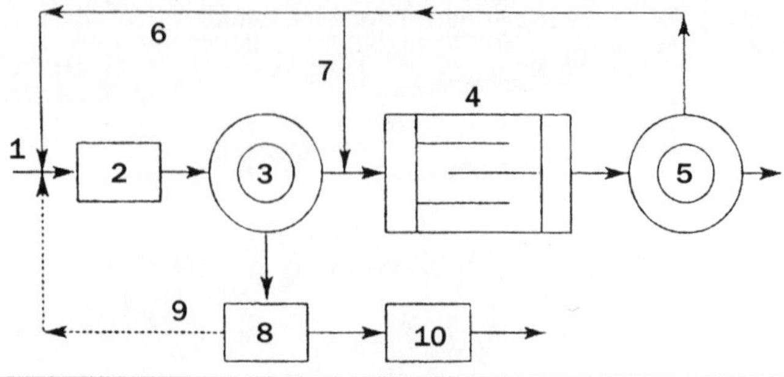

1. Wastewater Inlet
2. Preaeration tank
3. Primary settling tank
4. Aeration tank
5. Secondary settling tank
6. Activated sludge return line to preaerator
7. Activated sludge return line to aeration tank
8. Sludge mixture inlet for dewatering
9. Filtrate
10. Sludge preparation for utilization facility

Figure 3.16 Schematic of combined thickening.

impacts when supplied with either unstabilized or aerobically digested sludge. Advantages of facultative sludge lagoons include long-term storage of sludge, continuation of anaerobic sludge stabilization, low capital cost when land is readily available, low energy consumption, no chemical conditioning required, and the least operator attention and skill required. However, they are more land-intensive than other mechanical methods, may create odor problems, and produce a supernatant that contain 300 to 600 mg/L of TKN, mostly ammonia, which is a by-product of anaerobic autodigestion.

Facultative sludge lagoons are designed to maintain an aerobic surface layer free of scum or membrane-type film buildup. Surface mixers provide agitation and mixing of the surface layer. The surface layer is usually 0.3 to 0.9 m (1 to 3 ft) in depth and supports a dense population of algae. Dissolved oxygen is supplied to this layer by algal photosynthesis, by direct surface transfer from the atmosphere, and by the surface mixers. To maintain the aerobic top layer, organic loading rate to the lagoon must be at or below 1.0 ton of volatile solids per hectare per day (20 lb per 1000 ft^2 per day). The oxygen is used by the bacteria in the aerobic degradation of colloidal and soluble organic matter in the digested sludge liquor, while the solids settle to the bottom of the lagoon and continue their anaerobic decomposition. Supernatant from the lagoon is returned periodically to the treatment plant's liquid process stream. Decanting of supernatant allows initial sludge feed concentration of about 2% solids to thicken to a concentration of more than 6%. A part of the water is removed as a result of evaporation.

Emerging Technologies Many new technologies and equipment for sludge thickening being developed in Europe, Japan, and Russia include improved gravity thickeners, flotation thickeners, and belt thickeners. New materials for thickening, such as water-absorbing porous materials, stretched and elastic capillary materials, and nonwoven fibrous fabric materials, are being experimented with. Figure 3.17 is an experimental model of a *capillary thickener* with a water-absorbing porous material. Thickening of sludge occurs from water being absorbed by the capillary action of the porous material by the successive compression and decompression of the porous material. Thickened sludge formed on the surface of the material is removed by a scraper.

Figure 3.18 illustrates an experimental model of a *screw porofilter thickener* developed by Dr. Turovskiy. The thickener includes, in succession, a porous material, an electric heating zone, a filter cloth on a perforated sieve,

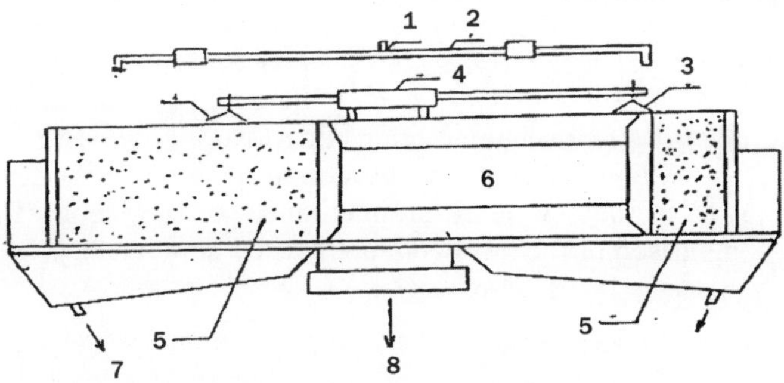

1 - Sludge feed, 2 - Distributor, 3 - Sludge scraper,
4 - Scraper drive, 5 - Porous media, 6 - Piston,
7 - Thickened sludge, 8 - Filtrate

Figure 3.17 Capillary thickener.

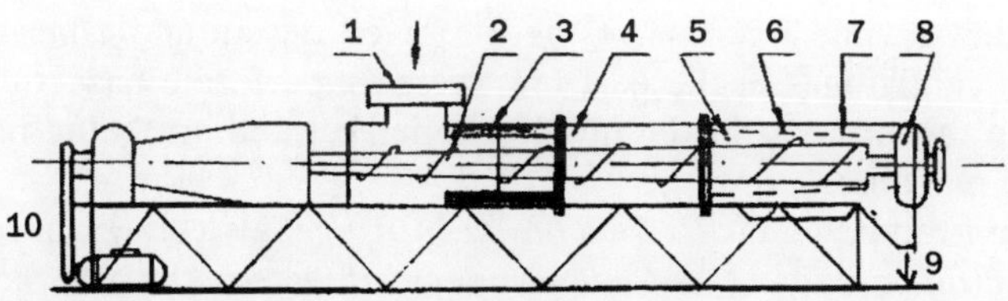

1 - Inlet, 2 - Screw conveyor, 3 - Porous media
4 - Heating zone, 5 - Final compression zone,
6 - Filter cloth, 7 - Perforated sieve, 8 - Locking
cone, 9 - Thickened sludge discharge, 10 - Drive.

Figure 3.18 Screw porofilter thickener.

and a locking cone to control back pressure. With this thickener, activated sludge with solids concentration of 0.4 to 0.8 has been thickened to 7 to 10% solids concentration.

3.4 DEWATERING

Dewatering is the physical operation of reducing the moisture content of sludge and biosolids to achieve a volume reduction greater than that achieved by thickening. Dewatering, because of the substantial volume reduction, decreases the capital and operating costs of subsequent handling of solids. Dewatering sludge and biosolids from a solids concentration of 4 to 20% reduces the volume to one-fifth and results in a nonfluid material.

The dewatering processes that are commonly used include mechanical processes such as centrifuges, belt filter presses, and pressure filter presses; and natural processes such as drying beds and drying lagoons. The main variables in any dewatering process are solids concentration and flow rate of the feed stream, chemical demand and solids concentration of dewatered sludge cake, and sidestream. The selection of particular process is determined by the type and volume of sludge to be dewatered, characteristics such as dryness required of the dewatered product, and space available. Table 3.13 presents a comparison of the most commonly used dewatering processes.

3.4.1 Centrifugal Dewatering

Dewatering of municipal sludge by centrifugation has been widely used in both the United States and Europe. Similar to its application in thickening, it is the process in which a centrifugal force of 500 to 3000 times the force of gravity is applied to sludge to accelerate the separation of the solids and the liquid.

Two basic categories of centrifuges are used for municipal wastewater sludge dewatering: imperforate basket and solid bowl. A third type, the disk nozzle centrifuge, has been used for thickening sludge but has seldom been used for dewatering. Because of the improved design and efficiency of the solid bowl centrifuges in the past few years, imperforate basket centrifuges have fallen out of favor in the municipal market and are being replaced by solid bowl machines.

The main components of a *solid bowl centrifuge* (also known as *continuous decanter scroll* or *helical screw conveyor centrifuge*) are the base, cover, rotating bowl, rotating conveyor scroll, feed pipe, gear unit, backdrive, and main drive (see Figure 3.11). The base provides a solid foundation to support the centrifuge. Vibration isolators below the base reduce the transmission of vibration from the centrifuge to its foundation. The cover that encloses the rotating bowl assembly completely serves as a safety guard. It also helps contain odors and dampens the noise.

TABLE 3.13 Comparison of Dewatering Processes

Process	Advantages	Disadvantages
Centrifuge	Relatively less space required Fast startup and shutdown capabilities Does not require continuous operator attention Clean appearance and good odor containment	Relatively high capital cost Consumes more direct power per unit of product produced Requires grit removal from feed sludge Requires periodic repair of scroll, resulting in long downtime Requires skilled maintenance personnel Moderately high solids concentration in centrate
Belt filter press	Relatively low capital, operating, and power costs Easier to shut down the system Easier to maintain	Very sensitive feed sludge characteristics Sensitive to polymer type and dosage rate Requires large quantity of belt wash water
Pressure filter press	High cake solids concentration Low suspended solids in filtrate A good dewatering process for hard-to-handle sludge Plates can be added to increase capacity without a significant increase in floor area	Batch operation High capital and labor costs Requires skilled maintenance personnel Often requires inorganic chemical conditioning that produces additional solids
Drying beds and drying lagoons	Low capital cost when land is readily available Low energy consumption Low to no chemical consumption Least operator attention and skill required	Large area requirement Requires stabilized sludge Design requires consideration of climatic effect Sludge cake removal is labor intensive Odor potential

The rotating bowl of the centrifuge consists of a cylindrical-conical design; the proportion of the cylindrical to conical shape varies depending on the manufacturer and the type of centrifuge. The conveyor scroll fits inside the bowl with a small clearance between its outer edge and the inner surface of the bowl. The conveyor rotates, but at a slightly lower or faster speed than the bowl. This difference in speed between the bowl and the conveyor scroll allows the solids to be conveyed from the zone of the stationary feed pipe where the sludge enters the bowl, to the dewatering beach, where the sludge cake is discharged. The dilute stream called *centrate* is discharged at the opposite end of the cake discharge port. The differential speed is controlled by the gear unit and the backdrive. Depending on the type of sludge, cake solids concentration varies from about 15 to 36%. Centrifuges are available with capacities as low as 40 L/min (10 gpm) to more than 3000 L/min (800 gpm).

Solid bowl centrifuges are available in both countercurrent and concurrent bowl designs (see Figure 3.19). In the countercurrent design, the sludge feed enters through the small-diameter end of the bowl, and the dewatered sludge cake is conveyed toward the same end. In the concurrent flow design, the sludge feed enters through the large-diameter end of the bowl, and the sludge cake is conveyed toward the opposite end.

Because of improvements in the design of the solid bowl centrifuges, cake solids concentrations in excess of 40% have been reported. These machines, known as *high-solids* (also called *high-torque*) *centrifuges* or *centripresses*, have a slightly longer bowl length, a reduced differential speed, higher torque, and a modified conveyor that presses the solids within the beach end of the centrifuge. These centrifuges may require higher polymer dosages to achieve higher cake solids concentrations.

Centrifuges can successfully dewater many different types of sludge. Operational variables that affect dewatering include the following:

- Feed flow rate
- Physicochemical properties of sludge, such as particle size and shape, particle density, temperature, and liquid viscosity
- Bowl/scroll geometry, such as bowl diameter, bowl length, conical bowl angle (beach), and conveyor shape and pitch
- Bowl speed (increasing bowl speed may decrease polymer use and increase cake dryness and solids recovery, but may shorten bearing life and increase equipment maintenance)
- Bowl/scroll differential speed
- Depth of liquid pool in the bowl
- Chemical conditioning

Design features of centrifuges substantially differ among centrifuge manufacturers. Therefore, the interrelationship of these variables will be different for each machine. The reliability of a centrifuge as a dewatering unit can only be evaluated by testing using a trailer-mounted test unit. From these test data, full-scale requirements of hydraulic and solids-loading capacities can be developed. Typical performance data for solid bowl centrifuges for different types of sludge are shown in Table 3.14.

Although many physical conditioning methods have been used for centrifugal dewatering, the most effective method has been polymer conditioning. Cationic polymers seem to be most effective for wastewater sludge. If aluminum and ferric salts are present, anionic polymers give a better result. Sludge can be dewatered using solid bowl centrifuges without the use of chemicals; however, it results in poor TSS capture.

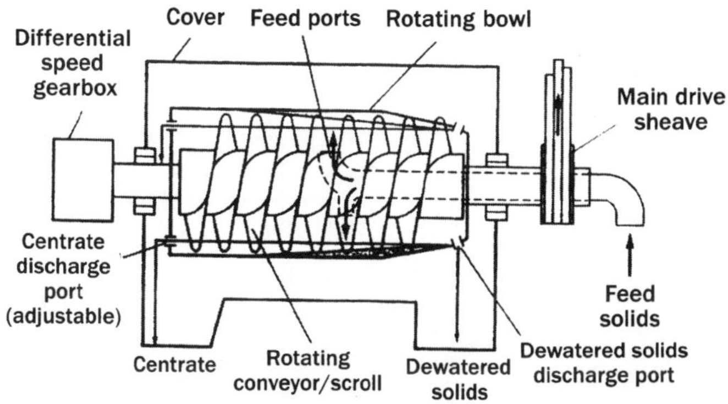

(a) Countercurrent

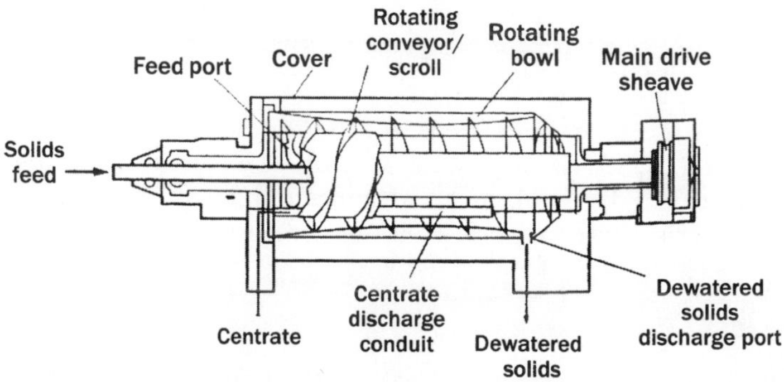

(b) Cocurrent

(c) Pictorial view of a typical installation

Figure 3.19 Solid bowl centrifuge dewatering. [Parts (a) and (b) from Metcalf & Eddy, 2003; Part (c) from Humboldt Decanker, Inc., Norcross, GA.]

TABLE 3.14 Operating Performance of Dewatering Solid Bowl Centrifuges

Sludge Type	Feed Solids Conc. (%)	Polymer Dosage g/kg dry solids	Polymer Dosage lb/ton dry solids	Cake Solids (%)	Solids Recovery (%)
Raw primary (PRI)	3–7	1–3	2–6	26–36	90–97
Raw WAS	0.5–2.5	4–8	8–16	8–20	85–94
Raw (PRI + WAS)	3–5	2–5	4–10	18–25	90–96
Anaerobically digested PRI	4–6	2.0–7.5	2–15	25–35	92–96
Anaerobically digested (PRI + WAS)	2–6	3–10	6–20	15–27	85–98
Aerobically digested WAS	1–3	1.5–5.0	3–10	8–12	88–91
Aerobically digested (PRI + WAS)	1.7–4.5	3.0–5.5	6–11	11–18	92–98

Source: Adapted in part from U.S. EPA, 1979.

Performance of a centrifuge is measured by the dewatered cake solids concentration and the TSS recovery (solids capture). The recovery is reported as the dewatered caked dry solids as a percentage of the feed dry solids. See equation (3.6) described for solids recovery for centrifugal thickening.

3.4.2 Belt Filter Press

A belt filter press (BFP) is a continuous-feed sludge dewatering machine with two porous moving belts that has a gravity drainage zone and mechanically applied pressure zones. Belt filter presses have been used in Europe first for dewatering paper pulp and later modified to dewater wastewater sludge. It was introduced in North America in the mid-1970s, mainly because of its ability to dewater secondary sludge economically, and its reduced energy requirements compared to centrifuges and vacuum filters. It is presently the most widely used dewatering equipment in the world.

Figure 3.20 shows schematics of belt filter presses commonly used in the United States and Europe. There are four basic processes in a BFP: conditioning, gravity drainage, low pressure compression, and high pressure compression. Polymer is injected into a sludge–polymer mixer placed in the feed line to the press. Flocculation occurs in a vertical column, a flocculation tank, or in a long stretch of pipe between the pipe mixer and the press. The conditioned sludge is then fed onto the gravity section of the belt via a feed chute. The solids–water mixture sits on the moving porous belt, allowing the water to drain through it. The water is collected in drain pans and routed to a sump. As the slurry moves on the belt, it is turned by plow blades. The plows greatly increase the gravity drainage process by clearing places for water to drain and by turning the solid mass on the belt. The slurry is stopped from running off the sides of the belt by restrainers and rubber seals. At the end of this gravity drainage section, the solids are usually a loosely structured cake.

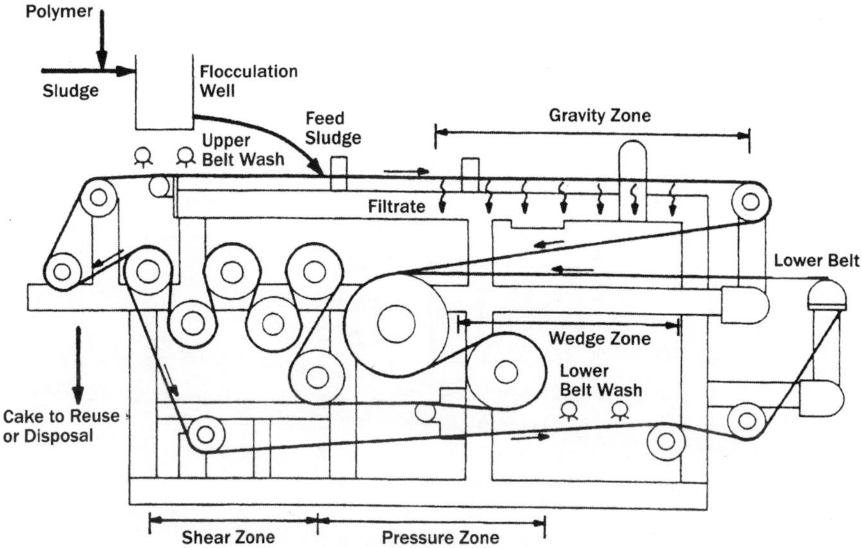

Figure 3.20 Schematic of a belt filter press. (From Ashbrook Corporation, 1992.)

From the gravity drainage section, the cake falls onto the bottom belt and begins to be compressed between the belts, forming a wedge. The belts compress the solids, thereby applying a low pressure. The sandwiched solid material then passes over a series of rollers applying pressure to the solids at a gradual rate to expel nearly all the free water from the slurry. At the last roller, the belts separate and the cake is removed from the belts by scrapers. The belts are washed, normally at two different locations, to remove the solids that have been forced into the pores of the weave. The filtrate and washwater are piped to the base of the machine for discharging into the sump.

A BFP can be used to dewater a variety of sludge of initial concentration as low as 1%. A solids capture efficiency greater than 95% can be achieved, and as high as 99% has been reported. Other advantages of the BFP include low capital and operating costs, low power requirements, and ease of maintenance. However, belt filter press dewatering is polymer dependent. Polymer addition of 1 to 10g/kg (2 to 20lb/ton) on a dry weight basis is required. Disadvantages of the BFP include being sensitive to incoming feed characteristics, hydraulically limited in throughput, and higher odor potential. Machines are presently available with covers to contain odors and ventilation through odor control equipment.

The most common size of BFPs is 2m in effective belt width, although machines as small as 0.5m and lately as large as 3.5m are available. Support systems for a BFP include sludge feed pumps, polymer feed equipment, sludge flocculation tank, sludge cake conveyor, and cake hopper. Figure 3.21 is a schematic of a belt filter press dewatering system.

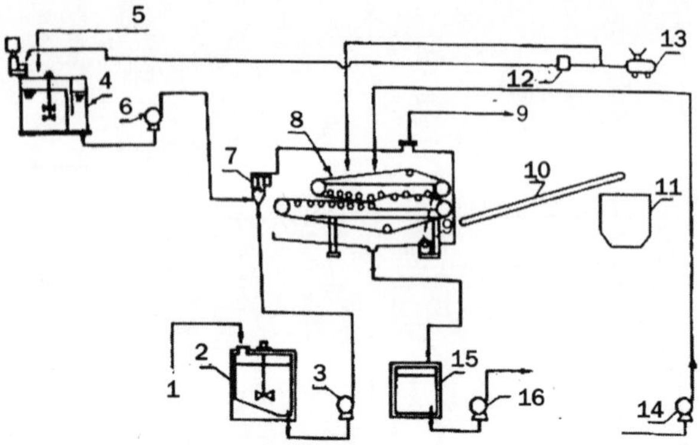

1 - Sludge Feed, 2 - Sludge day tank, 3 - Sludge feed pump,
4 - Polymer dilution tank, 5 - Polymer dilution water,
6 - Polymer feed pump, 7 - Flocculation tank, 8 - Belt filter
press, 9 - Air to odor control system, 10 - Cake conveyor,
11 - Cake dumpster, 12 - Air filter, 13 - Air compressor,
14 - Washwater pump, 15 - Filtrate tand, 16 - Filtrate pump.

Figure 3.21 Schematic of a belt filter press dewatering system.

Sludge loading rates for BFPs generally vary from 45 to 550kg/h·m (100 to 1200lb/hr-m), depending on the sludge type and feed concentrations. Hydraulic throughput generally ranges from 80 to 380L/min·m (20 to 100gpm/m). Polymer use of 1 to 10g/kg (2 to 20lb/ton) on a dry weight basis is required. Table 3.15 presents typical performance data for various type of sludge, and Table 3.16 presents the results of testing a BFP manufactured by a company in Russia for dewatering various types of sludge. The installation for this testing is shown in Figure 3.21.

Design Example 3.4 A wastewater treatment plant produces 50,000gpd (189,000L/d) of anaerobically digested combined primary and waste activated sludge. The digested sludge contains 2.8% solids. Design a belt filter press dewatering system for a normal 7-h/d, 5-d/wk operation. Assume the following operating parameters:

press loading rate:	600lb/m·h (272kg/m·h)
cake solids concentration:	22%
solids capture:	96%
wash water flow rate:	20gpm (75.6L/min)/m of belt width
specific gravity of feed sludge:	1.02
specific gravity of sludge cake:	1.05
specific gravity of filtrate:	1.00

TABLE 3.15 Typical Performance Data for a Belt Filter Press

Type of Sludge	Feed Solids (%)	Loading per Meter Belt Width				Polymer Dosage		Cake Solids (%)	
		L/min	kg[a]/hr	gpm	lb[a]/hr	g/kg[a]	lb/ton[a]	Typical	Range
Primary (PRI)	3–7	110–190	360–550	30–50	800–1200	1–4	2–8	28	26–32
WAS	1–4	40–150	45–180	10–40	100–400	3–10	6–20	15	12–20
PRI + WAS (50:50)	3–6	80–190	180–320	20–50	400–700	2–8	4–16	23	20–28
PRI + WAS (40:60)	3–6	80–190	180–320	20–50	400–700	2–10	4–20	20	18–25
PRI + Trickling filter	3–6	80–190	180–320	20–50	400–700	2–8	4–16	25	23–30
Anaerobically digested									
PRI	3–7	80–190	360–550	20–50	800–1200	2–5	4–10	28	24–30
WAS	3–4	40–150	45–135	10–40	100–300	4–10	8–20	15	12–20
PRI + WAS	3–6	80–190	180–320	20–50	400–700	3–8	6–16	22	20–25
Aerobically digested									
PRI + WAS unthickened	1–3	40–190	135–225	10–50	300–500	2–8	4–16	16	12–20
PRI + WAS (50:50) thickened	4–8	40–190	135–225	10–50	300–500	2–8	4–16	18	12–25
Oxygen-activated WAS	1–3	40–150	90–180	10–50	200–400	4–10	8–20	18	15–23

Source: Adapted from WEF, 1998.

[a] Dry solids.

TABLE 3.16 Results of Dewatering on a Belt Filter Press in Russia

Type of Sludge	Feed Solids (%)	Polymer Dosage g/kg dry solids	Polymer Dosage lb/ton dry solids	Belt Speed (m/min)	Cake Solids (%)	Solids Capture (%)
Primary (PRI)	4–7	2.0–7.0	4.4–14.0	1.7–2.2	25–30	99–99.6
WAS, thickened	2–3	8–12	16–24	2	18–24	92–96
PRI + WAS (50:50)	2.6–6.0	4–8	8–16	1.7–2.8	18–27	91–98
Aerobically digested WAS	2.0–4.4	3–8	6–16	2.6–3.0	18–20	96–97.5
Aerobically digested PRI + WAS	4–6	3.8–6.4	7.6–12.8	2.0–2.6	20–22	96–99.5

1. Weekly dry solids production rate

$$= (50{,}000\,\text{gpd})(8.34\,\text{lb/gal})(7\,\text{d/wk})(0.28)(1.02)$$
$$= 83{,}376\,\text{lb/wk}\ (37{,}825\,\text{kg/wk})$$

2. Hourly solids processing rate $= \dfrac{83{,}367\,\text{lb/wk}}{(5\,\text{d/wk})(7\,\text{h/d})}$

$$= 2382\,\text{lb/hr}\ (1081\,\text{kg/h})$$

Note: The 7-hour operation is one 8-hour shift less 1 hour for startup and cleanup.

3. BFP size required $= 2382\,\text{lb/hr}/600\,\text{lb/m}\cdot\text{h} = 3.97\,\text{m}$. Provide a total of three 2-m BFPs (two for duty, one for standby).
   Note: If space is limited, and also to save capital cost, two 2-m BFPs will be adequate if operation can be extended to two shifts a day if one of the two units is out of service. Operating time can also be extended as required for sustained peak sludge loads.
4. Check hydraulic loading rate

$$= \frac{(50{,}000\,\text{gpd})(7\,\text{d/wk})}{(5\,\text{d/wk})(7\,\text{hr/d})(60\,\text{min/h})(4\,\text{m total belt width})}$$
$$= 42\,\text{gpm}\ (159\,\text{L/min})/\text{m of belt width}$$

5. Sludge cake volume produced based on solids capture rate

$$= (2382\,\text{lb/hr})(7\,\text{hr/d})(0.96) = 16{,}007\,\text{lb/d}\ (7263\,\text{kg/d})$$

$$\text{sludge cake flow rate} = \frac{16{,}007\,\text{lb/d}}{(8.34\,\text{lb/gal})(0.22)(1.05)}$$
$$= 8309\,\text{gal/d}\ (31{,}440\,\text{L/d})$$

The specific weight of the sludge cake is about 50 lb/ft³. Therefore, the volume of sludge cake for storage or truck loading purposes is

$$\frac{16{,}007\,\text{lb/d}}{50\,\text{lb/ft}^3} = 320\,\text{ft}^3/\text{d} = 11.9\,\text{yd}^3/\text{d}\,(9.1\,\text{m}^3/\text{d})$$

6. Filtrate flow:

$$\begin{aligned}\text{daily sludge flow rate} &= (50{,}000\,\text{gal/d})(7\,\text{d/wk})(5\,\text{d/wk})\\ &= 70{,}000\,\text{gal/d}\,(264{,}978\,\text{L/d})\end{aligned}$$

$$\begin{aligned}\text{filtrate from sludge} &= (70{,}000 - 8309)\,\text{gal/d}\\ &= 61{,}696\,\text{gal/d}\,(233{,}544\,\text{L/d})\end{aligned}$$

$$\begin{aligned}\text{washwater flow} &= (20\,\text{gpm/m})(4\,\text{m})(60\,\text{min/hr})(7\,\text{hr/d})\\ &= 33{,}600\,\text{gal/d}\,(127{,}176\,\text{L/d})\end{aligned}$$

$$\begin{aligned}\text{total of filtrate and wash water flow} &= (61{,}696 + 33{,}600)\,\text{gal/d}\\ &= 95{,}296\,\text{gal/d}\,(360{,}733\,\text{L/d})\end{aligned}$$

7. TSS in filtrate = (2382 lb/hr) (7 hr/d) – 16,007 lb/d = 667 lb/d (303 kg/d)

$$\begin{aligned}\text{TSS concentration filtrate} &= \frac{667\,\text{lb/d}}{(8.34\,\text{lb/gal})(95{,}296\,\text{gal/d})}\times 100\\ &= 0.08\% = 800\,\text{mg/L}\end{aligned}$$

8. Filtrate flow can be computed assuming a suspended solids concentration in filtrate (normally, 500 to 1200 mg/L) instead of the solids capture rate, and then developing solids balance and flow rate equations, and solving the equations simultaneously. The following computations illustrate this. Assume the filtrate solids concentration to be 800 mg/L.

Solids balance equation:

solids in sludge feed = solids in cake + solids in filtrate

$$\begin{aligned}(2382\,lb/hr)(7\,h/d) &= (C\text{-gal/d})(8.34\,\text{lb/gal})(1.05\times 0.22)\\ &\quad + (F\text{-gal/d})(8.34\,\text{lb/gal})(0.0008)\end{aligned}$$

$$16{,}774 = 1.9265C + 0.0067F$$

where C is the cake flow rate (gal/d) and F is the filtrate flow rate (gal/d).

Flow rate equation:

sludge feed rate + washwater flow rate = $C + F$

$$\text{sludge feed rate} = (50{,}000\,\text{gal/d})(7/5) = 70{,}000\,\text{gal/d}\ (264{,}978\,\text{L/d})$$

$$\begin{aligned}\text{wash water flow rate} &= (20\,\text{gpm/m})(4\,\text{m})(60\,\text{min/hr})(7\,\text{hr/d})\\ &= 33{,}600\,\text{gal/d}\ (127{,}176\,\text{L/d})\end{aligned}$$

$$70{,}000 + 3600 = 103{,}600 = C + F$$

Solving the solids balance and flow rate equations simultaneously yields,

$$F = 95{,}276\,\text{gal/d}$$
$$C = 8324\,\text{gal/d}$$

$$\begin{aligned}\text{solids in filtrate} &= (95{,}276\ \text{gal/d})(8.34\,\text{lb/gal})(0.0008)\\ &= 636\,\text{lb/d}\ (288\,\text{kg/d})\end{aligned}$$

$$\begin{aligned}\text{solids capture} &= \frac{\text{solids in feed} - \text{solids in filtrate}}{\text{solids in feed}} \times 100\\ &= \frac{16{,}674 - 636}{16{,}674} \times 100\\ &= 96\%\end{aligned}$$

3.4.3 Pressure Filter Press

Pressure filter press dewatering is a batch process in which dewatering is achieved by forcing the water from the sludge under high pressure. It produces a cake that is drier than that produced by any other dewatering alternative. Another advantage is that the high solids capture results in good filtrate quality. Disadvantages include high capital cost, relatively high operation and maintenance costs, high chemical costs, and a large area requirement for the equipment in small wastewater treatment plants. The area requirements for large treatment plants are relatively small because presses are available with large plates that dewater several tons of sludge per hour. Additional plates can be added to a press without a significant increase in floor area.

Among the various types of pressure filter presses that have been used for dewatering sludge, the two most commonly used are the fixed-volume recessed plate filter press and the variable-volume recessed plate filter press.

Fixed-Volume Recessed Plate Filter Press The fixed-volume press consists of a series of recessed plates held in a frame and pressed together either hydraulically or electromechanically between a fixed end and a moving end (see Figure 3.22). Volume for the sludge to be dewatered is provided by the recesses on both sides of the plates. A filter cloth is mounted over the two surfaces of each plate. The plates have drainage ports between the filter cloths and the plate surfaces to drain the filtrate.

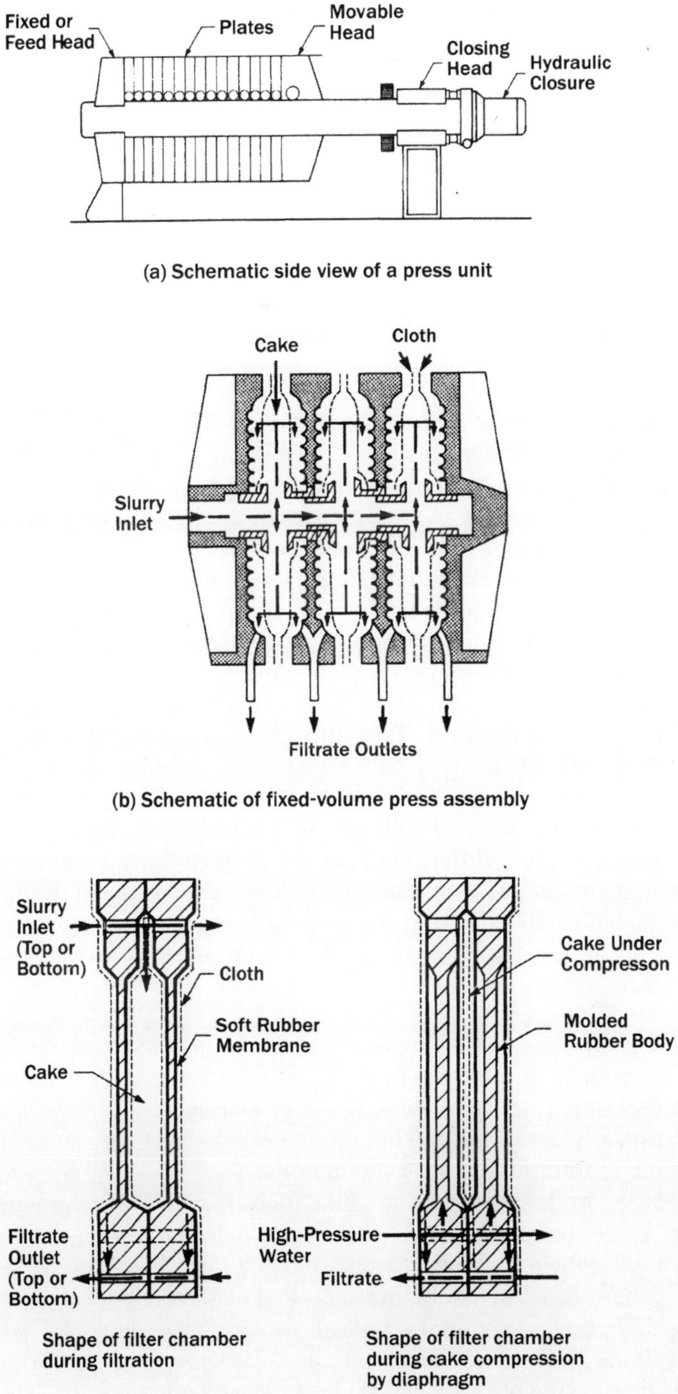

(a) Schematic side view of a press unit

(b) Schematic of fixed-volume press assembly

(c) Schematic of a variable volume press assembly

Figure 3.22 Recessed plate filter press.

A typical dewatering cycle begins with the closing of the plates. Chemically conditioned sludge is fed until the flow rate drops to 5 to 7%: usually, 20 to 30 minutes. The pressure at this point is generally the desired maximum, typically 226 to 1550 kN/m^2 (40 to 225 psi) and is maintained for 1 to 3 hours. During this time, more filtrate is removed and the desired cake solids concentration is achieved. The plates are then separated mechanically, and the cake is dropped from the chambers onto a truck or conveyor belt for removal. The cake thickness varies from 25 to 38 mm (1 to 1.5 in.) and solids concentration varies from 35 to 50%. Following cake removal, the filter press is washed and ready for the next cycle. Cycle time varies from 1.5 to 4 hours.

Variable-Volume Recessed Plate Filter Press The variable-volume press, commonly called the *diaphragm press*, is similar to the fixed-volume press except that a diaphragm is placed behind the media as shown in Figure 3.22. A dewatering cycle begins as conditioned sludge is fed into each recessed chamber at a sustained pressure of about 690 to 860 kN/m^2 (100 to 125 psi). Once the chambers are filled and the filter cake formation is started, the sludge feed pump is turned off automatically based on a set instantaneous sludge flow rate, filtrate flow rate, or time. Water or air under high pressure (normally, water, because of safety concerns) is then pumped into the space between the diaphragm and the plate, squeezing the already formed cake and releasing additional water from the cake. Typically, 15 to 30 minutes of constant pressure at 1380 to 2070 kN/m^2 (200 to 300 psi) are required to dewater the cake to the desired solids content. At the end of the cycle, the water is returned to a reservoir, plates are automatically opened, and sludge cake is discharged.

The diaphragm press differs significantly from the fixed-volume press in that the volumetric capacity is generally less, the cake is much thinner, there is greater flexibility in achieving the desired level of cake dryness, and the press is highly automatic. However, a diaphragm press requires considerable maintenance.

Performance Pressure filter press is an advantageous choice for sludge of poor dewaterability or for cases in which it is desirable to dewater sludge to solids content of greater than 35%. If sludge characteristics are expected to change drastically over a normal operating period, variable-volume units are recommended rather than fixed-volume units.

Pilot testing and scale-up from pilot to full scale is recommended for designing a dewatering system. All major manufacturers have small skid-mounted units that are simple to test and operate. Pilot testing generally results in cake solids 2 to 4% higher than that achieved in a full-scale unit.

Tables 3.17 and 3.18 contain typical performance data for fixed- and variable-volume units, respectively. Table 3.19 shows the performance of filter press dewatering of municipal sludge in Russia and Ukraine. These data indicate significant variations that reflect differences in filtration characteristics of different types of sludge.

TABLE 3.17 Performance Data for a Fixed-Volume Filter Press

Type of Sludge	Feed Solids (%)	Conditioning Chemical[a] FeCl$_3$ (%)	Lime[b] (%)	Cake Solids[c] (%)	Cycle Time[d] (h)
Primary (PRI)	5–10	5	10	45	2
Primary + WAS (50:50)	3–6	5	10	40–45	2.5
Primary + WAS (50:50)	1–4	6	12	45	2.5
Primary + trickling filter	5–6	6	20	38	2
Primay + ferric chloride (FC)[e]	4	—	10	40	1.5
Primary + WAS (FC)[e]	8	5	10	45	3
WAS	5	7.5	15	45	2.5
Primary + two-stage high lime	7.5	—	—	50	1.5
Digested (PRI)	8	6	30	40	2
Digested (PRI + WAS)	6–8	5	10	45	2
Digested (PRI + WAS) (FC)[e]	6–8	5	10	40	3
Digested (PRI + WAS) (50:50)	6–10	5	10	45	2
Digested (PRI + WAS) (50:50)	1–5	7.5	15	45	2.5

Source: Adapted from WEF, 1998.

[a] As % of dry solids.

[b] As CaO

[c] Includes conditioning chemicals.

[d] Length of time from initiation of feed to fed pump termination; excludes cake discharge time.

[e] Ferric chloride used as a coagulant aid in the secondary process.

TABLE 3.18 Performance Data for Variable-Volume Diaphragm Filter Press

Type of Sludge	Feed Solids (%)	Conditioning Chemical[a] FeCl$_3$ (%)	Lime (%)	Cake Solids (%)	Total Cycle Time (min)
Primary sludge(Metropolitan Wastewater Commision Minneapolis–St. Paul, MN)	12	2.9	13	48	21
Heat-treated primary + WAS (25:75) (Metropolitan Wastewater Commision Minneapolis–St. Paul, MN)	12–16	0	0	52	22

Source: Adapted from WEF, 1998.

[a] As % of dry solids.

TABLE 3.19 Performance of Filter Press Dewatering of Municipal Sludge in Russia and Ukraine

Type of Sludge	Feed Solids (%)	Conditioning Chemical[a] FeCl$_3$ (%)	Conditioning Chemical[a] Lime (%)	Filter Dry Solids Capacity (kg/m^2·cycle)	Cake Solids (%)
Raw primary (PRI)	3–6	2.4–4.2	7.2–10.8	12–16	35–42 (42–48)[b]
Raw WAS	1–2	4–8	12–24	6–10	25–30
Digested primary	3.5–6.0	3.6–4.8	9.6–12.0	12–17	35–40 (42–48)[b]
Digested WAS	2.0–3.5	4–8	10–24	6–10	25–30 (29–35)[b]
Mesophilically digested (PRI + WAS)	3–5	4.8–7.2	12–18	10–16	32–40 (36–42)[b]
Thermophilically digested (PRI + WAS)	3–5	4.8–7.2	12–18	7–13	30–38
Mesophilically digested (PRI + WAS)	3–4	Polymer 0.2–0.4		7–11	24–30

[a] As % of dry solids.
[b] Dewatered with variable diaphragm filter press.

Design Considerations The principal design elements in a filter press dewatering facility include chemical conditioning system, precoat system, sludge feed system, washing system, and cake handling. Standby capacity to overcome excessive downtime for equipment maintenance, adequate clear space around the machine, parts and material-handling equipment such as a bridge crane, adequate ventilation of the building, and a sludge grinder ahead of the conditioning tank are other elements that should be incorporated into the design.

For conditioning, most facilities use ferric chloride and lime added in batches to sludge contained in an agitation tank and the conditioned sludge is pumped from the tank to the filter press. The lime system requires a silo, a slaker, lime slurry tank, and feed pumps. Ferric chloride handling requires a storage tank and feed pumps.

Sludge with high biological content or some industrial sludge often has a tendency to stick to the filtration media. In such instances, a precoat system aids cake release from the filtration media and protects it from premature blinding. Precoat material can be fly ash, incinerator ash, diatomaceous earth, or cement-kiln dust and is pumped in slurry form to the chambers of the filter press before the start of each filtration cycle.

The sludge feed system should be designed to pump 1.9 to 126 L/s (30 to 2000 gpm) of a viscous, abrasive slurry at pressures of 276 to 1550 kN/m^2 (40 to 225 psi). Reciprocating ram high-pressure pumps with variable-speed drives are generally used in such a system. An alternative is to use a combination of pumps and pressure vessels.

Because recessed plate pressure filters operate at high pressures and because many use lime for conditioning, the filter cloth will require routine washing with high-pressure water as well as periodic washing with acid to remove the scale buildup from the clothes and from the walls of the plates. Practices vary according to the particular sludge and type of filter cloth. Some facilities wash with water after 20 cycles and wash with acid after 100 cycles.

Providing suitable cake breakers is a function of the structural properties of the cake and the ultimate solids disposal method. Cake dropped from the filter press is usually friable enough that use of breaker wires or bars beneath the press is sufficient, especially when cake is dropped directly onto trucks for hauling away. If further processing of sludge cake, such as incineration, is required, additional cake breaking equipment, such as a cake breaker screw conveyor, needs to be provided.

3.4.4 Drying Beds

Drying beds are the most widely used method of municipal wastewater sludge dewatering in the United States. They have been used for more than 100 years. Although the use of drying beds might be expected in small plants and in warmer, sunny regions, they are also used in several large facilities and in northern climates. In the United States, a majority of wastewater treatment plants with less than 5-mgd capacity use drying beds for biosolids dewatering. In Russia and other Eastern European countries, more than 80% of the plants use drying beds.

The main advantages of sludge drying beds are low capital cost, low energy consumption, low to no chemical consumption, low operator skill and attention required, less sensitivity to sludge variability, and higher cake solids content than that of most mechanical methods. Disadvantages include large space requirements, the need for prior sludge stabilization, consideration of climatic effects, odor potential, and the fact that sludge removal is usually labor intensive. Sludge drying beds may be classified as: (1) conventional sand, (2) paved, (3) artificial media, and (4) vacuum assisted.

Conventional Sand Drying Beds Conventional sand drying beds are the oldest and most commonly used type of drying bed. They are generally used for communities with populations of fewer than 20,000. Many design variations are possible, including the layout of drainage piping, the thickness and type of gravel and sand layers, and construction materials.

Figure 3.23 shows a typical sand drying bed construction. Current practice is to divide the bed into multiple rectangular cells, each with dimensions of 4.5 to 18 m (15 to 60 ft) wide by 15 to 47 m (50 to 150 ft) long. The cells are sized such that one or two beds will be filled in a normal loading cycle. The outer walls may be earth embankments or vertical walls constructed of concrete blocks or reinforced concrete. The interior partitions may be constructed of concrete blocks, reinforced concrete, or planks stretching between slots in concrete posts. The planks can be wood, but the preferable material is precast

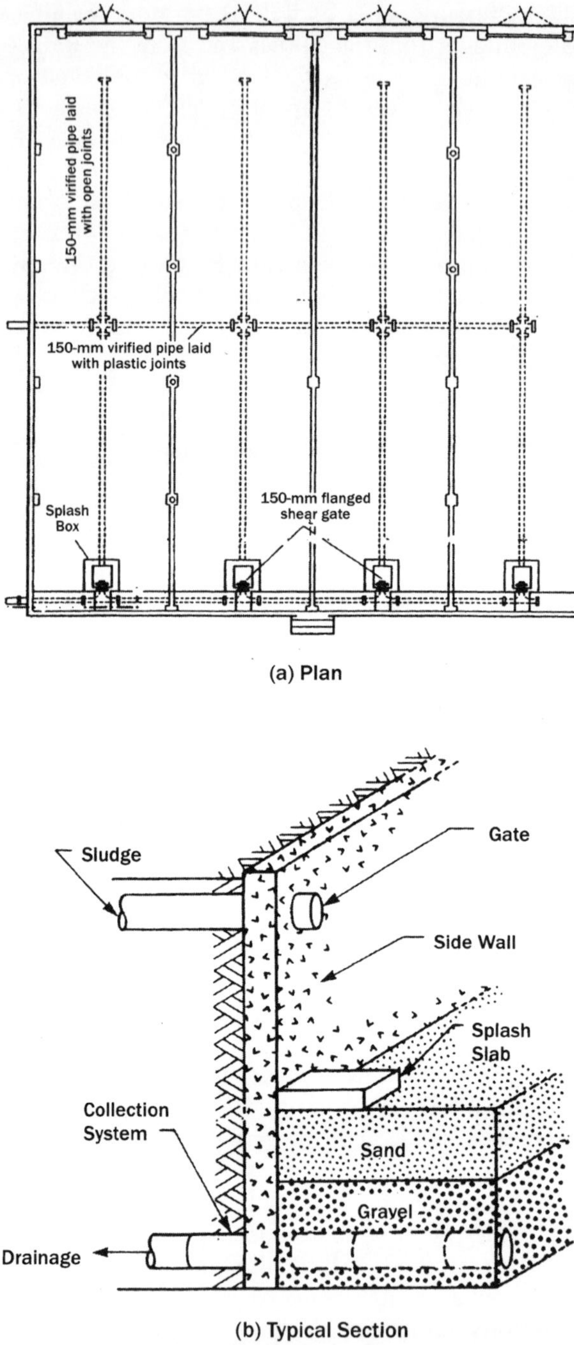

Figure 3.23 Conventional sand drying bed. (Plan from Metcalf & Eddy, 2003.)

concrete. If mechanical equipment such as a wheeled front-end loader is used for cake removal, at least one solid vertical wall in each cell against which the loader can push will speed bed cleaning.

Usually, 230 to 380mm (9 to 15in.) of sand is placed over 200 to 460mm (8 to 18in.) of graded gravel. A thicker sand layer secures a good filtrate and reduces the frequency of sand replacement caused by losses from cleaning operations. However, a deeper sand layer generally retards the draining process. The sand is usually 0.3 to 0.75mm (0.01 to 0.03in.) in effective diameter and has a uniformity coefficient of less than 4.0. Gravel is normally graded from 3 to 25mm (0.1 to 1.0in.) in effective diameter.

Underdrain piping is perforated plastic or vitrified clay pipe laid with open joints (without gaskets) and should take into account the type of sludge removal vehicles to be used to avoid damage to the pipes. The lateral underdrains pipes feeding into the main underdrain pipe should be spaced 2.4 to 6m (8 to 20ft) apart. The pipes should not be less than 100mm (4in.) in diameter and should have a minimum slope of 1%.

Sludge is applied to the cells of the sand drying bed through a pressurized pipe with a valved outlet to each cell or with a shear gate at the end of the outlet to each cell. Provisions should be made to flush the piping and, if necessary, to prevent it from freezing in cold climates. Sludge can also be applied through an open channel with an inlet to each cell controlled by a slide gate. With either type, a concrete splash block should be provided to receive the falling sludge and to prevent erosion of sand surface.

Sludge drying beds with greenhouse-type enclosures allow dewatering sludge throughout the year, regardless of the weather. They may also eliminate potential odor or insect problems. Sometimes a roof is placed over the top of the drying bed, leaving the sides open. Such a cover protects the bed from precipitation but provides little temperature control. Because of better temperature control, completely covered beds, require about 25 to 30% less area than open beds.

Sludge drying bed sizing criteria are given in unit area of bed required for dewatering on a per capita basis. These criteria are only valid for the characteristics of a particular wastewater and have no rational design basis. A better criterion for sizing the bed is the unit loading of kilograms of dry solids per square meter per year (pounds of dry solids per square foot per year) for a particular type of sludge. Table 3.20 shows the criteria for various types of biosolids. The criteria selected should take into consideration climatic conditions such as temperature, wind velocity, and precipitation; biosolids characteristics such as grit, grease, and biological content; and solids concentration.

The depths of application range from 200 to 400mm (8 to 16in.). The applied depth should result in an optimum solids loading of 10 to 15kg/m^2 (2 to 3lb/ft^2). The total drying time required depends on the desired cake dryness. In addition to the water draining through the sand bed, moisture is removed by evaporation also. The time require for evaporation of moisture

TABLE 3.20 Sand Drying Bed Design Criteria for Digested Sludge

Type of Biosolid	Uncovered Beds: Area		Uncovered Beds: Dry Solids Loading		Covered Beds Area	
	m^2/capita	ft^2/capita	$kg/m^2 \cdot yr$	lb/ft^2-yr	m^2/capita	ft^2/capita
Primary	0.09–0.12	1.0–1.3	120–150	25–30	0.07–0.09	0.8–1.0
Primary + chemical	0.18–0.23	1.9–2.5	100–160	20–33	0.09–0.17	1.0–1.8
Primary + trickling filter	0.12–0.17	1.3–1.8	90–120	18–25	0.09–0.14	1.0–1.5
Primary + WAS	0.16–0.23	1.7–2.5	60–100	12–20	0.12–0.16	1.3–1.7

Source: Adapted from U.S. EPA, 1979.

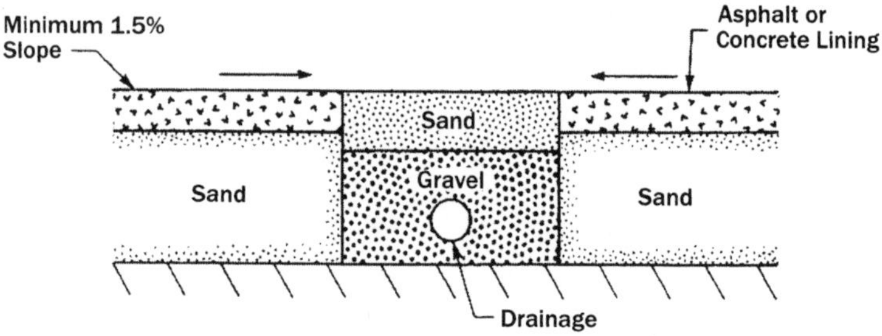

Figure 3.24 Cross section of a paved drying bed.

is considerably longer than that require for drainage. Therefore, the time the sludge must remain on the bed is determined by the amount of water that must be removed by evaporation. The drying time is shorter in regions that experience low rainfall and humidity and greater sunshine. Under favorable conditions, sludge may be dried to a solids content of about 40% after 10 to 15 days. As discussed previously, natural freezing and thawing in northern climates have been reported to improve dewaterability of sludge. Dried sludge has a coarse, cracked surface and is dark brown. At small treatment plants, sludge is usually removed by manual shoveling into wheelbarrows or trucks. At larger plants, a scraper or front-end loader, or special mechanical cake removal equipment, is used.

Paved Drying Beds Paved drying beds use concrete or asphalt lining. Normally, the lining rests on a 200- to 300-mm (8- to 12-in.)-thick built-up sand or gravel base. Figure 3.24 shows typical paved drying bed construction. The beds are normally rectangular in shape and are 6 to 15m wide (20 to 50ft) by 20 to 45 m (70 to 150 ft) long with vertical sidewalls. The lining should have a minimum 1.5% slope to a center drainage area that is not paved. The drainage area is 0.6 to 1 m (2 to 3 ft) wide. A minimum 100-mm (4-in.)-diameter perforated pipe would convey the drainage away. The main advantages of paved drying beds are that front-end loaders can be used for easy removal of sludge cake, augur mixing vehicles can speed up drying, and bed maintenance is reduced. However, for a given amount of sludge, paved drying beds require more area than do conventional sand drying beds.

Artificial Media Drying Beds Two types of artificial media can be used for drying beds: stainless steel wedgewire or high-density polyurethane panels. Wedgewire drying beds have been used successfully in England for about 50 years and in the United States for about 30 years. Figure 3-25 shows a typical cross section of a wedgewire bed. The bed consists of a shallow rectangular watertight basin fitted with a false floor of wedgewire panels. The slotted openings in the panels are 0.25 mm (0.1 in.) wide. An outlet valve to control the rate of drainage is located underneath the false floor.

The procedure used for dewatering sludge begins by introducing water, usually plant effluent, onto the surface of the bed to fill the septum and the wedgewire to a depth of approximately 25 mm (1 in.). This water serves as a cushion that permits the added sludge to float without causing upward or downward pressure across the wedgewire surface. When the bed is filled with sludge, the water permits the sludge to settle and initially compact against the screen so that the settled sludge acts as the filtration media. Next, the water is allowed to percolate at a controlled rate by controlling the outlet valve. After the free water has been drained, the sludge further concentrates by drainage and evaporation until it is ready for removal. In a high-density polyurethane media system, 300 mm (12-in.)-square panels with a built-in underdrain system are paved over a sloped slab. Each panel has an 8% perforated area for dewatering.

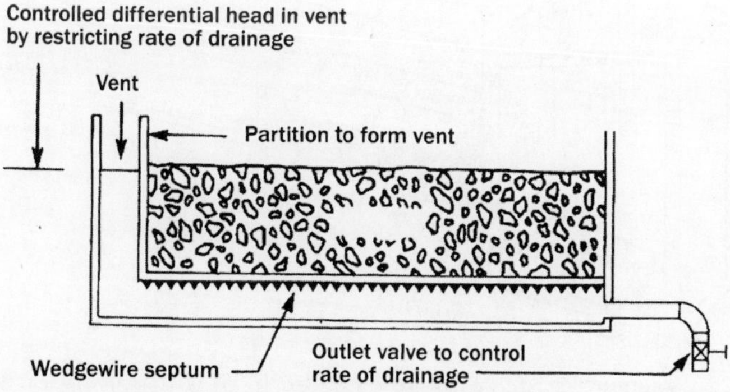

Figure 3.25 Cross section of a wedgewire drying bed.

Advantages of artificial media drying beds include (1) no clogging of the media, (2) constant and rapid drainage, (3) higher throughput rate than with sand beds, (4) easy bed maintenance, and (5) difficult-to-dewater sludges such as aerobically digested waste activated sludge can be dried. Drying beds with polyurethane panels have the added advantage of (1) dewatering dilute sludge, (2) low suspended solids in filtrate, and (3) easy removal of sludge cake possible with a front-end loader.

Artificial media beds can typically dewater 2.5 to 5.0 kg of solids per square meter (0.5 to 1.0 lb/ft^2) per each application. Most types of sludge dry to a concentration of 8 to 12% within 24 hours. However, such a sludge cake is still relatively wet and thus may complicate disposal.

Vacuum-Assisted Drying Beds Vacuum-assisted drying beds (see Figure 3.26) accelerate dewatering and drying from the application of vacuum to the underside of the porous filter plates. The bed is usually rectangular and has a reinforced concrete slab at the bottom. A layer of aggregate several millimeters thick is placed on top of the slab, which in turn supports a rigid multimedia porous filter top. The aggregate layer is also the vacuum chamber and is connected to a vacuum pump. The operating sequence is as follows:

- The sludge is preconditioned with a polymer.
- Sludge is introduced onto the drying bed by gravity flow at a rate of 9.4 L/s (9150 gpm) and to a depth of 300 to 750 mm (12 to 30 in.).

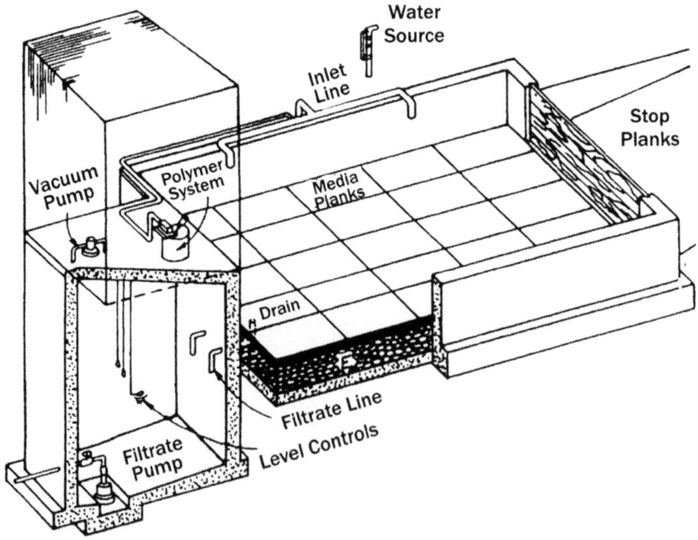

Figure 3.26 Vacuum-assisted drying bed. (Reprinted with permission from WEF, 1998.)

TABLE 3.21 Typical Performance Data for Vacuum-Assisted Sludge Drying Beds

Type of Sludge	Feed Solids (%)	Dry Solids Loading		Cycle Time (h)	Polymer Dosage		Cake Solids (%)
		kg/m³	lb/ft²		g/kg	lb/ton	
Anaerobically digested							
Primary	1–7	10–20	2–4	8–24	2–20	4–40	12–26
Primary + WAS	1–4	5–20	1–4	18–24	15–20	30–40	15–20
Primary + TF	3–10	15–30	3–6	18–24	20–26	40–52	20–26
Aerobically digested							
Conventional WAS	1–4	5–15	1–3	8–24	1–17	2–34	10–23
Oxidation ditch WAS	1–2	5–10	1–2	8–24	2–7	4–14	10–20

Source: Adapted from WEF, 1998.

- After the sludge is applied, it is allowed to drain by gravity for about 1 hour.
- At the end of the gravity drainage period, the vacuum system is started and it maintains a vacuum at 34 to 84 kPa (910 to 25 in.Hg).
- When the cake cracks and the vacuum is lost, the vacuum pump is shut off.
- The sludge is allowed to air dry for 1 to 2 days.
- The cake is removed from the bed using a front-end loader.
- The surfaces of the media plates are washed with a high-pressure hose.

Table 3.21 lists performance data for vacuum-assisted drying beds for various types of sludges. The principal advantages of this system are: the cycle time for sludge dewatering is short, thereby reducing the effects of weather on sludge drying; a smaller area is required; and the sludge cake is easily removable using a small front-end loader. Because of their small area requirements, these beds can be covered more easily for use in colder or wetter climates. The principal disadvantage is that polymer conditioning of sludge is required for successful operation.

3.4.5 Other Dewatering Methods

Vacuum Filters Since its first introduction in the United States in the mid-1920s, thousands of vacuum filters have been installed in municipal wastewater treatment plants for dewatering sludge. In vacuum filtration, a vacuum applied downstream of the media is the driving force on the liquid phase that moves it through the porous media. The medium can be natural or synthetic fiber cloth, woven stainless steel mesh, or coil springs. Figure 3.27 is a cross-sectional view of a rotary vacuum filter with a coil-spring filter. The unit consists of a horizontal drum that rotates, partially submerged, in a vat of conditioned sludge. The drum surface is divided into sections around its cir-

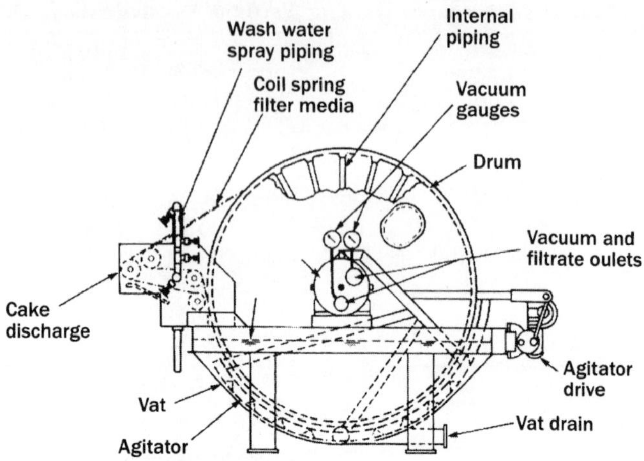

(a) Cross section rotary vacuum filter with coil spring media

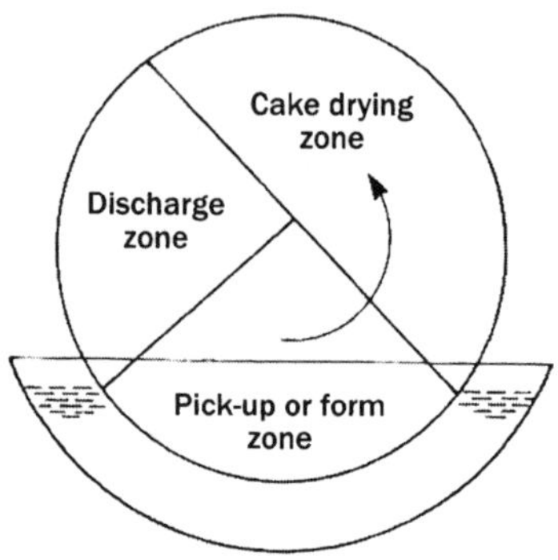

(b) Operation zones

Figure 3.27 Vacuum filter.

cumference. Each section is sealed from its adjacent section and the ends of the drum. A separate drain line connects each section to a rotary valve at the axis of the drum. As the drum rotates, the valve allows each segment to function in sequence as one of the three distinct zones: for cake forming, cake drying, and cake discharging. A vacuum is applied to each of the drum sections through drain lines. As the drum rotates, each section is carried successively through the cake-forming zone to the cake-drying zone. In the final zone, cake is removed from the coil-spring media. After cake discharge, the coils are washed.

Optimum performance is dependent on the type of sludge, solids concentration, type and quality of conditioning, and how the filter is operated. Ferric

TABLE 3.22 Typical Dewatering Performance Data for Vacuum Filters

Type of Sludge	Feed Solids (%)	Yield kg/m²·h	Yield lb/ft²-hr	Cake Solids (%)
Raw sludge				
Primary	4.5–9.0	20–50	4–10	25–32
WAS	2.5–4.5	5–15	1–3	12–20
Primary + trickling filter	4–8	15–30	3–6	20–28
Primary + WAS	3–7	12–30	2.5–6.0	18–25
Digested sludge				
Primary	4–8	15–34	3–7	25–32
Primary + trickling filter	5–8	20–34	4–7	20–28
Primary + WAS	3–7	17–24	3.5–5.0	20–28

Source: Adapted from U.S. EPA, 1979.

chloride/lime conditioning is the most common type of conditioning used. The selection of vacuum level, degree of drum submergence, type of media, and cycle time are all critical to optimum performance. Table 3.22 shows typical dewatering performance data for vacuum filters.

Vacuum filters consume the largest amount of energy per unit of sludge dewatered in most applications, and they require continuous operator attention. Because of the improvements to other dewatering devices and the development of new dewatering devices with lower operation and maintenance costs, the use of vacuum filter is declining.

Screw Presses A screw press is a dewatering device that employs a dewatering and conveying screw inside a screen. Figure 3.28 shows two types of screw presses, one with an inclined screw arrangement, the other with a horizontal installation with the added steam feed capability for additional drying of the cake.

In the inclined arrangement, the screw press consists of a wedge section basket with a 0.25-mm (0.01-in.) spacing. The slowly rotating screw, at variable speed, conveys the polymer-conditioned sludge upward through the inclined basket. The lower section of the basket serves as a predewatering zone, where free water drains by gravity. The upper section of the basket serves as the pressure zone. Here the sludge is compressed between narrowing flights of the screw (or progressively enlarging flights of the screw, as shown in Figure 3.28). The pressure in the pressure zone is controlled by the position of a cone at the discharge end of the basket. The dewatered sludge cake is driven through the gap between the cone and the basket and drops onto a conveyor or directly into a dumpster. The screw flights are provided with brushes for continuous internal cleaning of the wedge section basket. Spray nozzles are also provided for periodic cleaning of the basket from the outside with spray water. The basket is provided with a cover for housekeeping and odor containment.

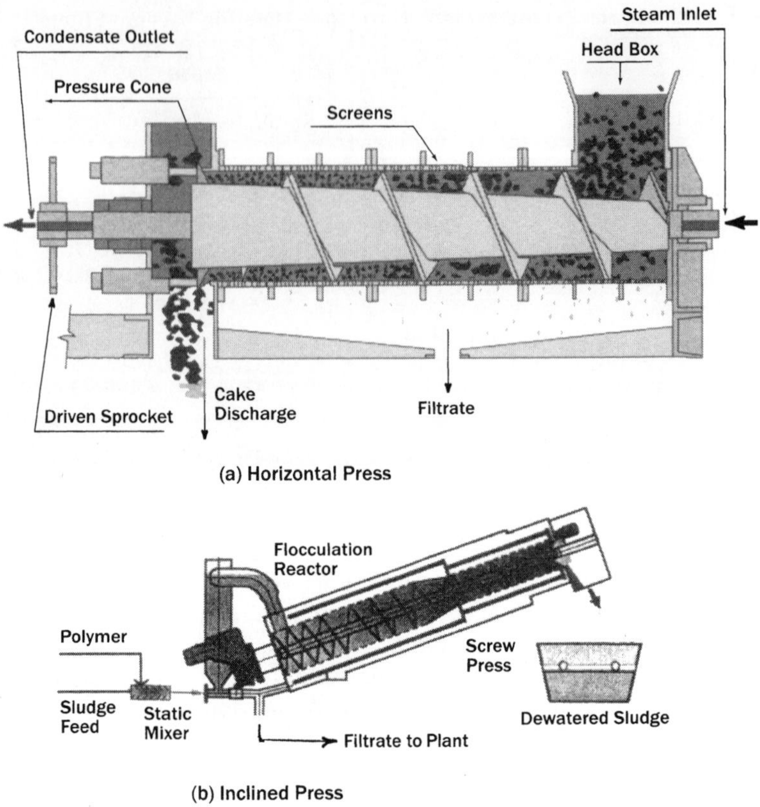

Figure 3.28 Screw presses. [Part (a) from FKC Co., Port Angles, WA; part (b) from Huber Technology Inc., Huntersville, NC.]

Advantages of screw presses include fewer space requirements and relatively low capital cost and power consumption. It is typically used in small wastewater treatment plants. The hydraulic capacity is about 10 m^3/h (45 gpm), and the solids capacity is about 275 kg/h (600 lb/hr). Cake solids concentration of 20 to 25% can be obtained with a polymer dosage of 4 to 6 g/kg (8 to 12 lb/ton) of dry solids. Solids capture rates of better than 95% have been reported.

Reed Beds The reed bed system for municipal sludge dewatering combines the action of conventional drying beds that have the effects of aquatic plants on water-bearing substrates. The system is constructed similar to drying beds with rectangular basins that have concrete sidewalls. The bottom of beds are provided with a 250-mm (10-in.)-thick layer of 20-mm (0.8-in.) washed gravel and perforated underdrain piping for filtrate removal. This bottom layer is topped with another 250-mm (10-in.) layer of 4 to 6 mm (0.16 to 0.25 in.) washed gravel, and a layer of filter sand of 100 to 150 mm (4 to 6 in.). A minimum freeboard of 1 m (3.3 ft) is provided above the sand. Reeds of the genus *Phragmites communis* are planted in the middle gravel layer on 300-mm (12-in.) centers. The plants provide a pathway for continuous drainage of water from the applied sludge. Additionally, the root system of the plants establishes a rich microflora that feeds on the organic content of the sludge. Degradation by the microflora is so effective that up to 97% of the organic content is converted to carbon dioxide and water, with a corresponding reduction in volume. The reeds are harvested once every year in the fall when they have become dormant.

The design solids loading rate is 30 to 60 kg/m^2·yr (6 to 12 lb/ft^2-yr). The basin is loaded over a 24-hour period, and a 1-week resting period is provided before the cycle is repeated. A bed can be operated for up to 10 years before the accumulated residues have to be removed. To remove the residues, the beds are first taken out of service for 6 months. This allows the uppermost layer to become mineralized and disinfected. When the solids are removed for disposal, the top gravel and sand layers are also removed and must be replaced. The use of reed beds is practical for wastewater treatment plants of capacities of less than 0.2 m^3/s (5 mgd).

Drying Lagoons Sludge drying lagoons are another method of dewatering stabilized sludge when sufficient land is available. They are similar to drying beds; however, the sludge is placed at depths three to four times greater than it would be in a drying bed. Dewatering occurs by evaporation and transpiration, of which evaporation is the most important dewatering factor. Sludge should be stabilized prior to discharging to the lagoons to minimize odor problems. The advantages and disadvantages of drying lagoons are listed in Table 3.13.

Drying lagoons are normally rectangular in shape, enclosed by earthen dikes 0.6 to 1.2 m (2 to 4 ft) high. Appurtenant equipment includes sludge feed lines, supernatant decant lines, and some type of mechanical sludge removal equipment. The removal equipment can be a bulldozer, dragline, or front-end loader. Stabilized sludge is pumped into the lagoon over a period of several months until a lagoon depth of 0.6 to 1.2 m (2 to 4 ft) is achieved. Supernatant is decanted, either continuously or intermittently, from the lagoon surface and returned to the treatment plant. Depending on the climate and the depth of the sludge, the time required for dewatering to a final solids content of 20 to 40% may be 3 to 12 months. After the sludge cake is removed, the cycle is repeated.

Proper design of sludge drying lagoons requires consideration of several factors, such as precipitation, evaporation, sludge characteristics, and volume. Solids loading criterion is 35 to 38 kg/m^3·yr (2.2 to 2.4 lb/ft^3-yr) of lagoon capacity. Per capita design criteria vary from 0.1 m^2/capita (1 ft^2/capita) with primary digested sludge in an arid climate to 0.3 to 0.4 m^2/capita (3 to 4 ft^2/capita) for activated sludge plants in areas where 900 mm (36 in.) of annual rainfall occurs.

REFERENCES

Albertson, O. E., and Walz, T. (1997), Optimizing Primary Clarification and Thickening, *Water Environment and Technology*, Vol. 9, No. 12.

Albertson, O. E., et al. (1991), *Dewatering Municipal Wastewater Sludges*, Noyes Data Corporation, Park Ridge, NJ, p. 189.

Agranonic, R. Y. (1985), *Technology of Wastewater Sludge Treatment with Centrifuges and Belt Filter Presses*, Stroyizdat, Moscow.

ASCE (1988), Belt Filter Press Dewatering of Wastewater Sludge, ASCE Task Force on Belt Press Filters, *ASCE Journal of the Environmental Engineering Division*, Vol. 115, No. 5, pp. 991–1006.

Ashbrook Corporation (1992), *Aquabelt Operation and Maintenance Manual*, Houston, TX.

Carmen, P. C. (1933, 1934), A Study of the Mechanism of Filtration, *Journal of the Society Chemical Industry London*, Vol. 52, p. 280 T; Vol. 53, pp. 159 T, 301 T.

Chang, L. W., Furst, A., and Nordberg, G. (1995), *Toxicology of Metals*, Vol. 1, July, p. 480.

Cheremisinoff, P. N. (1995), *Solids/Liquids Separation,* Technomic Publishing Co., Lancaster, PA.

Christensen, G. L., and Dick, R. I. (1985), Specific Resistance Measurements: Methods and Procedure, *ASCE Journal of the Environmental Engineering Division*, Vol. 111, No. 3, p. 258.

Citton, F. W., Jr., Adams, T. E., and Dohoney, R. W. (1991), Managing Sludge Through In-Vessel Composting, *Water Engineering and Management,* December, p. 21.

Coacley, P., Swenwic, V. D., Gabe, R. S., and Bascerville, R. C. (1956), *Water Pollution Research and Proceedings of the Institute for Sewage Purification*, Vol. 2.

Coker, C. S., et al. (1991), Dewatering Municipal Wastewater Sludge for Incineration, *Water Environment and Technology,* March.

Davis, J. (1986), Sludge Disposal Thinking Seminar in U.K. and U.S.A., *Water Engineering and Management,* Vol. 12, pp. 25–28.

Diaz, L. F., et al. (1993), *Composting and Recycling Municipal Solid Waste*, Water Environment Federation, Alexandria, VA, p. 320.

Dicht, N. (1986), Zweistufige Ferfahren der Shlämmstabkizierung, *Korrespondenza Abwasser*, Vol. 11, pp. 1055–1056.

Dick, R. I., and Ewing, B. B. (1967), Evaluation of Activated Sludge Thickening Theories, *ASCE Journal of the Environmental Engineering Division*, Vol. 93, No. 4, p. 9.

D. R. Sperry Co., Prospect, A vision for Tomorrow, North Aurora, IL.

Dvinskich, E. V., et al. (1991), Drying Beds, VNIPIE Lesprom, Moscow, p. 65.

Eckenfelder, W. W., and Santhanam, C. J. (1980), *Sludge Treatment,* Marcel Dekker, New York.

EPA Design Information Report (1987), The Original Vacuum Sludge Dewatering Bed, *Journal of the Water Pollution Control Federation*, Vol. 59, No. 4, pp. 228–234.

Epstein, E. (1997), *The Science of Composting*, Technomic Publishing Co., Lancaster, PA.

FDEP (1994), *Domestic Wastewater Residuals*, Chapter 62-640, Florida Department of Environmental Protection, Tallahassee, FL.

Foess, G. M., and Singer, R. B. (1993), Pathogen/Vector Attraction Reduction Requirement of the Sludge Rules, *Water Engineering and Management,* June, p. 25.

Frontier Technologies, Inc., Prospect, FT1 Belt Filter Press, Allegan, MI.

Garvey, D., Guairo, C., and Davis, R. (1993), Sludge Disposal Trends Around the Globe, *Water Engineering and Management*, December, p. 17.

Ghosh, S. (1987), Improved Sludge Classification by Two-Phase Anaerobic Digestion, *Environmental Engineer,* Vol. 6, pp. 1265–1284.

Goldfarb, L., Turovskiy, I., and Belaeva, S. (1983), *The Practice of Sludge Utilization*, Stroyizdat, Moscow.

Gulas, V., Benefield, L., and Randall, C. (1978), Factors Affecting the Design of Dissolved Air Flotation Systems, *Journal of the Water Pollution Control Federation,* Vol. 50, p. 1835.

Hallulen (1989), Thickening of the Activated Sludge, Symposium, Warsaw, Poland, April 24–29.

Hammer, M. J. (1975), *Water and Wastewater Technology*, New York.

Infilco Degremont, Inc. (1979), *Water Treatment Handbook*, 5th ed., Richmond, VA.

Jordan, V. J., and Scherer, C. H. (1970), Gravity Thickening Techniques at a Water Reclamation Plant, *Journal of the Water Pollution Control Federation,* Vol. 42, p. 180.

Karr, P. R., and Keinath, T. M. (1978), Influence of Particle Size on Sludge Dewaterability, *Journal of the Water Pollution Control Federation*, Vol. 50, p. 1911.

Komline-Sanderson, Inc., Prospect, GRS Series III Kompress, Peapack, NJ.

Lawler, D. F., and Chung, V. J. (1986), Anaerobic Digestion: Effect on Particle Size and Dewaterability, *Journal of the Water Pollution Control Federation,* Vol. 12, p. 1107.

Metcalf & Eddy, Inc. (2003), *Wastewater Engineering: Treatment and Reuse,* 4th ed., Tchobanoglous, G., Burton, F. L., and Stensel, H. D. (Eds.), McGraw-Hill, New York.

Ohara, G. T., Raksit, S. K., and Olson, D. R. (1978), Sludge Dewatering Studies at Hyperion Treatment Plant, *Journal of the Water Pollution Control Federation,* Vol. 50, p. 912.

Olson, R., Gendreau, A., and Cyr, S. (1999), *Biosolids Technical Bulletin*, March–April, pp. 5–9.

Parkin, G. F. (1986), Fundamentals of Anaerobic Digestion of Wastewater Sludges. *Environmental Engineer,* Vol. 5, pp. 867–920.

Popel, F. (1967), *Sludge Digestion and Disposal*, Stuttgart, Germany.

Randall, C. W. (1969), Are Paved Drying Beds Effective for Dewatering Digested Sludge? *Water and Sewage Works*, Vol. 116, p. 373.

Roberts, K., and Olsson, O. (1975), Influence of Collodial Particles on Dewatering of Activated Sludge with Polyelectrolyte, *Environmental Science and Technology*, Vol. 9, p. 945.

Schroeder, W. (1960), What Is Better—Raw or Digested Sludge? *Das Gas-und Wasserfach*, Vol. 101, No. 50, pp. 1298–1301.

Seabright Products, Inc., Prospect, Dewatering Solutions, Hopkins, MI.

Serna Technologies, Inc., Prospect, Belt Filter Press, Jasper, AL.

Siger, R. B. (1993), Practical Guide to the New Sludge Standards, *Water Engineering and Management,* November, p. 26.

———, and Hermann, G. (1993), Land Application Requirements of the New Sludge Rules, *Water Engineering and Management,* August, p. 26.

Sludge Treatment and Disposal, Vol. 1, *Sludge Treatment*, and Vol. 2, *Sludge Disposal* (1985), Stroyizdat, Moscow.

Spellman, F. R. (1996), *Wastewater Biosolids to Compost,* Technomic Publishing Co., Lancaster, PA.

——— (1997) *Dewatering Biosolids,* Technomic Publishing Co., Lancaster, PA.

Tchernova, N. M. (1966), *Zoological Characteristics of Compost,* Nauka, Moscow.

The Wave (1998), *U.S. Filter* magazine, Municipal Water and Wastewater, February, Vol. 2, No. 1.

The Wave (2000), *U.S. Filter* magazine, Municipal Water and Wastewater, March, Vol. 4, No. 1.

Turovskiy, I. S. (1986), *Design Handbook of Wastewater Systems, Methods of Wastewater Sludge Treatment,* Vol. 2, Sec. 10, Allerton Press, New York, pp. 531–610.

——— (1988), *Wastewater Sludge Treatment,* Stroyizdat, Moscow.

——— (1992), *Endickung, Ehtwasserung und Enseuhung von Abwasserschlämmen,* WAP 2.92, Munich, pp. 95–100.

——— (2000), *Dewatering of Wastewater Sludge*, Proceedings 75th Annual Florida Water Resources Conference, Tampa, FL.

Turovskiy, I. S., et al. (1970), *Thermal Treatment of Sludges*, Gos INTI, Moscow.

——— (1971), *Sludge Incineration,* Gos INTI, Moscow.

——— (1988), Sludge Dewatering by Belt Filter Presses, *Water Supply and Sanitary Technology*, Vol. 1, pp. 4–6.

——— (1989a), *Biothermal Treatment of Sludge*, ZBTI Minvodchoz, Moscow.

——— (1989b), *Systems with Press Filters,* ZINTICHIMNEFTE Mash, Moscow.

——— (1991), *Technology of Sludge Composting,* VNIPIEI Lesprom, Moscow.

U.S. EPA (1978), *Effects of Thermal Treatment of Sludge on Municipal Wastewater Treatment Cost,* EPA 600/2-78/073.

——— (1979), *Process Design Manual for Sludge Treatment and Disposal,* EPA 625/1-79/011.

——— (1987a), *Design Manual: Dewatering Municipal Wastewater Sludge,* EPA 625/1-87/014.

——— (1987b), *Innovation in Sludge Drying Beds: A Practical Technology.*

——— (1989), *Summary Report: In-Vessel Composting of Municipal Wastewater Sludge,* EPA 625/8-89/016.

U.S. Filter, Prospect, Belt Filter Presses, U.S. Filter, Dewatering Systems Group, Holland, MI.

Vesilind, P. A. (1979), *Treatment and Disposal of Wastewater Sludges,* Ann Arbor Science, Ann Arbor, MI.

——— (1996), Sludge Dewatering: Why Water Wins, *Industrial Wastewater,* Vol. 4 No. 2, pp. 43–46.

Wech, F. (1986), *Untersuchungen zur Optimiesierung der zweistufigen anaeroben Klarschlammstabilisierung,* GWF, *Wasser-Abwasser,* Vol. 3, pp. 109–117.

WEF (1980), *Sludge Conditioning,* Manual of Practice FD-1, Water Environment Federation, Alexandria, VA.

——— (1983), *Sludge Dewatering,* Manual of Practice 20, Water Environment Federation, Alexandria, VA.

——— (1987), *Operation and Maintenance of Sludge Dewatering Systems,* Manual of Practice OM-8, Water Environment Federation, Alexandria, VA.

——— (1988), *Sludge Conditioning,* Manual of Practice FD-14, Water Environment Federation, Alexandria, VA.

——— (1996), *Operation of Municipal Wastewater Treatment Plants,* 5th ed., Manual of Practice 11, Water Environment Federation, Alexandria, VA.

——— (1998), *Design of Municipal Wastewater Treatment Plants,* 4th ed., Manual of Practice 8 (ASCE 76), Water Environment Federation, Alexandria, VA.

Yakovlev, S. V. (1994), *Utilities in Building and Structures*, Stroyizdat, Moscow.

4

AEROBIC DIGESTION

Wastewater Sludge Processing, By Izrail S. Turovskiy and P. K. Mathai

Stabilization of wastewater sludge is processing the sludge for the purpose of eliminating the potential for putrefaction, inhibiting or reducing odor, and reducing pathogens, and thus converting it to a stable product for use or disposal. Although it is not practiced by all wastewater treatment plants, an overwhelming majority of plants, ranging in size from small to very large, use one of the following principal methods for stabilizing sludge:

1. Aerobic digestion
2. Autothermal thermophilic aerobic digestion
3. Anaerobic digestion
4. Alkaline stabilization
5. Composting
6. Thermal drying

In addition to the health and aesthetic reasons cited above, stabilization, except in the case of alkaline stabilization, reduces the volume of sludge solids to be used or disposed of. The stabilization processes listed above are described in this and subsequent chapters.

4.1 INTRODUCTION

Aerobic digestion is the most widely used sludge stabilization process in wastewater treatment plants that treat less than 20,000 m^3/d (5 mgd). It is a process of oxidation and decomposition of the organic part of the sludge by microorganisms in special open or enclosed tanks with the presence of oxygen. It is a suspended growth biological process that is similar to the extended aeration-type activated sludge process. The process produces a stable product, reduces mass and volume, and reduces pathogenic organisms. Advantages of aerobic digestion compared to anaerobic digestion include:

- Lower capital cost
- Odorless end product
- Volatile solids reduction only slightly less than that obtained in anaerobic digestion process

- Easier to operate
- Operational control of volatile acids–alkalinity relationship not necessary
- Lower BOD, TSS, and ammonia nitrogen in supernatant liquor
- Suitable for digesting nutrient-rich sludge solids while recovering more of the basic fertilizer values
- Safer to operate with no potential for gas explosion and less potential for odor problems

Major disadvantages of aerobic digestion compared with anaerobic digestion include:

- The operating cost is higher, in the form of the power cost for supplying the oxygen required.
- Methane gas, a useful by-product, is not produced.
- The process is dependent on temperature.
- The efficiency of the process is reduced during cold weather.
- The performance is affected by the concentration of solids, type of sludge, location, and type of mixing–aeration system

As discussed in Chapter 1, aerobic digestion is one of the processes defined in Part 503 regulations to meet the PSRP requirements. According to the regulations, the solids retention time in aerobic digestion must be at least 40 days at 20°C and 60 days at 15°C to meet the class B requirements. Most older aerobic digestion facilities have less than 40 days of SRTs. Therefore, if these facilities need to meet the class B requirements for pathogen reduction, additional capacities have to be added to the digesters, or thickeners have to be built to thicken the feed sludge. However, those facilities that do not meet these SRTs can meet the class B requirements by monitoring the process and demonstrating that the pathogen reduction criterion has been met. Monitoring is also required to demonstrate that the volatile solids reduction requirements to comply with the vector attraction reduction criterion have been met.

4.1.1 Process Theory

Aerobic digestion of excess activated sludge may be considered to be a continuation of the activated sludge process. Figure 4.1 illustrates this process graphically. When the soluble substrate (food) is completely oxidized by the microbial population, the microorganisms begin to consume their own protoplasm to obtain the energy for cell maintenance. This phenomenon of

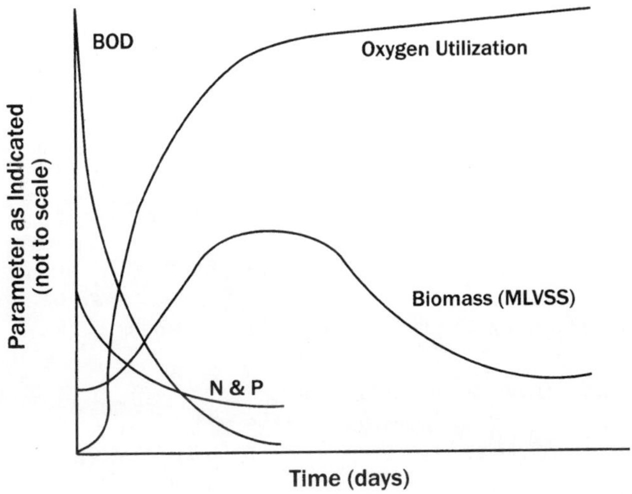

Figure 4.1 Relationship among BOD, MLVSS, oxygen utilization, and nutrients. (From Lue-Hing et al., 1998.)

obtaining energy from cell tissue, known as *endogenous respiration*, is the major reaction in aerobic digestion.

The cell tissue is oxidized aerobically to carbon dioxide, water, and ammonia. As the digestion process proceeds, the ammonia is subsequently oxidized to nitrate. In actuality, only about 65 to 80% of the cell tissue can be oxidized. The remaining 20 to 35% is composed of inert components and organic compounds that are not biodegradable. The material that remains after completion of the digestion process is at such a low-energy state that it is essentially biologically stable. The biooxidation of biomass results in the reduction of the volume of residual solids requiring disposal. However, this objective of volume reduction has not been fully realized in many facilities because of the problems with effective dewatering of the digested biosolids.

The first step of the aerobic digestion process, direct oxidation of biodegradable matter, can be illustrated by the equation

$$\text{organic matter} + O_2 \xrightarrow{\text{bacteria}} \text{cellular material} + CO_2 + H_2O \qquad (4.1)$$

The formula for the cell mass of microorganisms is $C_5H_7O_2N$. The second step, endogenous respiration, can be illustrated by the following equations (Enviroquip, 1997):

Destruction of biomass:

$$C_5H_7O_2N + 5O_2 \rightarrow 4CO_2 + H_2O + NH_4HCO_3 \qquad (4.2)$$

Nitrification of released ammonia nitrogen:

$$NH_4^+ + 2O_2 \rightarrow NO_3 + 2H^+ + H_2O \tag{4.3}$$

Overall equation with complete nitrification:

$$C_5H_7O_2N + 7O_2 \rightarrow 5CO_2 + 3H_2O + HNO_3 \tag{4.4}$$

Using nitrogen as an electron acceptor (denitrification):

$$C_5H_7O_2N + 4NO_3^- + H_2O \rightarrow NH_4^+ + 5HCO_3^- + 2NO_2 \tag{4.5}$$

With complete nitrification/denitrification:

$$2C_5H_7O_2N + 11.5O_2 \rightarrow 10CO_2 + 7H_2O + N_2 \tag{4.6}$$

With partial nitrification:

$$2C_5H_7O_2N + 12O_2 \rightarrow 10CO_2 + 5H_2O + NH_4^+ + NO_3^- \tag{4.7}$$

In the destruction of biomass [equation (4.2)], oxygen is used to oxidize cell mass to carbon dioxide and water. This reaction also produces ammonium bicarbonate. The important thing here is that it is NH_3 that is released, which then combines with some of the CO_2 that is produced to form ammonium bicarbonate.

In the second reaction [equation (4.3)], nitrification of released ammonia creates nitrate and 2 mol of acidity, which is shown as 2 mol of hydrogen ion (H^+). Therefore, the destruction of biomass produces 1 mol of alkalinity as ammonium bicarbonate, then nitrification destroys 2 mol of alkalinity. Complete oxidation and nitrification take 7 mol of oxygen, as shown in equation (4.4).

If the oxygen in the nitrate is used as the oxygen source just as is done in liquid stream processes, denitrification is possible. This is shown in equation (4.5): biomass plus nitrate producing ammonia, nitrogen gas, and alkalinity in the form of bicarbonate. So, taken together, biomass destruction and denitrification produce alkalinity, and nitrification consumes alkalinity.

Theoretically, approximately 50% of the alkalinity consumed by nitrification can be recovered by denitrification. If the dissolved oxygen is kept very low (less than 1 mg/L), nitrification will not occur. Therefore, cycling of the aerobic digester between aeration and mixing by mechanical means can be effective in maximizing denitrification while maintaining pH control. The complete nitrification–denitrification equation (4.6) shows a balanced stoichiometric equation. If all the ammonia released is nitrified and denitrified, there is balance. Biomass plus oxygen is converted to carbon dioxide, nitrogen gas, and water, with no net consumption of alkalinity.

What is often seen in aerobic digestion is in fact *partial nitrification*; that is, a portion of the nitrogen is being left as ammonia. The system will nitrify until the pH drops enough that it begins to inhibit the nitrifying bacteria. This is illustrated by equation (4.7). This is what is normally seen in many digesters, a mixture of both ammonia and nitrogen. This would be the case where there is only an inconsequential amount of alkalinity in the sludge that is fed to the digester. In situations where the buffering capacity is insufficient, resulting in pH depression below 5.5, it may be necessary to feed chemicals such as lime to maintain the desired pH.

When WAS is aerobically digested, the predominant phase that is maintained is endogenous respiration. However, if primary sludge is included in the process, the overall reaction can shift to a lengthy phase of direct oxidation of biodegradable matter. Most of the organic matter in primary sludge becomes the external food source for the active biomass in the biological sludge. Therefore, longer detention times are required to accommodate the metabolism and cellular growth that must occur before endogenous respiration conditions are achieved.

Theoretically, about 1.5 kg of oxygen is required per kilogram of active cell mass (1.5 lb/lb) in nonnitrifying systems, whereas about 2 kg of oxygen per kilogram of active cell mass (2 lb/lb) is required in nitrifying systems. In a system with complete nitrification–denitrification, it provides the opportunity to (1) reduce the oxygen requirements (a 17% reduction), (2) avoid alkalinity depletion because the alkalinity produced in denitrification is used to offset the existing alkalinity that is required for nitrification, and (3) and nitrogen is removed. The third benefit may or may not be of consequence. However, the first two result in savings in operating costs.

The oxygen requirements for mixed primary and activated sludge digestion are substantially greater than what is required simply for activated sludge digestion because of the longer time required to oxidize the organic matter in primary sludge. The oxygen uptake rates in the digestion process vary depending on the characteristics of the feed sludge. Thickened sludge has a very high oxygen uptake rate at the beginning of the digestion process. An adequate amount of air should be supplied to the digester to keep the solids in suspension and to avoid their settling to the bottom of the tank. An air supply rate of $1\,m^3/m^3 \cdot h$ (17 cfm per $1000\,ft^3$) is adequate to keep the solids in suspension, although some state standards might dictate much higher air requirements for mixing (Turovskiy, 2001).

4.2 CONVENTIONAL AEROBIC DIGESTION

The conventional aerobic digestion process is a continuous process that closely resembles the activated sludge process. The schematics shown in Figure 4.2 illustrate several variations of the aerobic digestion process. For wastewater treatment plants without primary settling tanks, Scheme 2a or 2b

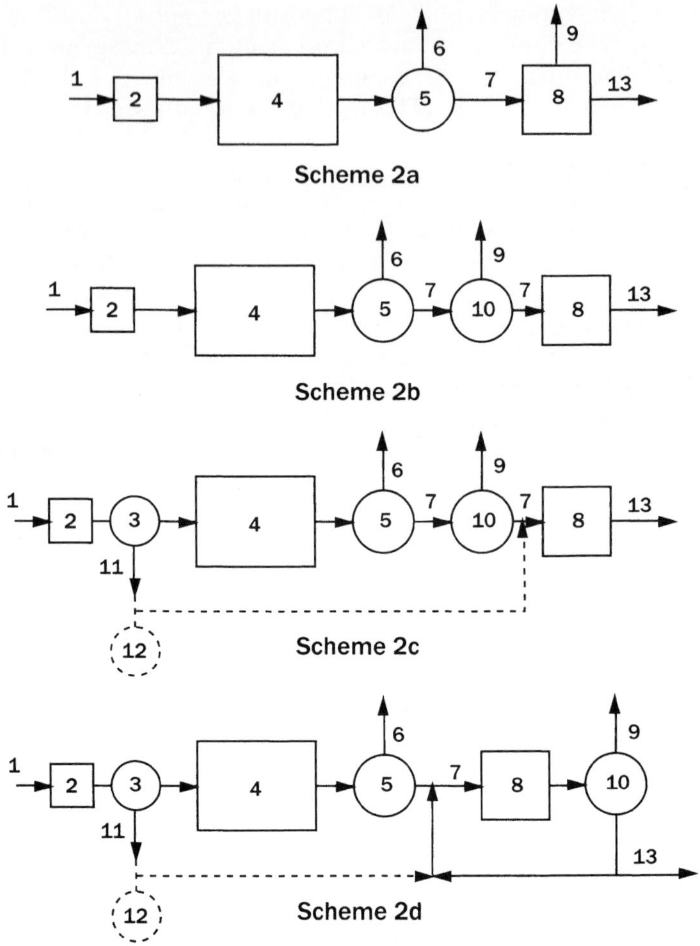

1- Wastewater inflow, 2 - Screen and grit removal,
3 - Primary settling tank, 4 - Aeration tank,
5 - Secondary clarifier, 6 - Treated effluent,
7 - Waste activated sludge, 8 - Aerobic digestor,
9 - Supernatant, 10 - Thickener, 11 - Primary sludge,
12 - Anaerobic digrestor, 13 - Digested sludge.

Figure 4.2 Aerobic sludge digestion process schemes.

is recommended. In Scheme 2a, waste activated sludge goes to the aerobic digester directly from the secondary clarifier. In Scheme 2a, the sludge goes to the digester after preliminary concentration in a sludge thickener. Schemes 2c and 2d are the most common processes in small to medium-sized wastewater treatment plants that include primary settling. In Scheme 2c, thickened secondary sludge is combined with primary sludge and discharged to the digester. In Scheme 2d, combined primary and unthickened secondary sludge is digested first and thickened in a thickener. For larger treatment plants, it is more practical to use the variations that digest primary sludge and WAS separately, as shown in Schemes 2c and 2d. The thickener–recycle system shown in the schematics can be incorporated into the digester tank design. A circular digester with the integral thickener–recycler is illustrated in Figure 4.3. It had been a common practice to pump the WAS directly from the clarifiers into the aerobic digester; however, this necessitated larger tanks to satisfy the SRT requirement. Modern practice is to thicken the WAS prior to discharging into the digester, thereby reducing the digester volume required.

Originally, aerobic digesters were designed as a semibatch process (see Figure 4.3). This concept is still functional at many facilities and is still

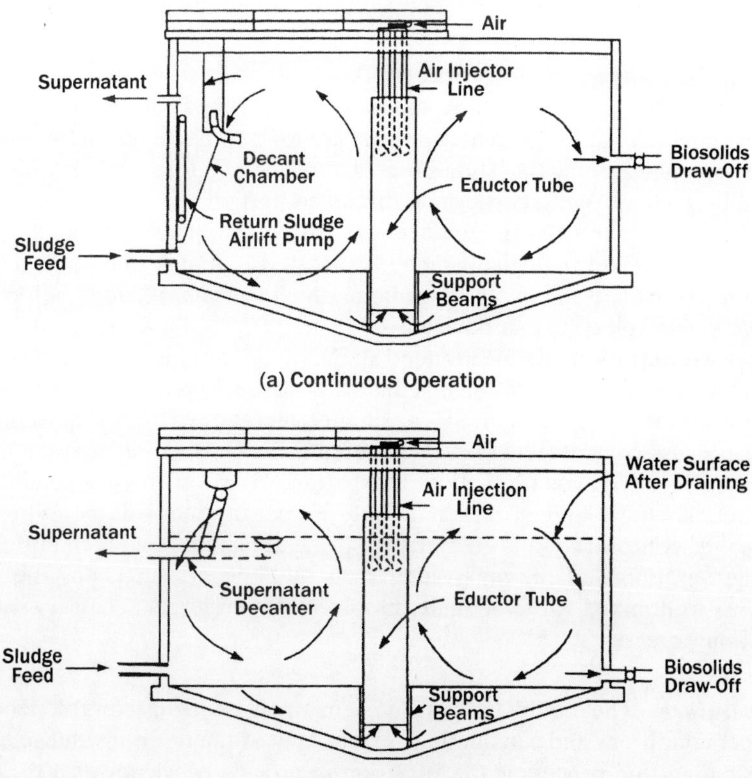

Figure 4.3 Circular aerobic digester. (Reprinted with permission from WEF, 1995.)

designed for smaller treatment plants. Sludge is pumped directly from the clarifiers, or after thickening, into the aerobic digester. During filling operation, the digester content is aerated continuously. When the tank is full, aeration continues for the required detention time. Aeration is then discontinued and the stabilized solids settled. The clear supernatant is then drawn off with a telescoping valve or similar decanting device and the thickened biosolids are removed. A sufficient amount of the biosolids is left in the digester to provide the necessary microbial population. The cycle is then repeated.

4.2.1 Process Design Considerations

Factors that govern the process design of conventional aerobic digesters include feed sludge characteristics, temperature, volatile solids reduction, oxygen requirements, and mixing. Other system design considerations and operational considerations are discussed later in this chapter.

Feed Sludge Characteristics Because the aerobic digestion process is similar to the activated sludge process, the same concerns, such as variations in influent characteristics and materials that are toxic to biological activities, are important. The aerobic digestion process is best suited for stabilizing biological solids such as WAS, because as discussed earlier, the process keeps the system predominantly in the endogenous respiration phase. If a mixture of primary sludge and biological sludge is digested, a longer detention time is required to oxidize the excess organic matter in primary sludge before the endogenous respiration can be achieved.

The concentration of solids in feed sludge to an aerobic digester is important in the design and operation of a digester. Advantages of higher feed solids concentration include longer SRTs, smaller digester volume requirements, easier process control (less or no decanting in batch-operated systems), and increased levels of volatile solids destruction. However, higher solids concentrations require higher oxygen input levels per digester volume. When the feed solids concentration is greater than 3%, care should be taken in designing the aeration and mixing system such that the system keeps the tank contents well mixed with adequate dissolved oxygen levels necessary for the digestion process.

Temperature The liquid temperatures in open-tank digesters depend on weather conditions and can fluctuate extensively. A major disadvantage of the aerobic digestion process is the variation in process efficiency as a result of the changes in operating temperatures. As with all biological systems, lower temperatures retard the process whereas higher temperatures speed it up. Because aerobic digestion is a biological process, the effects of temperature variations can be estimated by the equation

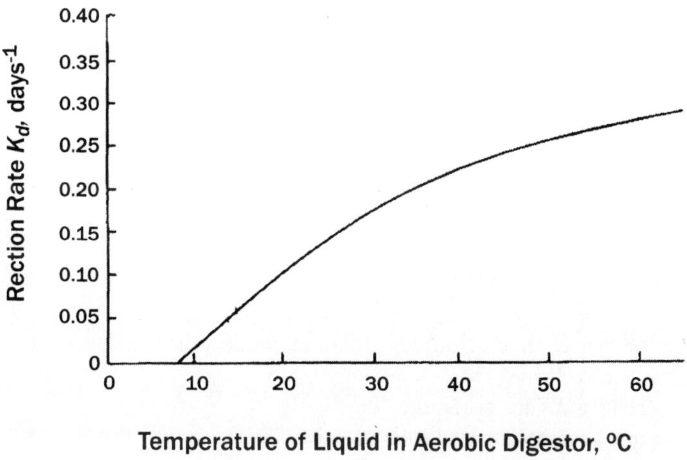

Figure 4.4 Reaction rate K_d versus digester liquid temperature.

$$(K_d)_T = (K_d)_{20}\, q^{T-20} \tag{4.8}$$

where

K_d = reaction rate constant, time

T = temperature, °C

$(K_d)_{20}$ = reaction rate constant at 20°C

q = temperature coefficient (ranges from 1.02 to 1.10, with an average of 1.05)

Figure 4.4 represents the change in reaction rate constant versus increasing operating temperature. An increase in the temperature of the system results in an increase in the reaction rate constant and implies an increase in digestion rate. In designing a digester, consideration should be given to minimizing heat losses by using concrete instead of steel, placing the tank below rather than above grade, and using subsurface instead of surface aeration. In extremely cold climates, consideration should be given to covering the tanks, heating the sludge, or both. The design should allow for the necessary degree of sludge stabilization at the lowest liquid operating temperature and should meet the maximum oxygen requirements at the maximum expected liquid operating temperature.

Volatile Solids Reduction The major objectives of aerobic digestion of sludge are to stabilize sludge, reduce pathogens, and reduce the mass of solids for disposal. The reduction in mass is possible only with the destruction of the biodegradable organic content of the sludge, although some studies (Randall, 1975; Benefield, 1978) have shown that there may be some destruction of nonorganics as well. Volatile solids reductions of 35 to 50% are attainable by aerobic digestion. To obtain the vector attraction reduction requirement of 40 CFR Part 503, a minimum of 38% volatile solids reduction or less than a specific oxygen uptake rate (SOUR) of 1.5 mg of O_2 per hour per gram of total sludge solids at 20°C has to be achieved.

The change in biodegradable volatile solids in a completely mixed aerobic digester can be represented by a first-order biochemical reaction as follows:

$$\frac{dM}{dt} = -K_d M \tag{4.9}$$

where

$\frac{dM}{dt}$ = rate of change of biodegradable volatile solids (M) per unit of time (Δ mass/time), MT^{-1}

K_d = reaction rate constant, T^{-1}

M = mass of biodegradable volatile solids remaining at time t in the aerobic digester

The time t in equation (4.9) is the sludge age or solids retention time (SRT) in the digester. Depending on how the aerobic digester is being operated, time t can be equal to or considerably greater than the theoretical hydraulic residence time. When there is no recycle and no decanting from the digester, SRT is equal to the hydraulic detention time. Use of biodegradable portion of the volatile solids in the equation recognizes that approximately 20 to 35% of the WAS from treatment plants with primary treatment systems, and 25 to 35% of the WAS from contact stabilization processes (no primary clarification), is nonbiodegradable.

The reaction rate constant K_d is a function of sludge type, temperature, and solids concentration. The reaction rate constant for WAS versus temperature may range from 0.06 d^{-1} at 15°C to 0.14 d^{-1} at 25°C, as represented in Figure 4.4. Because the reaction rate is influenced by several factors, it may be necessary to confirm decay coefficient values by bench- or pilot-scale studies.

Oxygen Requirements Oxygen requirements for aerobic digestion were discussed earlier in Section 4.1.1. Equation (4.2) indicates that, theoretically, 1.45 kg of oxygen is required to oxidize 1 kg (1.45 lb/lb) of cell mass with no nitrification. Similarly, according to equation (4.4), 1.98 kg of oxygen is theoretically required to oxidize 1 kg (1.98 lb/lb) of cell mass. The results of pilot- and full-scale studies indicate that actual oxygen requirement range from 1.74 to 2.07 kg of oxygen per kilogram of applied organic cell mass (1.74 to 2.07 lb/lb). Design experience indicates that a value of 2.0 is recommended. In a system with complete nitrification–denitrification, as discussed earlier, 17% less oxygen is required. For autothermal systems, which have temperatures above 45°C, nitrification does not occur and a value of 1.45 kg/kg (1.45 lb/lb) is recommended. Field studies have indicated that a minimum of 1 mg/L of oxygen be maintained in the digester under all operating conditions.

The inclusion of primary sludge in the digestion process requires an additional 1.6 to 1.9 kg of oxygen per kilogram (1.6 to 1.9 lb/lb) of volatile solids destroyed to convert the organic matter in primary sludge to cell tissue and to satisfy the endogenous respiration demand of the resulting cell mass.

Mixing Mixing is required in an aerobic digester to keep the solids in suspension. After the requirements for adequate mixing and oxygen transfer have been computed separately, the larger of the two requirements will govern the overall system design. Power levels of 20 to 40 kW per $10^3 m^3$ (0.75 to 1.5 hp per $10^3 ft^3$) of tank volume have been reported to be satisfactory for mechanical mixers. In diffused air mixing, air supply rates of 1.2 to 2.4 $m^3/m^3 \cdot h$ (20 to 40 cfm per $10^3 ft^3$) have been reported; the higher values are recommended for sludges of high solids concentrations. If polymers are used in the prethickening process, especially for centrifugal thickening, a greater amount of unit energy may be required for mixing. In cases where the air-mixing requirement exceeds the oxygen transfer requirement (e.g., when high-efficiency fine-bubble diffusers are used for aeration) supplemental mechanical mixing should be considered rather than overdesigning the oxygen transfer system. The increased capital cost for supplemental mixing should be balanced against power costs for more aeration to determine the optimum configuration.

4.2.2 System Design Considerations

Factors that must be considered in designing conventional aerobic digesters include method of operation, tank volume and detention time, tank design, and aeration and mixing equipment. Typical design criteria for aerobic digestion are presented in Table 4.1.

Method of Operation The two primary operational modes for conventional aerobic digesters are batch or continuous operation, referring to the manner in which supernatant is withdrawn from the process. Figure 4.3 shows cross sections of digesters with the two modes of operation. *Batch operation* is typically used for small-capacity digestion systems because of the relative simplicity of operation. The *continuous mode* allows regular operation without interrupting the oxygenation and mixing equipment. Although baffles and a stilling well can be incorporated inside the digester to separate the supernatant, as shown in Figure 4.3, a separate settling tank is recommended, as shown in Figure 4.2, because of the mixing-induced turbulence carrying into the stilling well. Design of separate settling basins is similar to the design of gravity thickeners or flotation thickeners for WAS. For gravity thickeners, surface loading rates ranging from 25 to 50 $kg/m^2 \cdot d$ (5 to 10 lb/ft^2-d) are used. For flotation thickeners, much higher loading rates of 50 to 100 $kg/m^2 \cdot d$ (10 to 20 lb/ft^2-d) are allowed.

TABLE 4.1 Design Criteria for Aerobic Digester

	SI Units		U.S. Customary Units	
Parameter	Value	Units	Value	Units
SRT[a]				
At 20°C	40	d	40	d
At 15°C	60	d	60	d
Volatile solids loading	1.6–4.8	$kg/m^3 \cdot d$	0.1–0.3	lb/ft^3-d
Oxygen requirements				
Cell tissue applied[b]	2.0	kg O_2/kg VSS	2	lb O_2/lb VSS
BOD in primary sludge destroyed	1.6–1.9	kg O_2/kg BOD	1.6–1.9	lb O_2/lb BOD
Energy requirements for mixing				
Mechanical aerators	20–40	$kW/10^3 m^3$	0.75–1.5	$hp/10^3 ft^3$
Diffused air mixing	1.2–2.4	$m^3/m^3 \cdot h$	20–40	$cfm/10^3 ft^3$
Dissolved O_2 residual in liquid	1–2	mg/L	1–2	mg/L
Reduction of VSS	38–50	%	38–50	%

Source: Adapted from Metcalf & Eddy, 2003.

[a] To meet pathogen reduction requirements (PSRP) of 40 CFR Part 503 regulations.

[b] With complete nitrification.

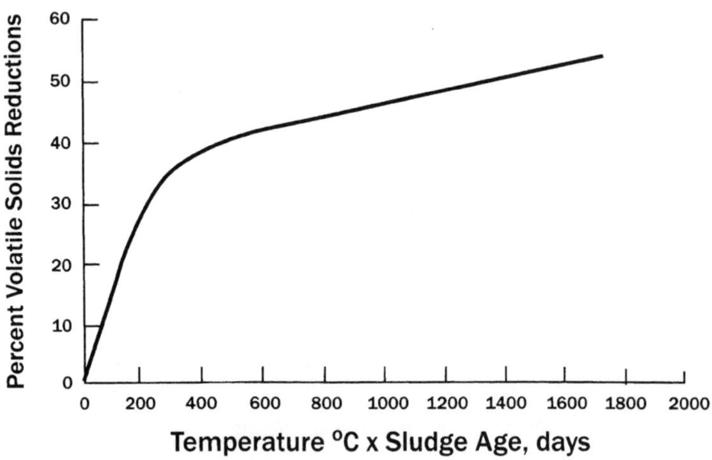

Figure 4.5 Volatile solids reduction versus liquid temperature and sludge age.

Tank Volume and Detention Time The tank volume of an aerobic digester is governed by the detention time necessary to achieve the desired volatile solids reduction. Solids destruction has been shown to be primarily a direct function of both liquid temperature in the digester and the SRT (sludge age), as illustrated in Figure 4.5. This figure is a plot of volatile solids reduction versus the parameter degree-days (temperature × sludge age) and is derived from data taken from both pilot- and full-scale studies on several types of municipal wastewater sludges. In the past, aerobic digesters were designed for a detention time ranging from 10 to 20 days to achieve a 38% volatile solids reduction. However, to meet the pathogen reduction requirements of 40 CFR Part 503 regulations, the SRT criteria of 40 days at 20°C and 60 days at 15°C (see Table 4.1) take precedence over the vector attraction criteria of 38% volatile solids reduction.

In some of the extended aeration wastewater treatment facilities with excessive detention time, 38% volatile solids reduction may not be achievable. In such instances, vector attraction reduction of aerobically digested sludge can be demonstrated, following the 503 regulations, by aerobically digesting a portion of the digested sludge from the digester having a solids concentration of 2% or less in a laboratory bench-scale unit for 30 days at 20°C. Vector attraction reduction is achieved if the volatile solids reduction is less than 15% from the beginning to the end of the 30-day period. In addition, vector attraction reduction, also based on Part 503 regulations, can be achieved by using the specific oxygen uptake rate (SOUR) criteria of less than 1.5 mg of oxygen per hour per gram of total solids at a temperature of 20°C.

The volume of an aerobic digester can be determined by using the equation (WEF, 1998)

$$V = \frac{Q_i(X_i + YS_i)}{X(K_d P_v + 1/SRT)} \tag{4.10}$$

where

V = volume of aerobic digester, m^3 (ft^3)
Q_i = average flow rate to digester, m^3/d (ft^3/d)
X_i = influent suspended solids, mg/L
Y = portion of influent BOD consisting of primary solids, %
S_i = influent BOD, mg/L
X = digester suspended solids, mg/L
K_d = reaction rate constant, d^{-1}
P_v = volatile fraction of digester suspended solids, %
SRT = solids retention time, d

The term YS_i can be disregarded if no primary sludge is included in the load to the digester. If the aerobic digestion process is operated in a staged configuration with two or three tanks in series, the total SRT should be divided approximately equal among the stages.

Tank Design Aerobic digesters can be designed with rectangular, circular, or annular geometry. In circular tanks where draft-tube air mixing is used, bottom slope in the tanks typically range from 1:12 to 1:4. Digesters where air diffusers are used for mixing and oxygenation, floors are generally flat, although a gentle slope may be provided to a small sump to facilitate biosolids removal or draining the tank if required. Side water depths are similar to those provided for activated sludge systems, but with freeboards in excess of 1 m (3 ft) to contain excessive foaming that may occur. At least two tanks should be provided to permit draining and equipment repair. Multiple units are especially important in batch operation systems to provide digester capacity during supernatant discharge and biosolids removal cycle.

Aerobic digesters are typically uncovered tanks of concrete construction. Steel tanks are normally used in small plants. Several recent installations have included covered tanks, especially in extremely cold regions, in consideration of the temperature-dependent nature of the process.

Aeration and Mixing Equipment Several types of aeration devices have been used successfully to provide the oxygenation and mixing requirements of aerobic digesters. These include diffused air, mechanical surface aeration, mechanical submerged turbines, draft-tube aeration, jet aeration, and combined systems.

The design of diffused air systems is similar to that used in aeration basins in conventional activated sludge systems. Coarse-bubble orifice-type diffusers can be located along one side of the tank near the tank bottom in rectangular tanks to produce a spiral or cross-roll pattern. Floor-mounted grid systems with coarse- or fine-bubble diffusers can also be used. Advantages of diffused air systems include the ability to control oxygen transfer by varying the air supply rate, and addition of heat from compressed air into the digester with a resulting increase in the rate of biological activity. However, diffused air systems can have recurring clogging problems, particularly in batch systems where solids are allowed to settle.

Mechanical surface aerators are used primarily in large tanks. They are typically floating, pontoon-mounted devices of either low- or high-speed design. Low-speed design is more common. Mechanical surface aerators have relatively high oxygen transfer efficiencies and are low-maintenance devices. The main disadvantages are their lack of control of the oxygenation rate and their potential ability to destroy the structure the solids flocs.

Mechanical submerged aerators combine several advantages and eliminate some disadvantages of the diffused air and surface aeration devices. Draft tube aeration devices are similar to the gas mixing devices used in anaerobic digesters. They are used in circular tanks. Jet aeration devices have somewhat higher oxygen transfer efficiency than submerged turbines. However, problems with plugging of the devices have occurred in the past where liquid flow paths have not been large enough to pass stringy solids commonly found in aerobic digesters. Combined systems of diffused air and submerged mixers may be required in digesters with high solids concentrations to keep the solids in suspension. Additionally, the submerged mixers in such a system can be operated as a mixer only, thereby promoting denitrification.

Design Example 4.1 Design a batch-operated aerobic digester system to treat waste activated sludge with 4000 lb/d (1814 kg/d) of solids to achieve a minimum 40% volatile solids reduction in winter. Compare it with a continuous-flow system. Assume the following operational parameters:

minimum digester liquid temperature in winter: 15° C

maximum digester liquid temperature in summer: 23° C

concentration of thickened WAS from gravity thickener: 2.5%

VSS in feed sludge: 78% of TSS in feed sludge

reaction rate constant K_d: 0.06 d^{-1} at 15° C

average solids concentration of liquid in digester:

batch–operation system: 70% feed sludge concentration

continuous operation system (with decanting and recycle): 3%

Neglect the specific gravity of sludge.

1. Thickened sludge flow rate

$$= \frac{4000\text{ lb/d}}{(8.34\text{ lb/gal})(0.025)} = 19{,}184\text{ gpd} = 2565\text{ ft}^3/\text{d}\ (73\text{ m}^3/\text{d})$$

2. SRT must be a minimum of 60 days to meet the pathogen reduction requirements (see Table 4.1); therefore,

$$\text{required digester volume} = (2565\text{ ft}^3/\text{d})(60\text{ d}) = 153{,}900\text{ ft}^3\ (4359\text{ m}^3)$$

Compute the required volume using equation (4.10).

$$\frac{(2565\text{ ft}^3/\text{d})(25{,}000\text{ mg/L})}{(25{,}000\text{ mg/L})(0.7)(0.06/\text{d})(0.78) + 1/60\text{ d}} = 57{,}736\text{ ft}^3\ (1635\text{ m}^3)$$

Use the larger of the two volumes, 153,900 ft^3 (4359 m^3). Provide two tanks of 76,950 ft^3 (2179 m^3) each, 70 ft (21.3 m) in diameter and 20 ft (6.1 m) in sidewater depth with 3 ft (0.9 m) of freeboard; or 62 ft (19.0 m) square and 20 ft (6.1 m) in sidewater depth with 3 ft (0.9 m) of freeboard.

3. VSS reduction: For winter conditions, degree-days = (15°C)(60 d) = 900 degree-days. From Figure 4.5, the VSS reduction for 900 degree-days = 45%. This exceeds the winter requirement of 40%, hence OK.
For summer conditions, degree-days = (23°C)(60 d) = 1380 degree-days. From Figure 4.5, the VSS reduction for 1380 degree-days = 49%.
4. Mass of VSS reduction:

$$\begin{aligned}
\text{total mass of VSS} &= (4000\,\text{lb/d})(0.78) \\
&= 3120\,\text{lb/d}\ (1415\,\text{kg/d}) \\
\text{VSS reduction in winter} &= (3120\,\text{lb/d})(0.45) \\
&= 1404\,\text{lb/d}\ (677\,\text{kg/d}) \\
\text{VSS reduction in summer} &= (3120\,\text{lb/d})(0.49) \\
&= 1529\,\text{lb/d}\ (693\,\text{kg/d})
\end{aligned}$$

5. Oxygen required: From Table 4.1, 2 lb (2 kg) O_2 per 1b (kg) of VSS applied is required.
Note: Some designers use 2.0 to 2.3 lb/lb (2.0 to 2.3 kg/kg) VSS destroyed instead of VSS applied. This will not, however, give adequate mixing energy. If primary sludge is also included in the feed to the digesters, additional air is required to oxidize the organic solids in the primary sludge. In such instances, the total air required should be compared to the maximum mixing energy requirement and should be adjusted accordingly.

$$\begin{aligned}
O_2 \text{ required} &= (3120\,\text{lb/d})(2\,\text{lb/lb-d}) \\
&= 6240\,\text{lb/d}\ (2831\,\text{kg/d})
\end{aligned}$$

6. Air required:

$$\begin{aligned}
\text{standard air} &= \frac{6240\,\text{lb/d}}{(0.075\,\text{lb/ft}^3)(0.232)} \\
&= 358{,}621\,\text{ft}^3/\text{d}\ (10{,}156\,\text{m}^3/\text{d})
\end{aligned}$$

After making corrections for plant elevation, ambient temperature, and α and β coefficients, assume an oxygen transfer efficiency of 10%:

$$\begin{aligned}
\text{airflow rate} &= \frac{358{,}621\,\text{ft}^3/\text{d}}{(0.1)(1440\,\text{min/d})} \\
&= 2490\,\text{cfm}\ (71\,\text{m}^3/\text{min}) \\
\text{airflow rate/unit volume} &= \frac{2490\,\text{cfm}}{153{,}900\,\text{ft}^3/1000} \\
&= 16\,\text{cfm}/10^3\,\text{ft}^3\ (0.02\,\text{m}^3/\text{m}^3\cdot\text{min})
\end{aligned}$$

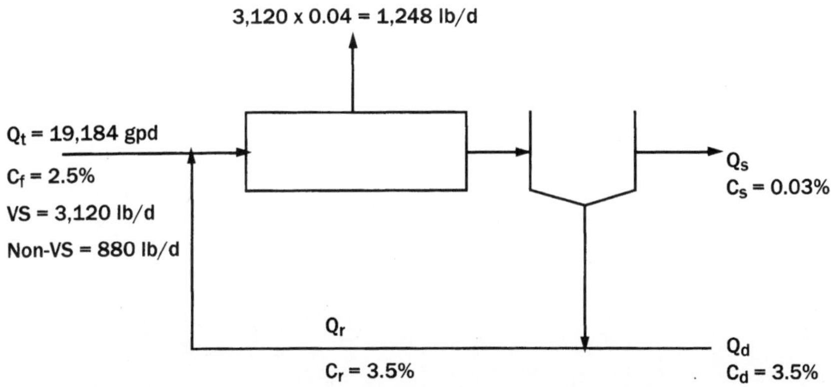

Figure 4.6 Mass balance with decanting and recycling.

Note: This is less than the minimum required that is listed in Table 4.1. Therefore, additional air or supplemental mixing devices must be provided to mix the digester content properly. Consult equipment manufacturers for the type of mixing devices and the horsepower required. Assuming 0.5 hp per $10^3 ft^3$ for supplemental mixing, the mixing energy required is 40 hp for each of the two digesters. An alternative is to provide mechanical aerators for oxygen transfer and mixing in lieu of diffused air. This problem is normally encountered when digesters are designed for the 60-day SRT at 20°C per the Part 503 regulations.

7. To design a continuous-flow digestion system, a mass balance with decanting and recycle (Figure 4.6) should be prepared as follows:

$$\text{non-VSS in feed sludge} = (4000 - 3120)\ \text{lb/d}$$
$$= 880\ \text{lb/d}$$
$$\text{total feed solids not destroyed} = (3120 \times 0.6 + 880)\ \text{lb/d}$$
$$= 2752\ \text{lb/d}$$

From the mass balance diagram,

$$Q_s + Q_d = Q_f$$
$$= 19{,}184\ \text{gpd} \quad (1)$$
$$(Q_s 8.34 C_s) + (Q_d 8.34 C_d) = \text{soids not destroyed}$$

Assuming a TSS value of 300 mg/L in supernatant and digested sludge concentration of 3.5% yields

$$(Q_s \times 8.34 \times 0.0003) + (Q_d \times 8.34 \times 0.035) = 2752$$
$$0.0025Q_s + 0.2919Q_d = 2752 \quad (2)$$

Multiplying formula (1) by 0.0025 gives

$$0.0025Q_s + 0.0025Q_d = 49 \tag{3}$$

Subtracting (3) from (2), we have

$$0.2669Q_d = 2703$$
$$Q_d = 10{,}127\,\text{gpd}$$
$$Q_s = 9057\,\text{gpd}$$

$$\text{fraction of feed solids not destroyed} = \frac{2752}{4000} = 0.67$$

Let the digester solids concentration = C, and let the recycle ratio $R = Q_r/Q_f$. Then

$$C = \frac{Q_f C_f\,(0.67) + Q_r C_r}{Q_f + Q_r}$$

$$3.0 = \frac{(2.5)(0.67) + R(3.5)}{1 + R}$$

$$R = 2.65$$

Therefore, the recycle of thickened biosolids should be 265%, or 35 gpm on a continuous basis, to maintain the 60-day SRT.
From Figure 4.1, for 40% VSS reduction,

degree-days required = 475

$$\text{SRT required in winter} = \frac{475}{15} = 32 \text{ days}$$

$$\text{SRT} = \frac{\text{total solids in digester}}{\text{total solids removed from the digester/day}}$$
$$= \frac{\text{total solids in digester}}{\text{total solids removed/day} + \text{solids lost in supernatant/day}}$$

$$60 = \frac{(\text{volume of digester } V)(\text{solids concentration in digester})}{[(10{,}127)(0.035)] + [(9057)(0.003)]}$$
$$= \frac{V(0.03)}{357}$$

$$V = 714{,}000\,\text{gal} = 95{,}455\,\text{ft}^3\ (2703\,\text{m}^3)$$

Provide two tanks of 47,728 ft^3 (1352 m^3) each, 55 ft (16.7 m) in diameter and a 20 ft (6.1 m) sidewater depth with 3 ft (0.9 m) of freeboard; or 50 ft (15.2 m) square and 20 ft (6.1 m) of sidewater depth with 3 ft (0.9 m) of freeboard.

Air required is same as computed before, which is 2490 cfm (35 m^3/m^3·min).

$$\text{airflow rate/unit volume} = \frac{2490\ \text{cfm}}{95{,}455\ \text{ft}^3}$$
$$= 26\ \text{cfm}/10^3\ \text{ft}^3\ (0.03\ \text{m}^3/\text{m}^3 \cdot \text{min})$$

Note: According to Table 4.1, the airflow rate is adequate for mixing.
Conclusion: The digester volume is reduced by 38% with a continuous-flow digester with recycle.

8. The 60-day requirement follows the Part 503 regulations for pathogen reduction requirements. However, an SRT of only 32 days is required for 40% VSS reduction at 15°C. The total volume require for this condition is 82,075 ft^3 (2324 m^3), which is only 53% of the volume for the 60-day SRT requirement. However, to meet the pathogen reduction requirements, testing is required as described in Section 4.1.

4.2.3 Operational Considerations

Compared to anaerobic digestion systems, an aerobic digestion process is relatively simple to operate. Similar to an activated sludge system, the aerobic digestion process is essentially self-sustaining. Operational considerations include pH reduction, foaming problems, supernatant quality, and dewaterability of digested sludge.

pH Reduction Decreases in pH and alkalinity have been observed in aerobic digesters at increasing detention times. The drop in pH is caused by acid formation that occurs during nitrification; the drop in alkalinity is caused by lowering of the buffering capacity of the sludge due to air stripping. It has been observed that the system will acclimate and perform well at a pH as low as 5.5. If the digester has separate aeration and mixing equipment (or to some extent is provided with mechanical aerators), it is possible to denitrify by operating only the mixing equipment (or turning off the mechanical aerators) during the fill cycle in a batch process. As discussed earlier, denitrification produces alkalinity. Filamentous growth may occur at low pH values. If the feed sludge has very low alkalinity and the pH in the digester continues to drop below 5.5, provisions may have to be made to increase the alkalinity by adding chemicals.

Foaming Problems Foaming may occur in aerobic digesters during warm-weather periods, due primarily to high organic loading rates. Foaming is more prevalent in digesters that use surface aeration devices. Water sprays are typically used to control the problems. Growth of filamentous organisms can also

TABLE 4.2 Characteristics of Supernatant from Aerobic Digesters

Parameter	Range
BOD (mg/L)	25–150
COD (mg/L)	20–300
Suspended solids (mg/L)	25–300
pH	6.0–7.8

cause foaming problems. Control methods that have yielded mixed results include chlorinating the digester feed to destroy filamentous organisms and turning off the aeration equipment to create temporary anaerobic conditions.

Supernatant Quality One of the advantages of aerobic digestion over anaerobic digestion is the better-quality supernatant, especially when the solids–liquid separation is carried out in a gravity thickener or by other mechanical means. Table 4.2 gives characteristics of supernatant from aerobic digesters. As the supernatant is normally returned to the head end of the treatment plant, the true loading to the aeration basins from the supernatant is represented by the soluble BOD, which is typically less than the organic strength of the wastewater. The suspended solids do not exert a high load to the aeration basin because the solids are in the endogenous stage of respiration. Nevertheless, a well-operated aerobic digester can produce a supernatant with less than 150 mg/L of suspended solids.

Dewatering Belt filter press dewatering of aerobically digested sludge produces cake with 14 to 22% solids. Dewatering characteristics show a definite deterioration with increasing SRT. Dewaterability of aerobically digested sludge is also affected by the degree of mixing provided during the digestion process because the high degree of mixing destroys the structure of the solids floc. Unless pilot plant data indicate otherwise, it is recommended that conservative criteria be used for designing mechanical sludge dewatering facilities.

4.3 PROCESS VARIATIONS

Several variations of the conventional aerobic digestion exist. These include high-purity oxygen digestion, low-temperature aerobic digestion, dual digestion, and mesophilic digestion. A short description of each follows. A fourth variation, autotheramal thermophilic aerobic digestion (ATAD), is described in detail later in the chapter. Studies on technological improvements of aerobic digestion of sludge are discussed at the end of the chapter.

4.3.1 High-Purity Oxygen Digestion

In this process, high-purity oxygen is used in lieu of air as the oxygen source. The process is typically carried out in a closed tank similar to the activated sludge process for wastewater using pure oxygen. High-purity oxygen atmosphere is maintained in the space above the liquid surface, and oxygen is transferred into the sludge via mechanical aerators. The process can also be performed in open tanks, in which case oxygen is introduced to the sludge in minute bubbles with special diffusers. The bubbles dissolve before they reach the liquid surface. Because of the high cost of generating pure oxygen, it is cost-effective only when used in conjunction with a pure oxygen activated sludge system. The principal advantage of the system is that it is relatively insensitive to changes in ambient temperatures when closed tanks are used because of the increased rate of biological activity and the exothermic nature of the process; consequently, it is particularly applicable in cold weather climates. High temperatures resulting from the exothermic process increase the rate of volatile solids destruction.

4.3.2 Low-Temperature Aerobic Digestion

Aerobic digesters in small package-type wastewater treatment plants have been studied to provide better operational control at temperatures lower than 20°C. Investigations at treatment plants in British Columbia, Canada have indicated that the SRT must be increased as operating temperatures decrease to ensure acceptable volatile solids reductions (Koers and Mavinic, 1977; Mavinic and Koers, 1979). At temperatures between 5 and 20°C, the system should operate at 250 to 300 degree-days (product of SRT in days and operating temperature in °C) to maintain an acceptable level of volatile solids ...duction. Heated air or wastewater can be used to decrease detention t... and to prevent freezing of digesters.

4.3.3 Dual Digestion

Dual digestion is aerobic thermophilic digesti... st stage followed by
mesophilic anaerobic digestion in a secon... s system has been used
extensively in Europe. Residence tim... oic digester is typically 18
to 24 hours at the temperature... 5°C. Residence time in the
anaerobic digester is about 1... sis in the aerobic digester results
in increased degradatio... ht anaerobic digestion and gas pro-
duction. Advantage... are (1) increased levels of pathogen
reduction, (2) i... lids reduction, (3) increased methane gas
production... ctor, (4) fewer odors produced by the stabi-
lized sl... d less tankage required than for a single-stage
ar...

4.3.4 Mesophilic Aerobic Digestion

Mesophilic aerobic digestion is an autoheating digestion process that includes sludge thickening followed by two or three stages of treatment in aerobic reactors. To achieve the balance of heat in the process, sludge has to be prethickened to 4 to 5% solids. In the first stage of digestion, the detention time is from 8 to 16 days, the dissolved oxygen is kept between 0.2 and 1.0 mg/L, and the temperature reaches 20 to 35°C. In this stage, ammonium bicarbonate alkalinity keeps the pH between 6.8 and 9.0. In the second stage, the detention time is from 10 to 17 days, the dissolved oxygen is kept between 0.2 and 1.0 mg/L, the temperature is kept between 15 and 30°C, and the pH is 6.5 to 7.5. A third stage may also be added with a detention time of 10 to 17 days. Detention time in all stages can be reduced by attaining temperature in the range 30 to 35°C in the first stage. Mesophilic aerobic digestion can qualify as a process to significantly reduce pathogens (PSRP); however, compared to conventional aerobic digestion, use of this process is limited, due to high capital and operating costs.

4.3.5 Autothermal Thermophilic Aerobic Digestion

Autothermal thermophilic aerobic digestion (ATAD) is a sludge digestion process that is capable of achieving a high degree of stabilization and pathogen reduction. The process is characterized by high reaction rates achieved at a thermophilic temperature of 40 to 70°C. The temperatures are attained by using the heat released by the exothermic microbial oxidation process. Approximately 15,000 kJ of heat is generated per kilogram of volatile solids destroyed. In a completely mixed and aerated environment, the thermophilic temperatures attained are sustained without the addition of supplemental heat (other than the heat introduced by aeration and mixing) by conserving the heat released during biological oxidation. Figure 4.7 illustrat... the heat balance in an ATAD reactor. If sufficient insulation, HRT, and a...quate solids concentrations and mixing are provided, the process can ... controlled at the thermophilic temperatures to achieve greater than 38% vo... solids destruction and sufficient pathogen reduction to meet the U.S. ...gulations for PFRPs or the 40 CFR Part 503 class A designation.

ATAD is a techn... ...ly applied in Europe since the 1970s and more recently in North Am... ...major advantages of ATAD over conventional aerobic digestion ...:

- Significantly reduced SRT ... achieve a volatile solids reduction of 40 to 50%
- Possibility of reducing pathoge... ...ia, viable helminth ova, and other parasites to below detec... ...eeting the pathogen reduction requirements of class A bi...

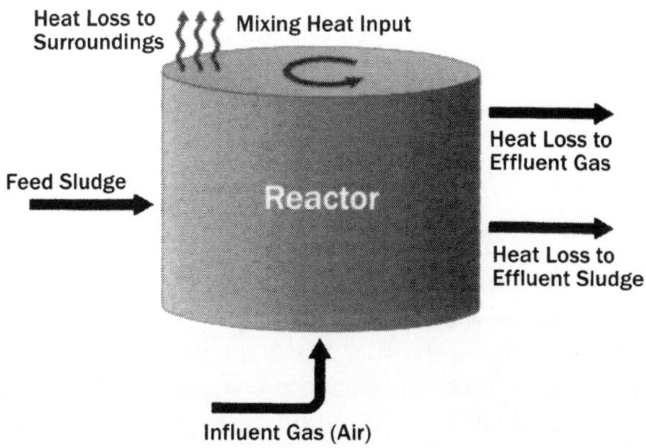

Figure 4.7 Heat balance in an Autothermal thermophilic aerobic digester. (From Fuchs, Cary, NC.)

- Destruction of all weed seeds, making the biosolids highly suitable as a soil amendment or fertilizer for lawns
- Approximately 25% lower oxygen requirement because few, if any, nitrifying bacteria exist in temperatures above 40°C
- Heat possibly recoverable for heating buildings

The disadvantages of ATAD are:

- High capital and operating costs
- Need for feed sludge to be thickened to a minimum solids concentration of 4% (preferably 5 to 6%)
- Requirement for extremely efficient aeration for systems using air instead of high-purity oxygen
- Objectionable odors
- Foam breakers required, due to the high degree of foaming in the reactors
- Poor dewatering characteristics of digested biosolids

Process Theory The aerobic destruction of volatile solids occurs in thermophilic aerobic digesters as illustrated by equation (4.2). Because of the destruction of nitrifying bacteria from the high temperatures in ATAD, the subsequent reactions outlined in equations (4.3) through (4.7) do not materialize. Equation (4.2) with the energy component added is as follows:

$$C_5H_7O_2N + 5O_2 \rightarrow 4CO_2 + H_2O + NH_4HCO_3 + \text{energy} \tag{4.11}$$

This equation indicates that approximately 1.5 kg of oxygen is required per kilogram of volatile solids destroyed. The system needs a hydraulic detention time of only about 5 to 8 days. The ammonia produced in the reaction reacts with water and carbon dioxide to form ammonium bicarbonate, resulting in increased alkalinity. Because nitrification does not occur, the pH will be in the range 8 to 9, higher than in conventional aerobic digesters. The energy produced is approximately 15,000 kJ/kg of volatile solids destroyed. As long as the system is well mixed and sufficient oxygen is provided, the temperature in the reactor will rise until a balance occurs (the heat lost equals the heat input from exothermic reaction and mechanical energy input). The temperature will continue to rise until the process becomes oxygen mass-transfer limited. Some of the ammonia nitrogen produced will be in the off-gas and in solution with concentrations of several hundred mg/L in each. Most of this ammonia will be returned to the treatment plant in the sidestreams from the off-gas odor control system in the biosolids dewatering facilities.

Process Design A typical ATAD system is shown in Figure 4.8. Key elements of this process are feed sludge characteristics, reactors, detention time, feed cycle, aeration and mixing, temperature and pH, foam control, digested biosolids storage and thickening, and odor control. Typical design criteria for the ATAD are given in Table 4.3.

Feed Sludge Characteristics The autothermal thermophilic digesters can effectively digest a mixture of primary and secondary sludge. Blending before feeding to an ATAD is optional. The solids concentration of the feed sludge should be at least 3%, the minimum necessary to attain and maintain thermophilic conditions; and a maximum of 6%, the upper limit for efficient aeration and mixing. The feed must contain a minimum volatile solids content of between 25 and 40 g/L COD. Fine screening or grinding and good grit removal are required to remove or grind the plastic and stringy materials and to minimize abrasion on aerators and mixers.

Reactors Typically, two rectors in series are installed. Concrete and steel have been used in the construction of the tanks. Steel tanks are less susceptible to heat stress and less costly to construct than concrete tanks. However, steel tanks require 140 mm (about 6 in.) of mineral insulation and are clad with ribbed aluminum sheeting on the sides and plane sheeting on the top to protect the insulation from the elements and for aesthetic purposes. Access hatches are provided on top of the tanks. The entire tank is constructed above grade on a concrete foundation. The height-to-depth ratios vary from 0.5 to 1.0. Actual ratios depend on the aerators used and their effectiveness for good mixing.

Heat exchange is not necessary for process requirements but has been incorporated into some facilities for energy recovery and for preheating the feed sludge before the first-stage reactor. Heat can be recovered during biosolids cooling with heat exchangers or by a water-cooling loop installed within

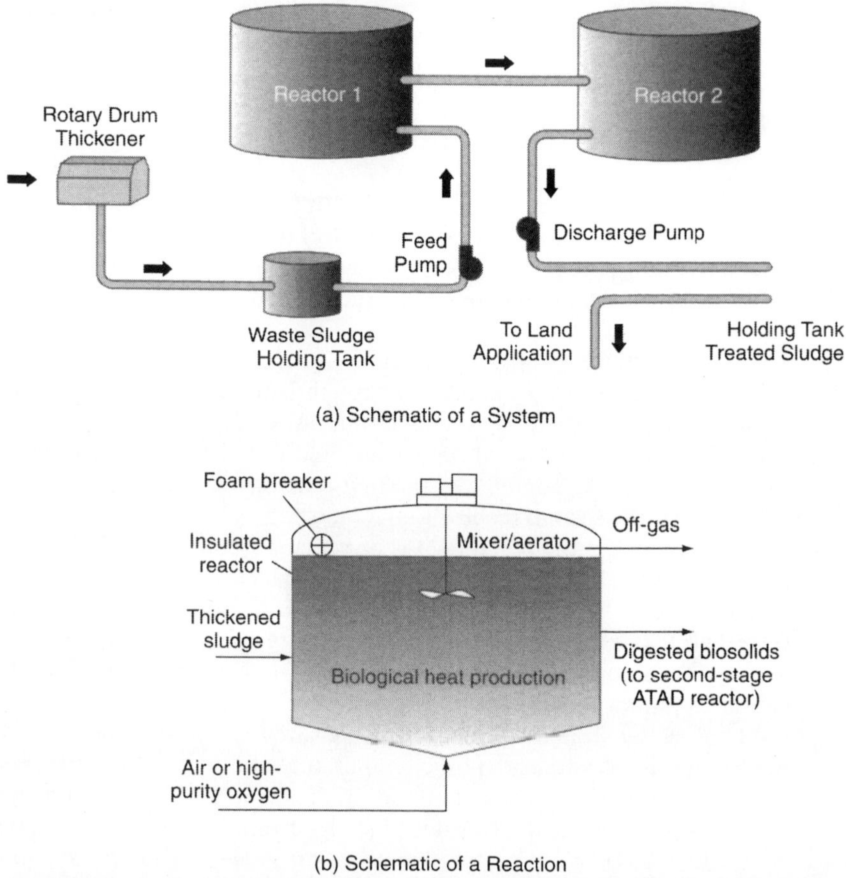

Figure 4.8 Autothermal thermophilic aerobic digestion system. [Part (a) from Fuchs, Cary, NC; part (b) from Metcalf & Eddy, 2003.]

TABLE 4.3 Design Criteria for Autothermal Thermophilic Aerobic Digestion

Parameter	SI Units		U.S. Customary Units	
	Value	Units	Value	Units
Hydraulic detention time	5–8	d	5–8	d
Feed sludge solids concentration	3–6	%	3–6	%
TSS loading	5–8	$kg/m^3 \cdot d$	320–500	$lb/10^3 ft^3$-d
VSS loading	3.2–4.2	$kg/m^3 \cdot d$	200–260	$lb/10^3 ft^3$-d
Temperature				
Stage 1	40–50	°C	104–122	°F
Stage 2	50–70	°C	122–158	°F
Aeration and mixing (aspirating mixer type)				
Air input	4	$m^3/h/m^3$		
Oxygen transfer efficiency	2	kg O_2/kWh	4.4	lb O_2/kWh
Energy requirement	130–170	W/m^3	5.0–6.4	$hp/10^3 ft^3$

Source: Adapted in part from Metcalf & Eddy, 2003.

the reactor shell. Heat can also be recovered from the off-gas. At the treatment plant in Salmon Arm, British Columbia, heat recovery of 1.2 J/s is used to heat the thickener building (WEF, 1998).

Detention Time Reactor hydraulic detention time (HRT) is from 5 to 8 days, with 2.5 to 4 days per reactor. About 40 to 50% of the volatile solids reduction takes place in the first reactor. German design standards include 5 to 6 days of HRT to comply with the pathogen destruction requirements.

Feed Cycle The influent sludge can be introduced into the first reactor continuously, intermittently, or in batches, although batch mode provides assurance in meeting class A pathogen reduction requirements.

In the continuous and intermittent feed modes, as sludge is introduced into the first reactor, the contents of the first reactor overflow to the second reactor, and from the second reactor to the biosolids holding tank. If aspirating aerators are used in the reactors, it is important to have a constant liquid level to ensure uniform and consistent oxygen transfer.

For batch mode, the system is designed to feed a one-day volume of the sludge to the reactor in less than 1 hour to ensure that the feed solids are exposed to the reactor temperature continuously for a minimum of 23 hours. This enhances pathogen destruction. Digested biosolids are withdrawn prior to introducing sludge from the first reactor to ensure that the effluent biosolids have been maintained at thermophilic temperatures for at least 24 hours and to minimize the possibility of contaminating the treated biosolids with partially treated sludge. After the transfer of sludge from the first reactor to the second reactor is completed, raw sludge is introduced again to the first reactor.

Aeration and Mixing The key to effective ATAD performance is aeration and mixing. The aeration system must be designed to (1) transfer sufficient oxygen to meet the high demand of the digestion process, (2) supply the oxygen required while minimizing the latent heat loss in the exhaust air, and (3) provide adequate mixing of the sludge to ensure complete stabilization. Aeration and mixing systems include aspirating aerators, Venturi aeration equipment, jet aeration, and immersible mechanical aerators.

Nearly all ATAD systems utilize aspirating aerators to introduce air and provide the mixing. Typical installations have a minimum of two aerators mounted on the side of each reactor. Larger installations may require a third unit on top of the roof at the center or additional wall-mounted units. The advantage of this type of aerator is that the motors and bearings are located outside the reactor. The design criteria for aspirating aerators are given in Table 4.3.

A combination recirculation pump and Venturi arrangement with air supplied to the Venturi has been used successfully. The main advantage of this aerator is that both the pump and the Venturi are located outside the reactor.

However, solids handling, pumps with corrosion-resistant impellers, and pump volute linings are required.

Temperature and pH Process operating temperature range in the first reactor is 40 to 50°C. During feeding, a drop in temperature will occur in this reactor. With an aspirating aeration system, the typical temperature recovery rate is 1°C/h (1.8°F/hr). The temperature in the first reactor should not be allowed to drop below 25°C (60°F), to avoid biological adaptation problems. Process operating temperature range in the second reactor is 50 to 70°C (102 to 158°F).

Typically, pH need not be controlled in an ATAD system because alkalinity is increased by the biological oxidation of cell mass, and no nitrification occurs to lower the alkalinity. Normally, pH in the first reactor is above 7 and in the second reactor is above 8.

Foam Control Substantial amounts of foam are generated in ATAD reactors because of cellular proteins, lipids, and oil and grease materials breaking down and releasing into solution. Control of the foam layer is important; however, the exact role of the foam layer has not been completely explained. The foam seems to improve oxygen utilization, provides insulation, and enhances biological activity. However, excessive foam inhibits air from entering the digesting sludge mass.

A freeboard of 0.5 to 1.0m (1.65 to 3.3ft) should be provided in the reactors to use as volume for foam development and control. Mechanical foam cutters suspended in the reactors at fixed elevations are used most commonly for foam control. Other methods include vertical mixers and spray systems. The design and operation of foam cutters are empirical and must consider the surface area of the reactors, solids concentration of sludge in the reactors, and the type and intensity of aeration.

Digested Sludge Storage and Thickening Cooling of the digested biosolids is necessary to achieve solids consolidation. A minimum of 20 days of detention is required for cooling and thickening. Detention time can be reduced substantially if heat exchangers are used for cooling the biosolids discharged from the reactors. Detention tanks are typically open-top, unmixed, and fitted with decant capability. Odor control is normally not required. Biosolids thickening by gravity thickeners can produce solids concentrations of 6 to 10%.

Odor Control Because there is no nitrification and because of the high temperatures in the ATAD systems, relatively high concentrations of ammonia are released. Reduced sulfur compounds, such as hydrogen sulfide, carbonyl sulfide, methyl mercaptan, ethyl mercaptan, dimethyl sulfide, and dimethyl disulfide, also result from the ATAD process.

Odors can be controlled if proper operating temperatures are achieved and the reactors are adequately mixed and aerated. Odor control systems may include wet scrubbers, biofilters, compost/soil filters, and diversion of off-gas to the activated sludge process.

Performance and Operation To meet the class A biosolids requirements of the Part 503 regulations, the requirement that needs to be demonstrated is (1) fecal coliform densities are less than 1000 MPN/g of total solids on a dry weight basis, or (2) *Salmonella* sp. bacteria concentrations are below detention limits of 3 MPN/g of total solids on a dry weight basis. Batch mode operation is better suited for meeting these requirements. In the continuous feed mode, it is possible that some pathogens pass through the system. Two or more reactors are required to ensure that all solids in the reactors are subject to the time and temperature requirements.

Part 503 regulations' vector attraction reduction requirement is a minimum 38% volatile solids reduction, or a SOUR of less than 1.5 mg of oxygen per gram per hour on a dry weight basis at a temperature of 20°C. Limited data are available on the ability of the ATAD process to meet these requirements. Volatile solids reduction is influenced by the feed sludge characteristics, HRT, operating temperature, and reactor loading. Data show volatile solids reduction ranging from 30 to 60%. Therefore, the ability of the process to meet the minimum requirement of 38% volatile solids reduction depends on a properly designed and operated system.

4.3.6 Technological Improvements

The principal author of this book, together with the Medical Hygiene Institute in Russia and several wastewater treatment plants, conducted research on aerobic sludge digestion (Turovskiy, 2001). The object of this research was to obtain sanitarily harmless biosolids with an effective sludge digestion technology that was also cost-effective.

Thickening Experiments with activated sludge thickening were conducted with samples from treatment plant thickeners using simulators. The thickening of activated sludge decreases its dewaterability significantly. The longer the activated sludge thickening process is, the higher the specific resistance of the thickened sludge, as shown in Figure 4.9. During the process of waste activated sludge thickening, the concentration of dry solids increased from 0.2% to 2.0%, and the specific resistance increased severalfold. However, when concentrations increased from 2.0% to 3.0%, the volumes were reduced only one and a half times, while the specific resistance rose from 1000 m/kg to 3500 m/kg (see Figure 4.9). When thickening lasts longer than 8 hours, organic putrefaction takes place. In addition, microorganisms perish due to lack of oxygen, the amount of colloids increases, and part of the free water

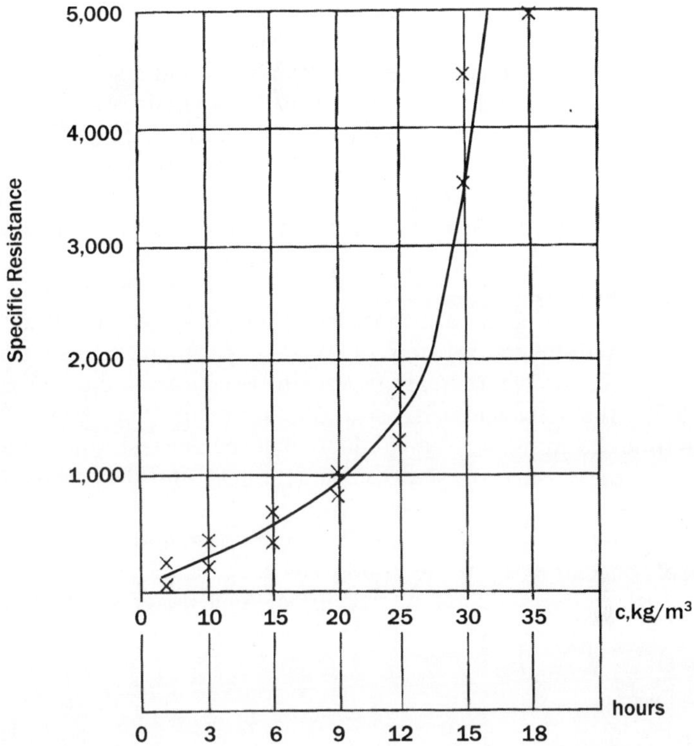

Figure 4.9 Specific resistance of activated sludge versus thickening time.

transforms into bound water with solids. Unthickened activated sludge usually has lower specific resistance and better dewatering capabilities than those of thickened sludge. On the other hand, digesting unthickened activated sludge does not make sense, due to the large volume and low initial concentration of solids. The kinetics of activated sludge thickening allows determining the rational time of thickening and concentration of solids.

Detention Time The duration of the volatile phase of solids oxidation depends on the food/microorganism ratio, the sludge temperature, and the intensity and quantity of air supplied. The wastewater composition also plays a role. The process of aerobic digestion of a sludge's organics and biomass was described by equations (4.1) through (4.7). At the beginning of the aerobic digestion process, oxidation of organic contaminants takes place. This process is followed by its mineralization, and finally, by self-oxidation and disintegration of the biomass. In digestion process studies, volatile solids were reduced 5 to 50%, fat 65 to 75%, and protein 20 to 30%. Activated sludge needs 7 to 10 days to stabilize; primary sludge takes 20 to 30 days at 20°C. At 8 to 10°C, stabilization takes 2 to 2.5 times longer. Thickened activated sludge needs a longer detention time to stabilize volatile solids. Figure 4.10 shows how volatile solids decrease in activated sludge digestion. If the initial volatile solids concentration is equated to 100%, after 7 to 10 days of digestion at a solids concentration of 10 to 18 kg/m^3, the solids content is 32 to 37% (line 1 in Figure 4.10); at 18 to 24 kg/m^3, the solids content is 25 to 33% (line 2 in Figure 4.10); and at 24 to 30 kg/m^3, the solids content is 15 to 23% (line 3 in Figure 4.10).

Table 4.4 illustrates changes in specific resistance during the aerobic digestion of activated sludge. A long detention time decreases the dewaterability of digested sludge. However, in several experiments, the specific resistance of digested sludge decreased. An effective process may be achieved by thickening the activated sludge 4 to 6 hours before digestion and with a digestion time of 3 to 5 days. Some reduction of specific resistance takes place when the organic part of sludge disintegrates and the solids become heavier.

Detention time also relates to air consumption. A high food/microorganism ratio of concentrated activated sludge needs more oxygen at the beginning of the digestion process: up to 2 m^3/h of air per cubic meter of activated sludge

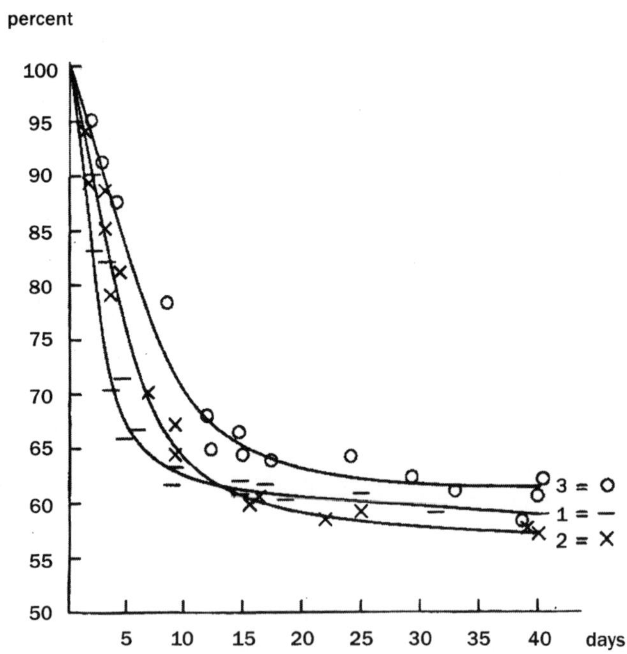

Figure 4.10 Volatile solids reduction versus detention time in activated sludge system.

TABLE 4.4 Changes in Specific Resistance (m/kg) During Aerobic Digestion of Activated Sludge

Dry Solids (%)	Detention Time (days)						
	0	1	3	5	10	15	40
1.0	300–400	160–270	70–180	230–390	400–690	—	—
1.5	500–700	360–530	220–370	410–640	540–870	—	—
2.0	800–1000	810–1100	1150–1480	1290–1850	4140–5510	4030–4690	1900–2600
2.5	1200–1800	1360–2100	1480–2600	1670–4500	5780–6250	4970–5800	2810–3960
3.0	3500–4500	3800–4100	4090–4920	5300–6170	6190–7020	5910–6690	4300–5720

at 20°C. When volatile solids decrease, the air consumption also decreases. This fact can be put into practical use by reducing the aeration to lower and lower rates as detention time increases. However, auxiliary mixing should be provided to prevent the solids from settling to the bottom of the digester. The oxygen requirement for a mixture of waste activated sludge and primary sludge at the beginning of the digestion process is 5 to 10 times more than that needed for activated sludge alone. Compared to activated sludge, the aerobic digestion of sludge from primary clarifiers needs more oxygen, a longer detention time, and increases specific resistance.

Disinfection Aerobic digestion of activated sludge with a detention time of 40 days and at a temperature of 20°C. leads to relatively safe levels of coliforms and pathogenic viruses. Reduction in indicator organisms and viruses from 70 to 99% takes place in 10 days at 20°C. A larger microbial population can be found in agricultural soils (Lue-Hing et al., 1998). One of the possible causes of pathogen destruction is the high Eh-potential of the digestion process (200 to 700 mV).

Aerobic digestion destroys only part of the helminth ova. These eggs number several hundreds in 1 kg of aerobically digested sludge, and they can survive a long time. Experiments showed that helminth ova could be destroyed by heating digested sludge at 50°C for 2 hours, at 60°C for a few minutes, and at 70°C for a few seconds (Turovskiy, 1999). After mechanically dewatering and heating sludge to 65°C, aerobically digested sludge revealed no presence of the intestinal typhoid group of bacteria on Wilson–Bleaur, Ploshiryov, or Miller Kaufman medium, or on media with different inhibitors. Studies also showed that because of the extreme changeability of the *colon bacillus* (revealed in the process of reactivation), there should be no fear of livability or virulence of pathogenic microbes during the utilization of dewatered heated biosolids.

Technological Schemes The technological schemes developed by the senior author are shown collectively in Figure 4.11. The technology includes thickening, aerobic digestion, disinfection of sludge, and converting the sludge to

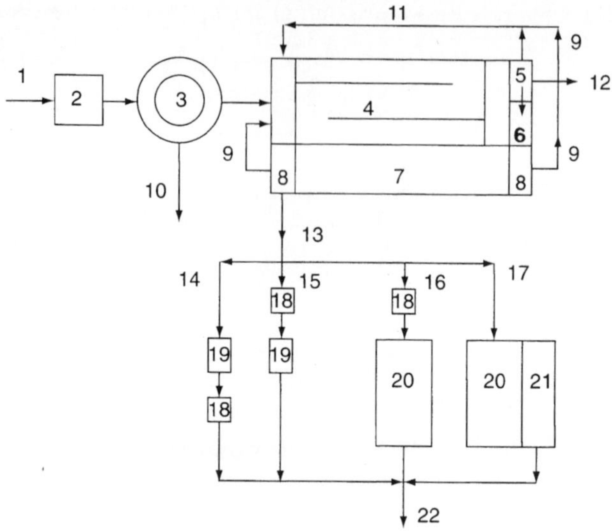

1- Wastewater inflow, 2 - Screening and grit removal, 3 - Primary clarifier, 4 - Aeration tank, 5 - Secondary clarifier, 6 - Sludge thickener, 7 - Aerobic digester, 8 - Thickening zone, 9 - Supernatant, 10 - Primary sludge to treatment, 11 - Return activated sludge, 12 - Treated effluent, 13 - Digested sludge, 14, 15, 16, 17 - Variations of disinfecting and dewatering of digested sludge, 18 - Sludge heater, 19 - Mechanical dewatering, 20 - Drying beds, 21 - Composting, 22 - Disinfected biosolids as fertilizer.

Figure 4.11 Technical improvements for aerobic digestion and disinfection of sludge.

useful biosolids. For small municipal wastewater treatment plants treating up to $1500\,m^3/d$ (0.4 mgd) and with BOD up to 150 mg/L, a scheme without primary clarifiers can be used. Waste activated sludge is thickened in a special zone inside the digester. The duration of thickening should be 3 to 5 hours, with a concentration of thickened sludge at 10 to 13 g/L. In this scheme, the detention time of activated sludge digestion should be 5 to 8 days with an aeration rate of $1\,m^3/h$ of air per cubic meter of sludge. Sedimentation time for digested sludge is 1.5 to 3 hours. Supernatant with a BOD value of about 100 mg/L is recycled to the aeration tank. After sedimentation, the digested sludge with solids concentration of 1.5 to 2.5% goes to a heater, where it is heated to 70°C and then discharged to drying beds. The loading rate of drying beds in a region with a mean annual ambient temperature of 4 to 6°C and annual precipitation of up to 500 mm can be $3\,m^3/m^2$ of drying bed per year. The resulting biosolids have moisture content of 75 to 80% and can be used as agricultural fertilizer.

For treatment plants treating 1500 to 5000 m^3/d (0.4 to 1.3 mgd), a scheme that includes primary clarifiers and aerobic digestion of combined primary sludge and thickened waste activated sludge can be used. The duration of thickening waste activated sludge should be 5 to 6 hours to produce a solids concentration of 13 to 15 g/L. The detention time in the digester should be 10 to 15 days at 20°C, and the aeration rate should be 6 m^3/h of air per cubic meter of combined sludge. This technology has been implemented at several treatment plants in Russia.

Experience has also shown that in northern regions in winter, when the temperature drops to –30°C and the digester temperature is about 3°C, the efficiency of this technology can be improved by discharging to the digesters a mixture of mixed liquor from the aeration tanks and the waste activated sludge, and by warming the primary sludge to 60°C or by heating the air supply. Two to four hours of thickening of digested sludge yields a dry solids concentration of 3 to 4%. Thickened digested sludge is normally pumped to the drying beds. The drying bed loading rate is 2 m^3/m^2 of drying bed per year. Dewatered sludge from drying beds with a moisture content of 70 to 78% is composted with bulking materials such as sawdust, wood chips, and compost. The composted biosolids are a good organic fertilizer.

For wastewater treatment plants with a capacity of more than 5000 m^3/d (1.3 mgd), an effective scheme includes separate treatment of waste activated sludge and primary sludge (e.g., aerobic digestion of waste activated sludge and anaerobic digestion of primary sludge). In this scheme, waste activated sludge is thickened 3 to 8 hours to a solids concentration of 10 to 18 g/L and discharged to the aerobic digester. Digester detention time is 7 to 10 days with air consumption being 2 m^3/h per cubic meter of sludge at a temperature of 20°C. Digested sludge is concentrated for 3 to 5 hours to 2.5 to 3.0% dry solids. The thickened digested sludge is dewatered using belt filter presses or centrifuges. Due to the low specific resistance, relatively less polymer is required for dewatering. For disinfecting, the sludge is heated to 70°C before or after dewatering. The dewatered sludge may be used as organic fertilizer.

Overall, the studies proved that heating of digested sludge allows digestion of waste activated sludge of low concentrations at reduced detention time, which results in smaller thickeners and digesters. In addition, the scheme of separate treatment of waste activated sludge and primary sludge is very efficient because of the decrease in detention time and air consumption and the improved dewaterability of digested sludge. The technological improvements produce biosolids that meet class A or class B requirements of the 40 CFR Part 503 regulations.

REFERENCES

Ahlberg, N. R., and Boyko, B. I. (1972), Evaluation and Design of Aerobic Digesters, *Journal of the Water Pollution Control Federation*, Vol. 44, p. 634.

Benefield, L. D., et al. (1978), Design Relationships for Aerobic Digestion, *Journal of the Water Pollution Control Federation*, Vol. 50, p. 518.

Benefield, L. D., and Randall, C. W. (1980), *Biological Process Design for Wastewater Treatment*, Prentice-Hall, Englewood Cliffs, NJ.

Chu, A., and Mavinic, D. S. (1998), The Effects of Macromolecular Substrates and a Metabolic Inhibitor on Volatile Fatty Acid Metabolism in Thermophilic Aerobic Digestion, *Water Science and Technology*, Vol. 38, No. 55.

Deeney, F. B., et al. (1991), Autothermal Thermophilic Aerobic Digestion, *Water Environment and Technology*, Vol. 65.

Enviroquip (1997), *Aerobic Digestion Workshop*, Vol. 1, Enviroquip, Inc., Austin, TX.

——— (2001), *Aerobic Digestion Workshop*, Vol. III, Enviroquip, Inc., Austin, TX.

Eyma, R., et al. (1999), Thermophilic Aerobic Digestion to Achieve Class A Biosolids, *Florida Water Resources Journal*, December, pp. 24–28.

Farrell, J. (1999), Summary of Designs, in *Aerobic Digestion Workshop*, Vol. III, Enviroquip, Inc., Austin, TX.

Federal Register (1993), 40 CFR Part 503, *Standards for the Use or Disposal of Sewage Sludge*.

Fuchs ATAD Systems, *Autothermal Thermophilic Aerobic Digestion*, I. Kruger, Inc., Cary, NC.

Ganczarczyk, J., Hamoda, M. F., and Wong, H. L. (1980), Performance of Aerobic Digestion at Different Sludge Solids Levels and Operating Patterns, *Water Research*, Vol. 14, No. 11, pp. 627–633.

Hartman, R. B., et al. (1979), Sludge Stabilization Through Aerobic Digestion, *Journal of the Water Pollution Control Federation*, Vol. 51, p. 2353.

Jewell, W. J., and Kabrick, R. M. (1980), Autoheated Aerobic Thermophilic Digestion with Air Aeration, *Journal of the Water Pollution Control Federation*, Vol. 52, p. 512.

Kelly, H. G. (1991a), Autothermal Thermophilic Aerobic Digestion: A Two-Year Appraisal of Canadian Facilities, *Proceedings of the Environmental Engineering Specialty Conference*, ASCE, Reno, NV.

——— (1991b), Autothermal Thermophilic Digestion of Municipal Sludge: Conclusion of a One-Year Full-Scale Demonstration Project, presented at the 64th Annual Conference of the Water Pollution Control Federation, Ontario, Canada.

———, et al. (1993), Autothermal Thermophilic Aerobic Digestion: A One-Year Full-Scale Demonstration Project, *Water Environment Research*, Vol. 65, p. 849.

Koers, D. A., and Mavinic, D. S. (1977), Aerobic Digestion of Waste Activated Sludge at Low Temperatures, *Journal of the Water Pollution Control Federation*, Vol. 49, No. 3, p. 460.

Krishnamoorthy, R., and Loehr, R. C. (1989), Aerobic Sludge Stabilization: Factors Affecting Kinetics, *ASCE Journal of the Environmental Engineering Digestion*, Vol. 115, pp. 283–301.

Lue-Hing, C., Zery, D. R., and Kuchenither, R. (Eds.) (1998), *Water Quality Management Library*, Vol. 4, *Municipal Sewage Sludge Management: A Reference Text on Processing, Utilization and Disposal*, Technomic Publishing Co., Lancaster, PA.

Mavinic, D. S., and Koers, D. A. (1979), Performance and Kinetics of Low Temperature, Aerobic Sludge Digestion, *Journal of the Water Pollution Control Federation*, Vol. 51, p. 2088.

Maxwell, M. J., et al. (1992), Impact of New Sludge Regulations on Aerobic Digester Sizing and Cost-Effectiveness, *Proceedings of the Water Environment Federation 65th Annual Conference and Exposition*, New Orleans, LA.

Metcalf & Eddy, Inc. (2003), *Wastewater Engineering: Treatment and Reuse*, 4th ed., Tchobanoglous, G., Burton, F. L., and Stensel, H. D. (Eds.), McGraw-Hill, New York.

Murray, K. C., et al. (1990), Thermophilic Aerobic Digestion: A Reliable and Effective Process for Sludge Treatment at Small Works, *Water Science and Technology*, Vol. 22, p. 225.

Pride, C. (2002), ATADs, Odor and Biofilters, *Florida Water Resources Journal*, April, pp. 18, 20, 25, 26.

Porteous, J. (1998), Controlled Aerobic Digestion of Thickened Sludge, *Water Engineering and Management*, August, pp. 26–28.

Randall, C. W., et al. (1975), Temperature Effects on Aerobic Digestion Kineties, *ASCE Journal of the Environmental Engineering Division*, Vol. 101, p. 795.

Reynolds, T. D. (1973), Aerobic Digestion of Thickened Waste Activated Sludge, *Proceedings of the 28th Purdue Industrial Waste Conference*, Purdue University, Lafayette, IN, pp. 12–37.

Roediger, M., and Vivona, M. A. (1998), Process for Pathogen Reduction to Produce Class A Solids, *Proceedings of the 71st Annual Conference and Exposition*, Water Environment Federation, Alexandria, VA, pp. 137–148.

Turovskiy, I. S. (1998), New Technology for Wastewater and Sludge Treatment in Northern Region, *Water Engineering and Management*, May, pp. 40–43, 56.

——— (1999), Beneficial Use of Wastewater Sludge in Russia, *Florida Water Resources Journal*, May, pp. 23–26.

——— (2001), Technological Improvements for the Aerobic Digestion of Sludge, *Water Engineering and Management*, August, pp. 33–36.

U.S. EPA (1974), *Process Design Manual for Upgrading Existing Wastewater Treatment Plants*, EPA 625/1-71/004a.

——— (1979), *Process Design Manual for Sludge Treatment and Disposal*, EPA 625/1-79/011.

——— (1989), *Design Manual: Fine Pore Aerator Systems*, EPA 625/9-89/023.

——— (1990), *Autothermal Thermophilic Aerobic Digestion of Municipal Wastewater Sludge*, EPA 625/10-90/007.

——— (1992), *Control of Pathogen and Vector Attraction in Sewage Sludge*, EPA 625/R1-87/014.

——— (2003), *Control of Pathogen and Vector Attraction in Sewage Sludge*, EPA 625/R1-92/013, revised July 2003.

WEF (1987), *Aerobic Sludge Digestion*, Manual of Practice 16, Water Environment Federation, Alexandria, VA.

——— (1995), *Sludge Stabilization*, Manual of Practice FD-9, Water Environment Federation, Alexandria, VA.

——— (1996), *Operation of Wastewater Treatment Plants*, 5th ed., Manual of Practice 11, Water Environment Federation, Alexandria, VA.

——— (1998), *Design of Municipal Wastewater Treatment Plants*, 4th ed., Manual of Practice 8 (ASCE 76), Water Environment Federation, Alexandria, VA.

Wolinski, W. K. (1985), Aerobic Thermophilic Sludge Stabilization Using Air, *Water Pollution Control*, p. 433.

5

ANAEROBIC DIGESTION

Wastewater Sludge Processing, By Izrail S. Turovskiy and P. K. Mathai

5.1 INTRODUCTION

Anaerobic digestion is one of the oldest and most widely used processes for wastewater sludge stabilization for plants with average flows greater than 20,000 m^3/d (5 mgd). The process transforms organic solids in sludge, in the absence of oxygen, to gaseous end products such as methane and carbon dioxide and to innocuous substances. A net reduction in the quantity of solids, and destruction of pathogenic organisms are also accomplished in the anaerobic digestion process.

5.1.1 Advantages and Disadvantages

Anaerobic digestion offers several advantages over the other methods of sludge stabilization, which include:

- The methane gas produced is a source of usable energy. In most cases the energy produced exceeds the energy required to maintain the temperature for sludge digestion. Excess methane can be used for heating buildings, running engines for aeration blowers, or generating electricity.
- Reduction in total sludge mass through the conversion of organic matter primarily to methane, carbon dioxide, and water. Commonly, 30 to 65% of the raw sludge solids are destroyed. This can significantly reduce the cost of sludge disposal.
- The digested solids are generally free of objectionable odors.
- The digested biosolids contain nutrients such as nitrogen and phosphorus, and organic matter that can improve the fertility and texture of soils.
- A high rate of pathogen distribution can be achieved, especially with the thermophilic digestion process.

The principal disadvantages of anaerobic sludge digestion are the following:

- The capital cost is high because large closed digestion tanks fitted with systems for feeding, heating, and mixing the sludge are required.
- Large reactors are required to provide the hydraulic detention time in excess of 10 days to stabilize the sludge effectively. This slow digestion process also limits the speed with which the system can adjust to changes in waste loads, temperature, and other environmental conditions.
- Microorganisms involved in anaerobic digestion are sensitive to small changes in the environment. Therefore, the process is susceptible to upsets. Monitoring of performance, and close process control are required to prevent upsets.
- The process produces a poor-quality sidestream. Supernatants often have a high oxygen demand and a high concentration of suspended solids, nitrogen, and phosphorus. These flows may require additional treatment before recycling to the influent flows in plants that are required to remove nitrogen and phosphorus from the wastewater.

5.1.2 Theory of Anaerobic Digestion

Anaerobic digestion involves several successive stages of chemical and biochemical reactions involving enzymes and a mixed culture of microorganisms. The process comprises three general degradation phases: hydrolysis, acidogenesis, and methanogenesis. Figure 5.1 is a simplified representation of the reactions involved in anaerobic digestion.

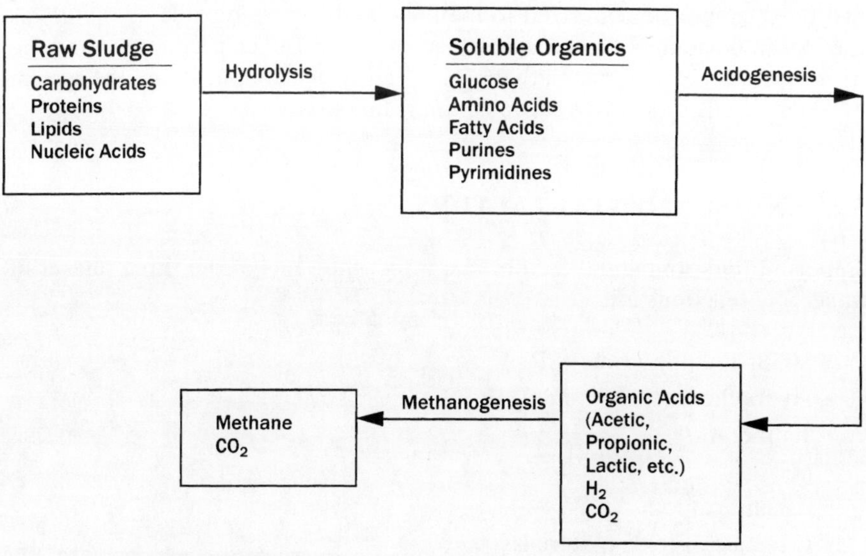

Figure 5.1 Schematic of reaction in anaerobic digestion.

In the first phase of anaerobic digestion, *hydrolysis*, complex organisms such as carbohydrates, proteins, and lipids are converted to their soluble forms and hydrolyzed further to simple monomers. In the second phase, *acidogenesis* (also known as *fermentation*), acid-forming bacteria convert the products formed in the first phase to short-chain organic acids: primarily, acetic, propionic, and lactic acids, and hydrogen and carbon dioxide. In the third phase, *methanogenesis*, methanogens convert the volatile acids to methane and carbon dioxide.

Hydrolysis is the rate-limiting step in the acid-forming phase (Eastman and Ferguson, 1981). After the solubilized organics are formed, they are converted immediately to volatile acids. The acid formers are primarily facultative bacteria. They are relatively tolerant to changes in pH and temperature. Facultative bacteria can also use dissolved oxygen during metabolism. Therefore, they can protect the methanogens, which are strict anaerobes, from the dissolved oxygen in the feed sludge to digesters.

In an anaerobic digestion system, the acidogenesis and methanogenesis are in dynamic equilibrium; that is, after the organics are converted to volatile acids and hydrogen, they are converted to methane and carbon dioxide at the same rate at which they are formed. As a result, volatile acid and hydrogen levels are low in a digester that is working properly. However, methanogens are inherently slow growing, with doubling times measured in days. They can also be affected adversely by even small changes in pH and temperature. In contrast, acid formers, with doubling times measured in hours, can function over a wide range of environmental conditions. Therefore, when a digester is stressed by shock loads or temperature fluctuations, methane production lags behind acid and hydrogen production. When this happens, volatile acids and hydrogen cannot be converted to methane and carbon dioxide as rapidly as they form. Once the balance is upset, acids and hydrogen accumulate and the pH drops. As a result, the methanogens are further inhibited, and the system eventually fails unless corrective action is taken.

5.2 ENVIRONMENTAL FACTORS

Important environmental factors that affect the rates of the three phases of anaerobic reactions are:

- Solids retention time (SRT)
- Hydraulic retention time (HRT)
- Temperature
- pH
- Alkalinity
- Presence of toxic materials

5.2.1 Solids and Hydraulic Retention Times

The most important factor in sizing the anaerobic digester is that the bacteria be given sufficient time to reproduce and metabolize volatile solids. The key parameters in providing sufficient time are the solids retention time (SRT), which is the average time the solids are held in the digester, and the hydraulic retention time (HRT), which is the average time the liquid sludge is held in the digester. They can be defined operationally as follows:

- SRT, in days, is equal to the mass of solids in the digester (kg) divided by the mass of solids withdrawn daily (kg/d).
- HRT, in days, is equal to the volume of sludge in the digester (m^3) divided by the volume of digested sludge withdrawn daily (m^3/d).

For digestion systems without recycle, HRT can be calculated based on either the sludge feeding rate or the removal rate. For such a system, SRT and HRT are equal. The three reactions in an anaerobic digestion system are directly related to SRT (or HRT). An increase in SRT increases the extent of reactions. Similarly, a decrease in SRT decreases the extent of reactions. Because a portion of the bacterial population is removed with each withdrawal of digested sludge, the rate of cell growth must at least match cell removal to maintain the system in steady state. Otherwise, the population of bacteria in the digester declines and the process eventually fails. Therefore, a minimum SRT is essential to ensure that bacteria are being produced at the same rate at which they are withdrawn daily.

Figure 5.2 illustrates the relationship between SRT and the performance of a lab-scale anaerobic digester fed with primary sludge. The figure shows how the production of methane and the reduction of degradable proteins, carbohydrates, lipids, COD, and volatile solids are related to SRT. As the SRT is reduced, the concentration of each component in the effluent gradually increases until the SRT reaches a value beyond which the concentration increases rapidly. This is the breakpoint SRT (or critical SRT) at which washout of bacteria begins; that is, the point where the rate at which bacteria leave the system exceeds their rate of reproduction. Calculations based on process kinetics predict a critical SRT of 4.2 days for the anaerobic digestion of wastewater sludge at 35°C (95°F).

5.2.2 Temperature

Temperature has an important effect on bacterial growth rates and, accordingly, changes the relationship between SRT and digester performance. Figure 5.3 illustrates the effect of temperature on methane production and volatile solids reduction. The figure shows that stabilization is slowed at lower temperatures, with 20°C (68°F) appearing to be the minimum temperature at

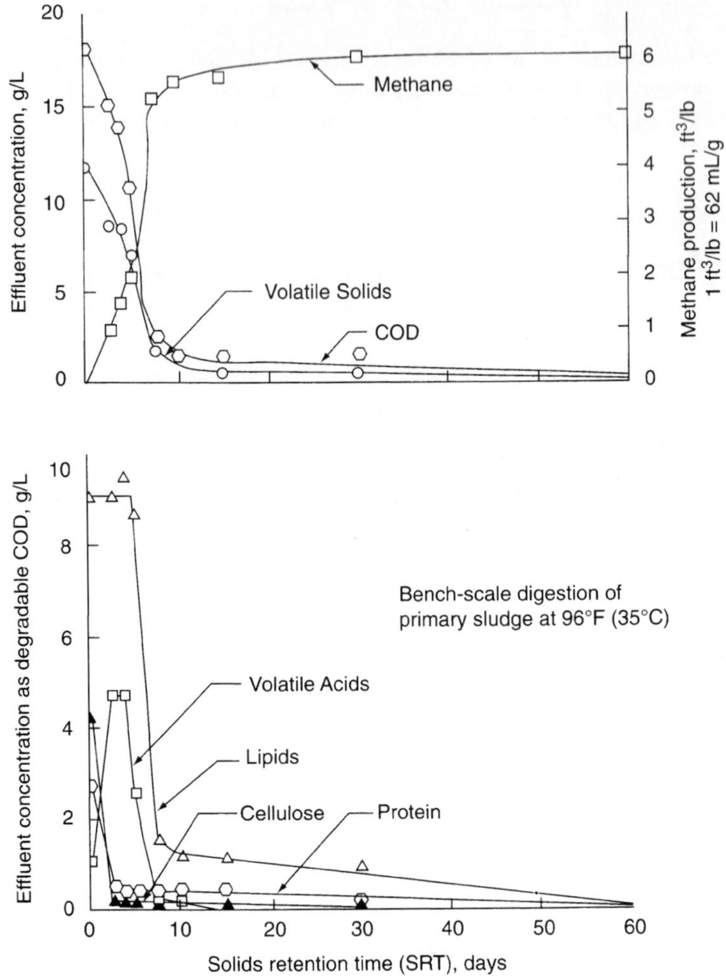

Figure 5.2 Effects of SRT on the relative breakdown of degradable components and methane production. (From U.S. EPA, 1979.)

which sludge stabilization can be accomplished with a practical SRT. Most anaerobic digesters are designed to operate in the mesophilic temperature range 30 to 38°C (85 to 100°F), 35°C (95°F) being the most common. Some digesters are designed to operate in the thermophilic range 50 to 57°C (122 to 135°F).

It is important that a stable operating temperature be maintained in the digester. Sharp and frequent fluctuations in temperature affect the bacteria, especially methanogens. Process failure can occur at temperature changes

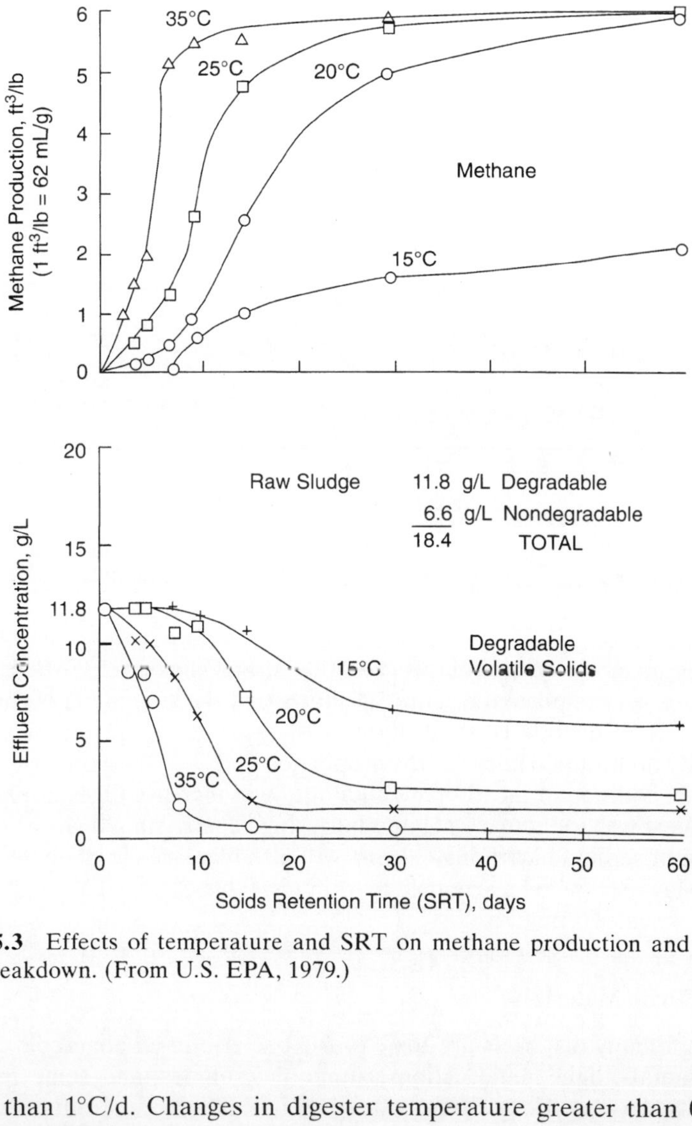

Figure 5.3 Effects of temperature and SRT on methane production and volatile solids breakdown. (From U.S. EPA, 1979.)

greater than 1°C/d. Changes in digester temperature greater than 0.6°C/d should be avoided.

5.2.3 pH and Alkalinity

Methane-producing bacteria are extremely sensitive to pH. Optimum pH for methane formers is in the range 6.8 to 7.2. Volatile acids produced in the acid-forming phase tend to reduce the pH. The reduction is normally countered by methane formers, which also produce alkalinity in the form of carbon dioxide, ammonia, and bicarbonate.

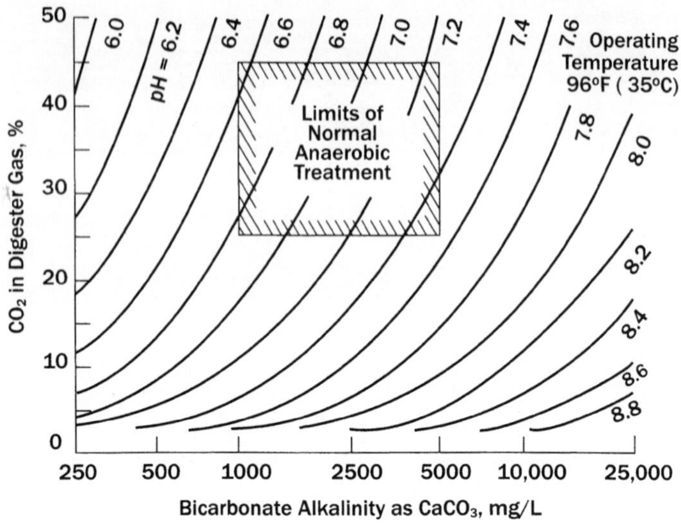

Figure 5.4 Relationship between pH and bicarbonate concentration at 35°C. (From U.S. EPA, 1979.)

In the anaerobic digestion process, the carbon dioxide–bicarbonate relationship is very important. Figure 5.4 shows that the system pH is controlled by the carbon dioxide concentration of the gas phase and the bicarbonate alkalinity of the liquid phase. If the proportion of CO_2 in the gas phase remains the same, addition of bicarbonate alkalinity will increase digester pH.

The best way to increase pH and buffering capacity in a digester is by the addition of sodium bicarbonate. Lime will also increase bicarbonate alkalinity but may react with bicarbonate to form insoluble calcium carbonate, which promotes scale formation.

5.2.4 Toxic Materials

Although many materials are toxic to the bacteria in an anaerobic digester, heavy metals, light metal cations, ammonia, sulfides, and some inorganic materials are of concern. Toxic conditions normally occur from overfeeding and excessive addition of chemicals. Toxic conditions can also occur from industrial wastewater contributions with excessive toxic materials to the plant influent. Tables 5.1 and 5.2 present toxic and inhibitory materials of concern.

Heavy metal toxicity has frequently been cited as the cause of anaerobic digestion failures, although trace amounts of most heavy metals are necessary for cell synthesis. Domestic wastewater sludge normally has low concentrations of light metal cations (sodium, potassium, calcium, and magnesium). However, significant contributions can come from industrial discharges and

TABLE 5.1 Selected Toxic and Inhibitory Inorganic Materials in Anaerobic Digestion

Substance	Moderately Inhibitory Concentration (mg/L)	Strongly Inhibitory Concentration (mg/L)
Na^+	3,500–5,500	8,000
K^+	2,500–4,500	12,000
Ca^{2+}	2,500–4,500	8,000
Mg^{2+}	1,000–1,500	3,000
Ammonia nitrogen, NH_4^+	1,500–3,500	3,000
Sulfide, S^{2-}	200	200
Copper, Cu^{2+}		0.5 (soluble) 50–70 (total)
Chromium		
Cr^{6+}		3.0 (soluble) 200–250 (total)
Cr^{3+}		2.0 (soluble) 180–420 (total)
Nickel, Ni^{2+}		30 (total)
Zinc, Zn^{2+}		1.0 (soluble)

Source: WEF, 1998.

TABLE 5.2 Selected Toxic and Inhibitory Organic Materials in Anaerobic Digestion

Compound	Concentration Resulting in 50% Reduction in Activity (m*M*)
1-Chloropropene	0.1
Nitrobenzene	0.1
Acrolein	0.2
1-Chloropane	1.9
Formaldehyde	2.4
Lauric acid	2.6
Ethyl benzene	3.2
Acrylonitrile	4
3-Chlorol-1,2-propanediol	6
Crotonaldehyde	6.5
2-Chloropropionic acid	8
Vinyl acetate	8
Acetaldehyde	10
Ethyl acetate	11
Acrylic acid	12
Catechol	24
Phenol	26
Aniline	26
Resorcinol	29
Propanol	90

Source: WEF, 1998.

from the addition of alkaline material for pH control. Ammonia, produced during the anaerobic digestion of proteins and urea, may reach toxic levels in highly concentrated sludge. Ammonia nitrogen concentrations of more than 1000 mg/L can be highly toxic.

When wastewater sludge contains high concentrations of sulfide, it can cause a problem in anaerobic digestion because the sulfate-reducing bacteria reduce sulfate to sulfide, which is toxic to methanogens at concentrations over 200 mg/L. This can be controlled by precipitating the sulfide as iron sulfide by adding iron salts to the digesters at controlled amounts.

5.3 PROCESS VARIATIONS

Experimentation over the years has yielded four basic variations of anaerobic sludge digestion: low-rate digestion, high-rate digestion, two-stage digestion, and two-phase digestion. In addition, anaerobic digestion can also be operated in two temperature regimes: mesophilic at 30 to 38°C (86 to 100°F) and thermophilic at 50 to 57°C (122 to 135°F). Variations of two-phase digestion are also being used.

5.3.1 Low-Rate Digestion

Low-rate digestion is the oldest and simplest type of the anaerobic sludge digestion process. The basic features of this system are illustrated in Figure 5.5.

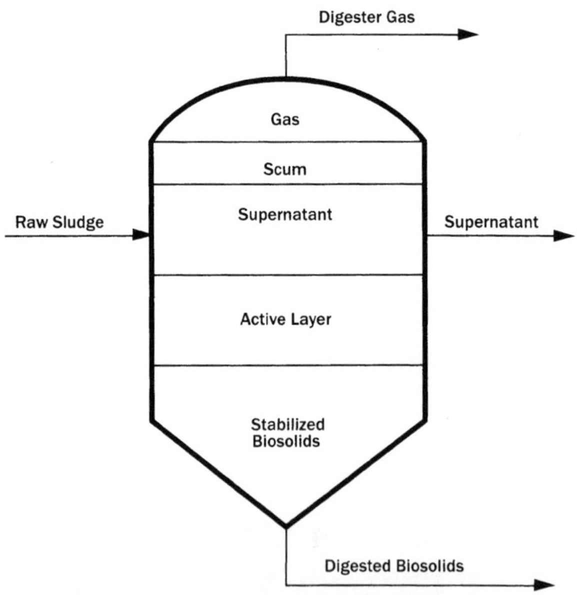

Figure 5.5 Low-rate anaerobic digestion.

Essentially, it is a large cylindrically shaped tank with a sloping bottom and a flat or domed roof. An external heat source may or may not be provided. No mixing is provided. Although the gas generated and its rise to the surface provide some degree of mixing, the stabilization results in a stratified condition within the digester, as shown in Figure 5.5. Supernatant is drawn off and recycled to the treatment plant influent. Stabilized biosolids, which accumulate and thicken at the bottom of the tank, are periodically drawn off for removal. Methane gas collects above the liquid surface and is drawn off through the cover. Low-rate digestion is characterized by a long detention time, 30 to 60 days, and is sometimes considered for small plants of less than 1.0 mgd (3800 m^3/d); however, they are seldom built today.

5.3.2 High-Rate Digestion

In the 1950s, several improvements to low-rate digestion were developed, resulting in a high-rate anaerobic digestion system. Heating, auxiliary mixing, thickening the raw sludge, and uniform feeding, the essential elements of a high-rate digestion system, act together to create a uniform environment. As a result, the tank volume is reduced and the stability and efficiency of the process are improved. Figure 5.6 shows the basic layout of this system.

Heating Heating is essential because the rate of microbial growth, and therefore the rate of digestion and gas production, increase with temperature. Most commonly, high-rate digestion operates in the mesophilic range 30 to 38°C (86 to 100°F). The contents of the digester are heated and maintained

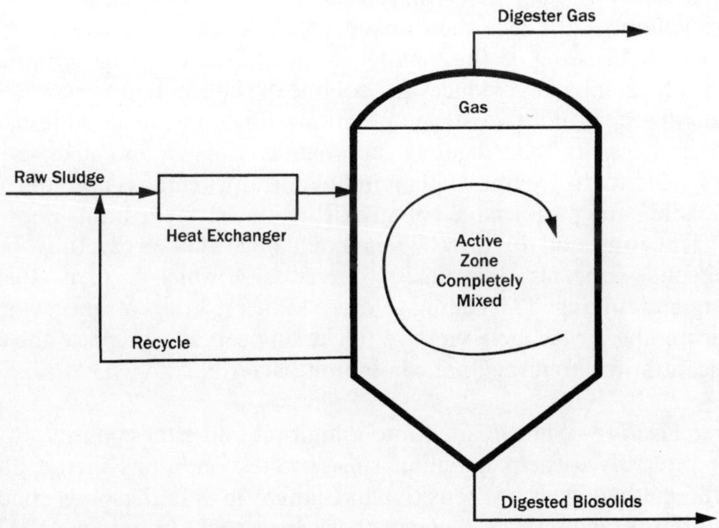

Figure 5.6 Single-stage high-rate anaerobic digestion.

consistently within 0.5°C (1°F) of design temperature because anaerobic bacteria, especially methanogens, are easily inhibited by even small changes in temperature. Heating by external heat exchangers is the most common method of heating because of their flexibility and ease of maintenance. Other heating methods include internal heat exchangers and steam injection. Heating methods are discussed further in Section 5.5.4.

Auxiliary Mixing Auxiliary mixing of digester contents is beneficial for (WEF, 1998):

- Reducing thermal stratification
- Dispersing the raw sludge for better contact with the active biomass
- Reducing scum buildup
- Diluting any inhibitory substances or any adverse pH and temperature feed characteristics
- Increasing the effective volume of the reactor
- Allowing reaction product gases to separate more easily
- Keeping in suspension more inorganic material, which has a tendency to settle

The three types of mixing typically used are mechanical draft tube mixing, pumping, and gas recirculation. Methods of mixing are discussed further in Section 5.5.3.

Thickening In the early 1950s, (Torpey, 1954), thickening raw sludge before feeding it to the digester was found to be beneficial. Thickening reduces the biomass volume and hence the digester volume. It also reduces the heating requirements because of the smaller volume of raw sludge entering the digester. Thickening also reduces the volume of supernatant to be withdrawn from the digester (in a two-stage operation). Since thickener supernatant is of far better quality than digester supernatant, it has a less adverse impact when recycled to the treatment plant influent. If thickening is used to enhance solids loading and performance of an existing digester, the limits imposed by pumping, heating, and mixing systems should be evaluated carefully. Increasing the solids concentration affects the viscosity, which in turn affects the pumping and mixing. Thickening to more than 7% concentration can cause digester mixing problems. Excessive thickening can also increase the chemical concentrations to levels that can inhibit bacterial activity.

Uniform Feeding Sludge is fed into a high-rate digester continuously or at regular intervals to help maintain steady-state conditions in the digester. Since the methanogens are sensitive to changes in volatile solids concentrations, uniform feeding can alleviate shock loadings. Multiple feed points in the digester can also alleviate or reduce shock loadings. Shock loading a digester affects its temperature and dilutes the alkalinity necessary for the buffering against pH in the digester.

5.3.3 Thermophilic Digestion

Most high-rate digesters are operated in the range 30 to 38°C (86 to 100°F). Bacteria that grow in this temperature range are called *mesophilic*. Another group of microorganisms, called *thermophilic* bacteria, grow in the temperature range 50 to 57°C (122 to 135°F). Thermophilic anaerobic digestion has been studied since the 1930s, at both the laboratory and plant scales (U.S. EPA, 1979).

In general, advantages claimed for thermophilic anaerobic digestion over mesophilic digestion are:

- Faster reaction rates, which permit increased volatile solids destruction
- Increased destruction of pathogens

Disadvantages of thermophilic anaerobic digestion include:

- Higher energy requirements for heating
- Lower-quality supernatant, containing large quantities of dissolved materials
- Higher odor potential
- Poorer process stability because thermophilic bacteria are more sensitive than mesophilic bacteria to temperature fluctuations
- Poor dewaterability

U.S. federal regulations controlling land application of sludge classify thermophilic digestion, along with mesophilic digestion, as a process to significantly reduce pathogens (PSRP). That is, although there may be greater reduction of pathogens levels in thermophilic digestion, it is not classified as a process to further reduce pathogens (PFRP). Therefore, single-stage thermophilic digestion is limited in its application.

5.3.4 Two-Stage Digestion

Two-stage digestion is carried out in a high-rate digestion tank coupled in series with a second tank (sometimes called a *secondary digester*), as shown in Figure 5.7. The second tank is neither heated nor mixed. Its main function is to allow gravity concentration of digested solids and decanting of supernatant liquor. Decanting reduces the volume of digested sludge, requiring further processing and disposal. A secondary digester fitted with a floating cover can also provide storage for digested sludge and digester gas. Very little solids reduction and gas production take place in the second tank. In some

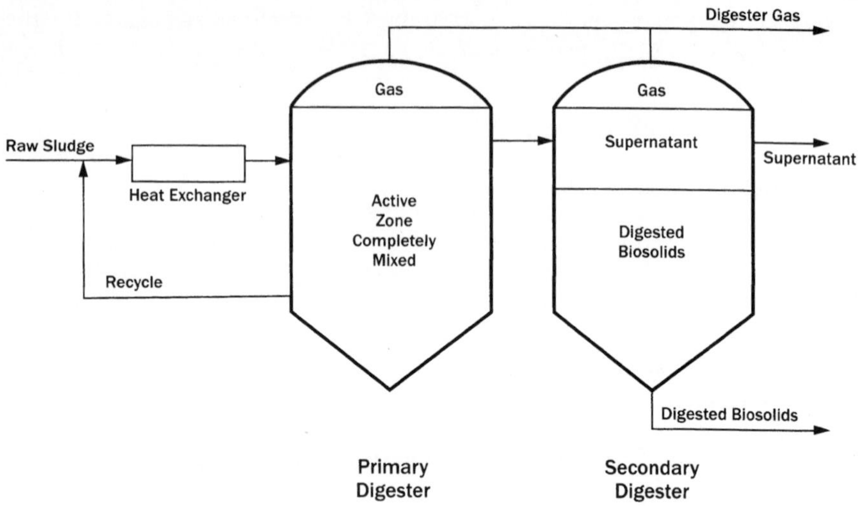

Figure 5.7 Two-stage high-rate anaerobic digestion.

instances, the second tank is similar in design to the first, with heating and mixing capabilities to provide standby digester capacity.

Many secondary digesters have performed poorly as thickeners, producing dilute sludge and high-strength supernatant. This is because some gas will come out of the solution in small bubbles if there is incomplete digestion in the primary digester or if the sludge transferred from the primary digester is supersaturated with gas. These bubbles attach to the sludge particles and provide a buoyant force that hinders settling. Another reason for poor settling is the fine-sized solids produced during digestion by both mixing and the natural breakdown of particles through biological decomposition (U.S. EPA, 1979). The problem is compounded when the digester is fed with secondary and tertiary solids, especially when those have been flocculated and broken up during digestion. Therefore, two-stage digestion, used frequently in the past, is seldom used in newer facilities.

5.3.5 Two-Phase Digestion

As discussed in Section 5.1.2, anaerobic digestion involves two major phases: hydrolysis and acid formation together, and methane production. In the three preceding high-rate digestion processes, both phases take place in a single reactor. In two-phase digestion, the two major phases are divided into separate tanks coupled in series. Pilot studies (Ghosh et al., 1975; Lee et al., 1989) and data from two full-scale facilities (Ghosh et al., 1995) have shown that better sludge digestion can be achieved by optimizing the two phases separately.

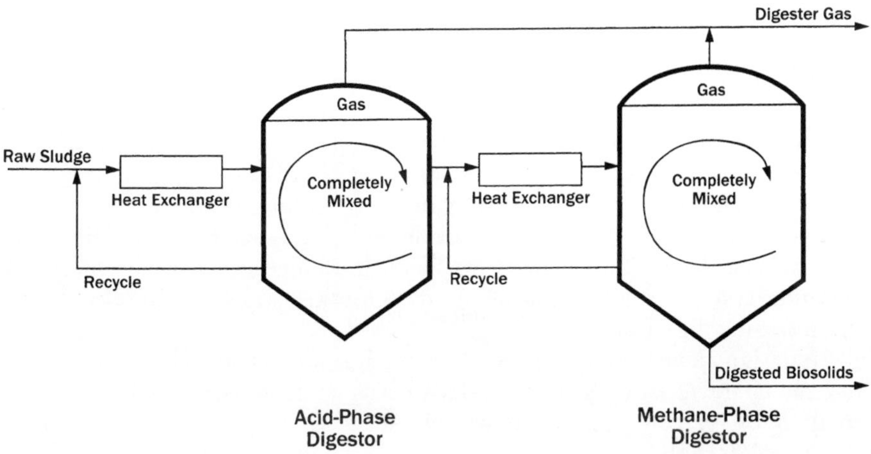

Figure 5.8 Two-phase anaerobic digestion.

Figure 5.8 illustrates a two-phase anaerobic digestion system. The first reactor, known as an *acid-phase digester*, is for hydrolysis and acidogenesis and is designed for a 1- to 2-day detention time. This phase can be operated in either the mesophilic or thermophilic regime. pH in the reactor is between 5.5 and 6.5. Methane generation is negligible in this reactor. The second reactor, known as a *methane-phase digester*, is designed for about 10 days of detention time and operates in the mesophilic temperature range. Advantages cited for two-phase anaerobic digestion compared to single-phase digestion include:

- Higher volatile solids reduction because it allows the creation of an optimum environment for the acid formers
- Increased production of gases
- Higher content of methane in the final product gas
- Higher pathogen reduction
- Fewer foaming problems
- Better stability of the digestion process

Recent studies (Schafer and Farrell, 2000b) have produced several variations of the two-phase anaerobic digestion system, including staged mesophilic digestion, temperature-phased digestion (TPAD), acid/gas-phase digestion (as described above), and staged thermophilic digestion. These variations use different solids retention times in the two reactors and operate either reactor in the mesophilic or thermophilic regimes. Some of these processes are reported to be capable of meeting class A biosolids requirements.

5.4 PROCESS DESIGN

Determination of digester tank volume is the first important consideration in the design of an anaerobic digestion system. Various methods have been used to determine the size of the digester, including per capita basis, solids loading, solids retention time, volumetric loading, and volatile solids destruction. Other design considerations, such as mixing, heating, energy recovery, and chemical additions, should also be evaluated in the design of the digestion system. The discussion below is for mesophilic digestion, as it is the most common system in use. Some of the design criteria are also applicable for thermophilic digestion.

To design an anaerobic digestion system, it is important to know the quantity and characteristics of the feed sludge. The characteristics include percent total solids, percent volatile solids, and the ratio of primary to secondary sludge solids. Solids production can be calculated based on the solids mass balance of the treatment plant (see Chapter 2) or from the existing plant operating data. If there are significant contributions of industrial wastewater to the treatment plant, the feed sludge should be analyzed for the toxic materials described in Section 5.2.4.

5.4.1 Per Capita Basis

Empirical loading criteria have traditionally been used for determining anaerobic digester volume. The oldest and simplest of these criteria is the per capita volume allowance, that is, based on population served by the treatment plant. Table 5.3 lists typical per capita design values along with other design

TABLE 5.3 Typical Design Criteria for Mesophilic Anaerobic Sludge Digesters

	SI Units		U.S. Customary Units	
Parameter	Units	Value	Units	Value
Volume criteria				
Primary sludge	m^3/capita	0.03–0.06	ft^3/capita	1.3–2.0
Primary sludge + trickling filter sludge	m^3/capita	0.07–0.09	ft^3/capita	2.6–3.3
Primary sludge + waste activated sludge	m^3/capita	0.07–0.11	ft^3/capita	2.6–4.0
Solids loading rate	kg VSS/m^3·d	1.6–3.2	lb VSS/ft^3-d	0.1–0.02
Solids retention time	d	15–20	d	15–20
Sludge concentration				
Primary sludge + biological sludge feed	%	4–7	%	4–7
Digested sludge draw-off[a]	%	4–7	%	4–7

Source: Adapted from U.S. EPA, 1979 and WEF, 1998.

[a] Lower values for single-stage digesters (no supernatant draw-off).

criteria. Per capita loading factors should be used only for initial sizing estimates because it implicitly assumes a value for such important parameters as per capita waste load, solids removal efficiency in treatment, and digestibility of the sludge. These parameters vary widely from one community to another and cannot be predicted accurately. The values for per capita design criteria shown in Table 5.3 should be increased on a population-equivalent basis if the plant influent includes considerable industrial waste loads.

5.4.2 Solids Loading

A more direct design criterion in determining digester volume is the volatile solids loading rate. It is the mass of volatile solids fed to the digester per unit volume and time. A typical range of loading rates is shown in Table 5.3. Design loading criteria generally are based on sustained loading conditions: typically, peak month solids production. It should be recognized that excessively low volatile solids loading criteria would result in a large digester system, which can be expensive to operate because it may not produce sufficient gas to provide the energy needed to heat the digester content to its desired temperature.

5.4.3 Solids Retention Time

Volatile solids digestion is time dependent. Therefore, the most important consideration in sizing a digester is that the bacteria be given sufficient time to reproduce. To provide sufficient time, the key design parameter in determining the digester volume is the solids retention time (SRT), which is the average time the microbial mass is retained in the digester. It can be defined operationally as the total mass of solids in the digester divided by the mass of solids withdrawn daily. In anaerobic digesters without recycle or supernatant draw-off (e.g., single-stage high-rate digestion), the SRT is equivalent to the hydraulic detention time.

The relationship between SRT and process performance, and critical SRT, which is the lowest SRT beyond which the process fails, have been described in Section 5.2.1. The design SRT should be selected with care. A margin of safety should be provided because the critical SRT values shown in Figure 5.2 were determined in bench-scale digester studies maintained with ideal conditions of temperature, complete mixing, and uniform feeding and withdrawal rates. A minimum safety factor of 2.5 is recommended.

The recommended minimum design SRT is 10 days for systems operating at 35°C (95°F). Values for systems operating at other temperatures are shown in Table 5.4. These design values must be met at all conditions expected, including peak sustained loading, and must account for additional volume for grit accumulations and for differences in the rate of feeding and withdrawal. Taking all these into account, SRT values in the range 15 to 20 days are recommended for design.

TABLE 5.4 Suggested Solids Retention Time for a High-Rate Digester

Operating Temperature		Critical SRT	Minimum Design SRT
°C	°F	(days)	(days)
18	65	11	28
24	75	8	20
30	85	6	14
35	95	4	10
40	105	4	10

5.4.4 Volatile Solids Reduction

It is usually assumed that solids reduction takes place only in the volatile portion of the sludge solids. Therefore, the common measure of the degree of stabilization is the percent volatile solids destroyed. Volatile solids reduction in high-rate anaerobic digesters usually ranges from 50 to 65%. Volatile solids reduction achieved in any particular application depends on both the characteristics of the sludge and the operating parameters of the digestion system. For a high-rate digestion system, the following empirical equation (Liptak, 1974) applies:

$$V_d = 13.7 \ln(\text{SRT}_d) + 18.9 \tag{5.1}$$

where V_d is the volatile solids reduction (%) and SRT_d is the design SRT (days). Figure 5.9 shows volatile solids reduction as a function of SRT and is developed using equation (5.1).

5.4.5 Gas Production

Generation of digester gas is a direct result of the destruction of volatile solids. Specific gas production for wastewater sludge generally ranges from 0.8 to 1.1 m^3/kg (13 to 18 ft^3/lb) of volatile solids destroyed. Specific gas production values will be closer to the high end of this range if the sludge contains a higher percentage of fats and grease as long as adequate SRT is provided for these slow-metabolizing materials. A healthy digestion process produces a gas with about 65 to 75% methane, 30 to 35% carbon dioxide, and very low levels of nitrogen, hydrogen, and hydrogen sulfide. The heat value of digester gas is approximately 24 MJ/m^3 (640 Btu/ft^3), compared to 38 MJ/m^3 (1010 Btu/ft^3) for methane.

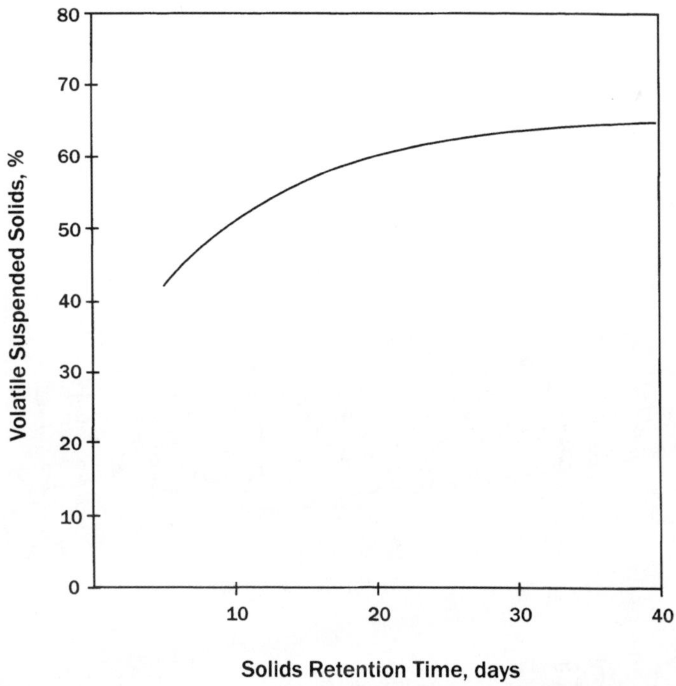

Figure 5.9 Volatile solids reduction as function of solids retention time.

5.5 SYSTEM COMPONENT DESIGN

Typical characteristics of high-rate digesters are auxiliary heating and mixing. Equipment for heating and mixing varies by manufacturer. High-rate two-stage systems have been the choice of anaerobic digestion process since the 1950s. However, modern digesters are designed as single-stage high-rate systems fed with sludge of high solids concentrations (5 to 7%) and no supernatant draw-off.

5.5.1 Tank Design

Anaerobic digesters are rectangular, cylindrical, or egg-shaped. Rectangular tanks, used in the past, are no longer used because of the difficulty in mixing the tank contents uniformly. Figure 5.10 shows simplified sketches of cylindrical and egg-shaped design.

The most common tank design in the United States is a low vertical cylinder 6 to 38 m (20 to 125 ft) in diameter with a sidewall depth of 6 to 14 m (20 to 46 ft). Tanks are usually made of concrete, although steel tank designs are not uncommon in smaller sizes. Tank floors are usually conical with slopes

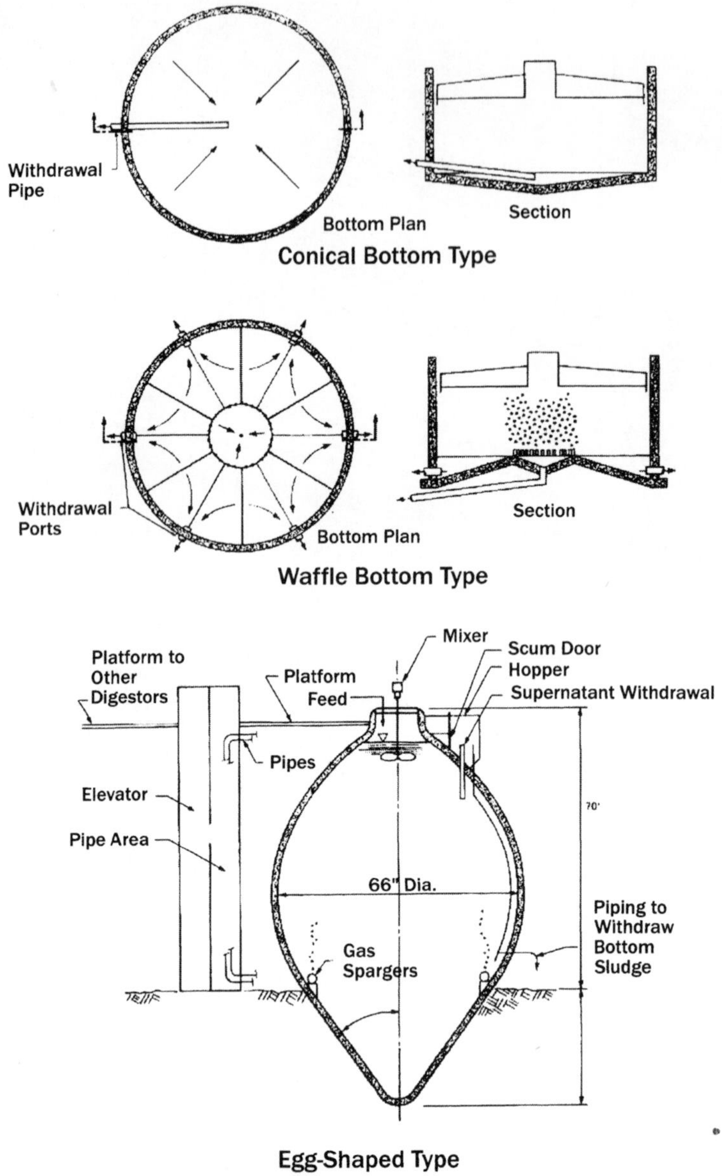

Figure 5.10 Anaerobic digester tank design.

varying between 1:3 and 1:6. Digested sludge is withdrawn from the low point in the center of the tank. Some digesters have been designed with "waffle bottoms" as shown in Figure 5.10, to minimize grit accumulation and eliminate the need for cleaning. However, construction of waffle-bottom floors is more complex and costly.

Egg-shaped digesters, originated in Germany over 50 years ago, are becoming popular in North America. The steeply sloped bottom of the tank eliminates grit accumulation; therefore, cleaning is not required. The liquid surface area at the top is small, so the scum accumulated there can be kept fluid with a mixer and removed through a scum door. Egg-shaped digesters can be built with steel or concrete. Steel construction is more common because concrete construction requires complex formwork and special construction techniques. The outer surface is always insulated and clad with aluminum for aesthetic purpose. Tanks are relatively tall. Digesters at the Deer Island plant in Boston, the largest system currently in operation, are over 40m (130ft) in height.

5.5.2 Digester Covers

Cylindrical digesters are covered to contain odors, maintain operating temperature, maintain anaerobic condition, and collect digester gas. Covers can be classified as either fixed or floating. Figure 5.11 shows several types of fixed and floating covers.

Fixed covers are either dome-shaped or flat and are fabricated from reinforced concrete, steel, or fiberglass-reinforced polyester. Concrete roofs are susceptible to cracking and therefore are sometimes lined with polyvinyl chloride (PVC) or steel plate to contain gas. Generally, fixed-cover digesters are operated so as to maintain a constant water level in the tank by compensating digested sludge withdrawals with the addition of raw sludge. This eliminates the possibility of air entering the tank and producing an explosive mixture of gas and oxygen.

Floating covers float on the surface of the digester contents and allow the volume of the digester to change without allowing air to enter the digester and mix with gas. Floating covers are normally used for single-stage digesters and in the second stage of two-stage digesters. A variation of the floating cover is the floating gas holder, which is a floating cover with an extended skirt (see Figure 5.11) to allow storage of gas during periods when gas production exceeds demand. A recent development in gas-holder covers is the membrane cover shown in Figure 5.11. The cover consists of flexible air and gas membranes under a support structure. As gas storage volume decreases or increases in the space between the liquid surface and the gas membrane, the space between the membranes is pressurized or depressurized accordingly, using an air-blower bleed-valve system.

In egg-shaped digesters, the volume available for gas storage is small. Therefore, supplemental external gas storage is required.

5.5.3 Mixing

Mixing of digester contents is essential for creating uniformity throughout the digester, preventing the formation of a surface scum layer and preventing the deposition of suspended matter on the bottom of the tank. A certain

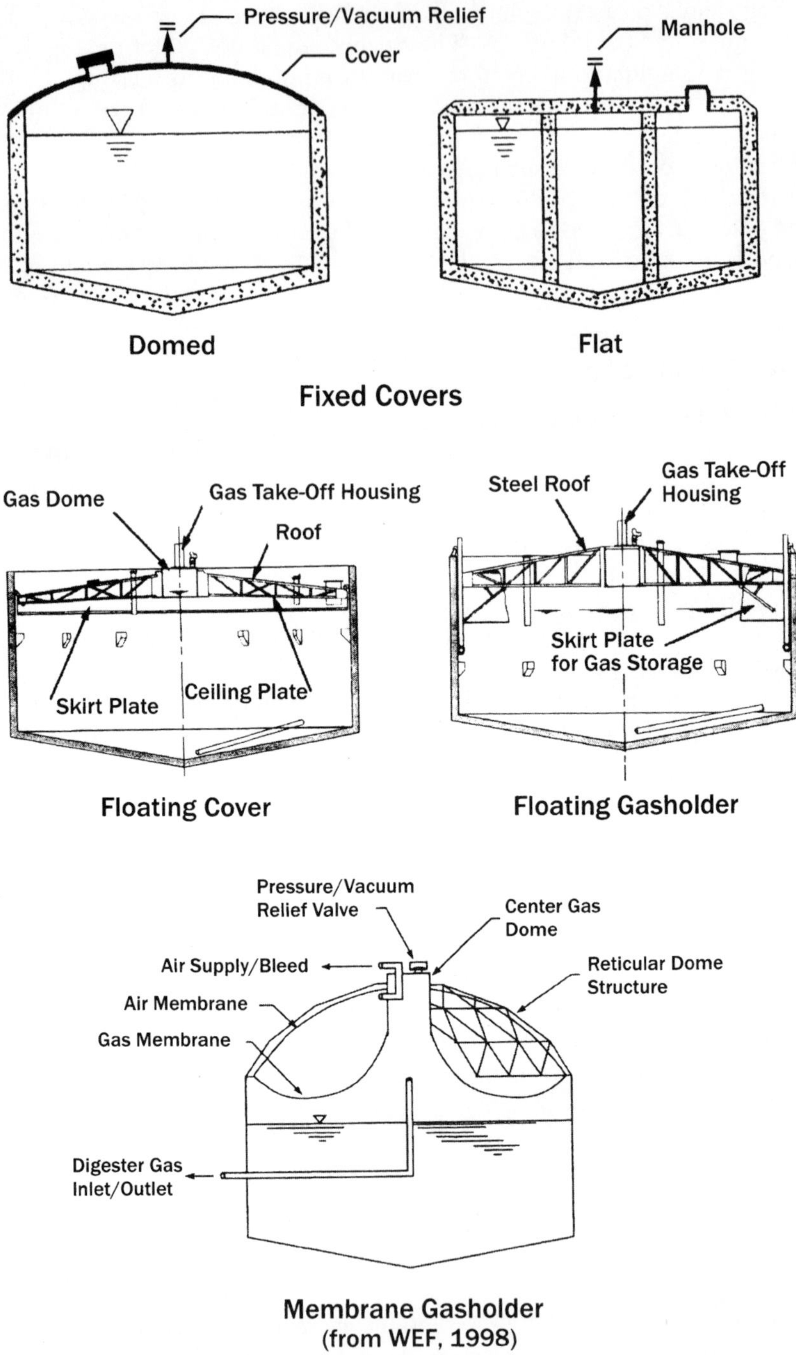

Figure 5.11 Anaerobic digester covers.

amount of natural mixing occurs in the digester from the rise of gas bubbles and the thermal convection currents created by the addition of heated sludge. However, auxiliary mixing is required to optimize performance. Methods used for mixing include external pumped recirculation, mechanical mixing, and gas mixing. Figure 5.12 shows various methods used for auxiliary mixing.

Pumped Recirculation In pumped recirculation systems, externally mounted pumps withdraw sludge from the tank and reinject it through nozzles at the bottom of the tank and near the surface to break up scum accumulation. Pumps are easier to maintain and provide better mixing control. Pumped recirculation also allows external heat exchangers for heating the digester and uniform blending of raw sludge with heated circulating sludge. Disadvantages include impeller wear from grit in the sludge, plugging of pumps by rags, and bearing failures. High-flow low-head solids handling pumps such as axial flow, mixed flow, or screw centrifugal pumps are used for recirculation. Pumps are sized for 20 to 30 minutes of turnover time, and the power required is 0.005 to 0.008 kW/m^3 (0.2 to 0.3 hp per $10^3 ft^3$) of digester volume.

Mechanical Mixing Mechanical mixing is performed by low-speed flat-blade turbines or high-speed propeller mixers. In one design a propeller drives sludge through a draft tube to promote vertical mixing. Draft tubes can be either internally or externally mounted (see Figure 5.12). Mechanical mixing provides good mixing efficiency and break up of any scum layer. However, rags in sludge can foul impellers. Efficient mechanical mixing can be effected with about 0.007 kW/m^3 (0.25 hp per $10^3 ft^3$) of digester volume.

Gas Mixing The four variations of gas mixing of anaerobic digesters are the following:

- Confined injection of gas bubbles intermittently at the bottom of pistons to create piston pumping action and surface agitation
- Confined release of gas within a draft tube positioned inside a tank
- Unconfined injection of gas through a series of lancers suspended from the digester cover
- Unconfined release of gas through a ring of spargers (diffusers) mounted on the floor of the digester

In gas pistons, gas bubbles rise up the tubes and act like pistons, pushing the sludge to the surface. This confined gas mixing method generally has a low power requirement and is effective against scum buildup.

In a draft tube system, the tube acts as a gas lift pump, causing the flow of sludge entering the bottom of the tube to exit at the top. This confined method induces bottom currents and prevents or at least reduces accumulation of settleable material. Unit gas flow requirement for confined mixing systems is 0.005 to 0.007 $m^3 \cdot min$ (5 to 7 ft^3 per $10^3 ft^3$-min).

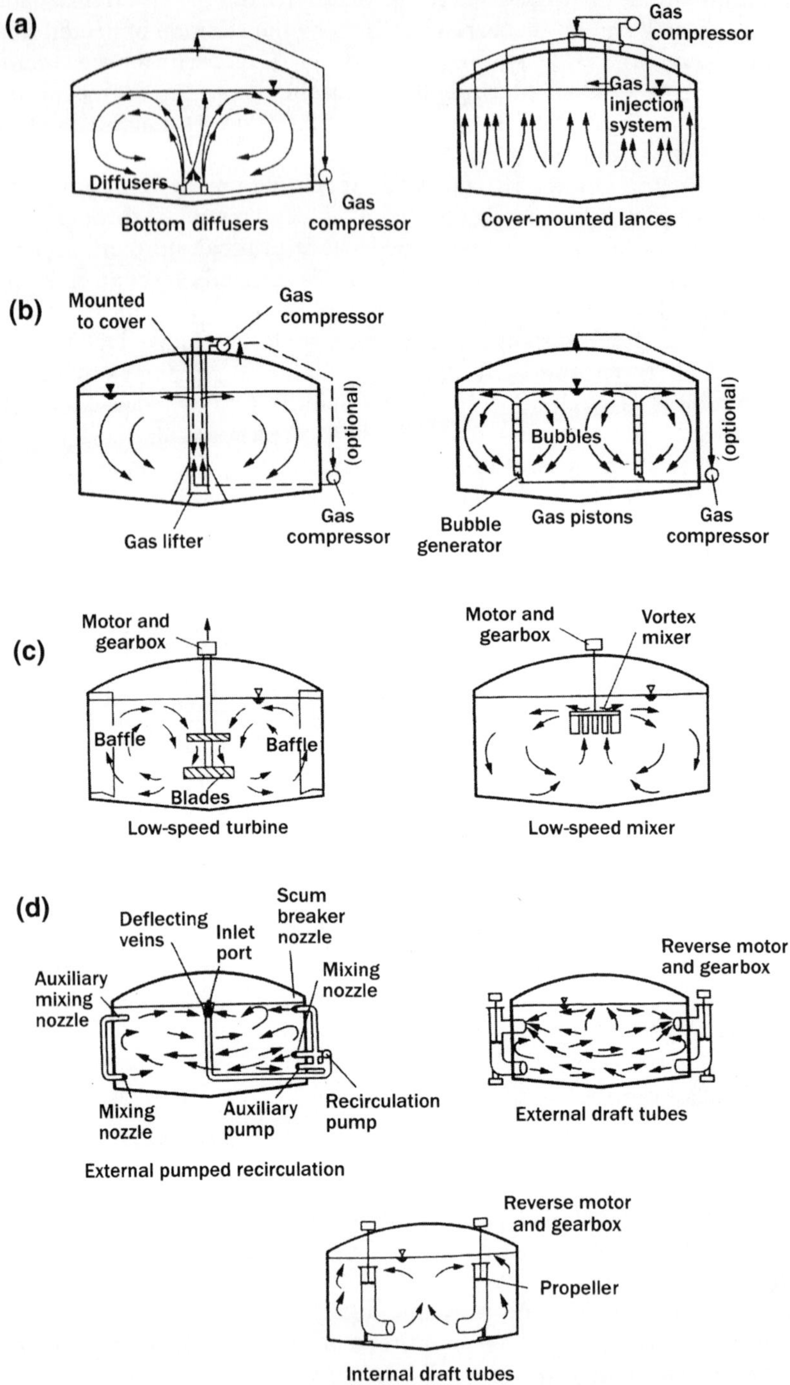

Figure 5.12 Mixing systems for anaerobic digesters. (Reprinted with permission from WEF, 1987.)

In an unconfined lance gas recirculation system, gas injection pipes are located throughout the tank. Gas is discharged continuously through all the lances (gas injection pipes) or sequentially by the use of a rotary valve and preset timers. This system is effective against scum buildup; however, a greater chance of solids deposits results, due to the less efficient mix regime.

In an unconfined diffuser mixing system, gas is discharged continuously through floor-mounted diffusers. This system is effective against solids deposits. However, it does not provide good top mixing, which results in scum buildup. There is also a potential for diffuser plugging, which requires the digester to be drained for maintenance. The unit gas flow requirement for unconfined mixing systems is 0.0045 to 0.005 $m^3/m^3 \cdot min$ (4.5 to 5.0 ft^3 per $10^3 ft^3$-min).

5.5.4 Heating

Maintaining a constant temperature [typically, 35°C (95°F) for mesophilic organisms and 55°C (130°F) for thermophilic organisms] is important for the efficient operation of an anaerobic digester. Heating is required to raise the feed sludge to the operating temperature of the digester and to compensate for heat loss through the walls, floor, and roof of the digester.

Heating Requirements The amount of heat required to raise the temperature of the sludge is given as

$$Q_1 = W_f C_p (T_2 - T_1) \tag{5.2}$$

where

Q_1 = heat required, J/d (Btu/d)
W_f = feed sludge weight, kg/d (lb/d)
C_p = specific heat of sludge (assumed to be same as water), 4200 J/kg · °C (1 Btu/lb-°F)
T_2 = design operating temperature of digester, °C (°F)
T_1 = temperature of feed sludge, °C (°F)

The amount of heat required to make up for the heat loss from the digester is given by

$$Q_2 = UA(T_2 - T_3) \tag{5.3}$$

where

Q_2 = heat loss, J/s (Btu/h)
U = heat transfer coefficient, $J/m^2 \cdot s \cdot °C$ (Btu/ft^2-hr-°F)
A = surface area of digester through which heat losses occur, m^2 (ft^2)
T_2 = temperature of sludge in digester, °C (°F)
T_3 = temperature outside the digester, °C (°F)

Various values of heat transfer coefficients for different wall, floor, and roof construction are given in Table 5.5. Coefficients, hence heat losses, can be

TABLE 5.5 Heat Transfer Coefficients for Anaerobic Digester Components

Digester Component	SI Units W/m^2·°C	U.S. Customary Units Btu/ft^2-°F-hr
Plain concrete walls (above ground)		
300 mm (12 in.) thick, not insulated	4.7–5.1	0.83–0.90
300 mm (12 in.) thick with air space plus brick facing	1.8–2.4	0.32–0.42
300 mm (12 in.) thick with insulation	0.6–0.8	0.11–0.14
Plain concrete wall (below ground)		
In dry earth	0.57–0.68	0.10–0.12
In moist earth	1.1–1.4	0.19–0.25
Plain concrete floors		
300 mm (12 in.) thick in dry earth	0.85	0.15
300 mm (12 in.) thick in moist earth	0.7	0.12
Floating covers		
With 35-mm (1.5-in.) wooden deck, built-up roofing, and no insulation	1.8–2.0	0.32–0.35
With 25-mm (1-in.) insulating board installed under roofing	0.9–1.0	0.16–0.18
Fixed concrete covers		
100 mm (4 in.) thick, covered with built-up roofing, and not insulated	4.0–5.0	0.70–0.88
100 mm (4 in.) thick, covered, and insulated with 25-mm (1-in.) insulating board	1.2–1.6	0.21–0.28
225 mm (9 in.) thick, not insulated	3.0–3.6	0.53–0.63
Fixed steel covers, 6 mm (0.25 in.) thick	4.0–5.4	0.70–0.95

Source: Adapted from U.S. EPA, 1979, and Metcalf & Eddy, 2003.

reduced by insulating the cover and the exposed walls of the digester. Common insulating materials are glass wool, insulation board, urethane foam, and dead air space. A facing is placed over the insulation for protection and to improve aesthetics. Common facing materials are brick, metal siding, and stucco.

In computing heat losses, various surfaces should be considered separately (roof, walls exposed to air, walls below ground, and floor) to develop the total heat loss. Walls below ground level are exposed to different temperature regimes depending on groundwater level and frost penetration. An average temperature is generally assumed for the entire wall below grade. If the groundwater level is not known, it is assumed that the digester floor is in saturated earth.

Heating Equipment External heat exchangers are the most commonly used heating method. Other heating methods, such as steam injection directly into the tank, direct flame heating by passing hot combustion gases through the sludge, and heat exchanger coils placed inside the tank, have been used in the past but have been discontinued because of the better heat transfer efficiency of the external heat exchangers. Another advantage of external heat exchangers is that recirculating digester sludge can be blended with raw sludge feed before heating in order to seed the raw sludge with anaerobic organisms. Cold sludge should never be added directly to the digester because the thermal shock will be detrimental to the anaerobic bacteria.

Three types of external heat exchangers are commonly used: water bath, jacketed pipe, and spiral. In a water bath heat exchanger, boiler tubes and sludge piping are located in a common water-filled container (see Figure 5.13). Hot water is pumped in and out of the bath to increase heat transfer. In a jacketed pipe heat exchanger (also known as a tube-in-tube heat exchanger), hot water is pumped countercurrent to the sludge flow through a concentric pipe surrounding the sludge pipe. The spiral heat exchanger (see Figure 5.13) is also a countercurrent flow design; however, the sludge and the

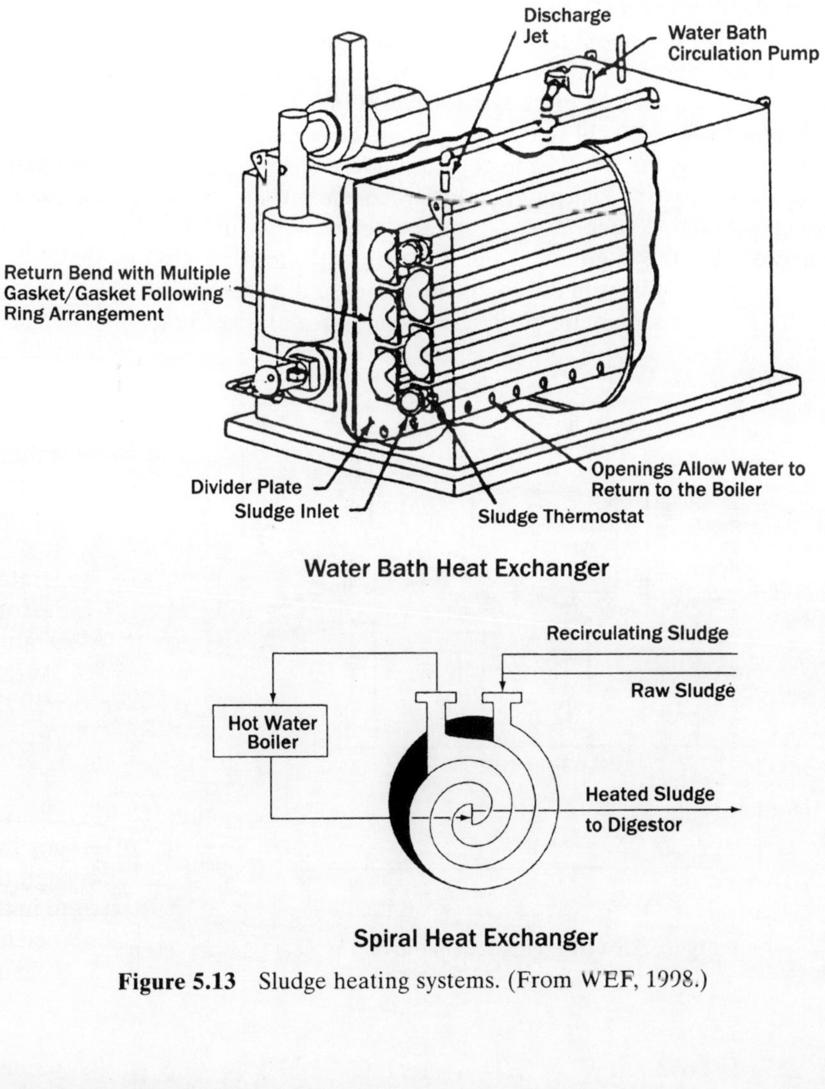

Figure 5.13 Sludge heating systems. (From WEF, 1998.)

water passageways are cast in a spiral. Heat transfer coefficients for external heat exchangers range from 0.85 to 1.6 kJ/m^2·°C·s (150 to 280 Btu/ft^2-°F-hr).

Hot water used in a heat exchanger is most commonly generated in a boiler fueled by sludge gas. Provisions for burning an auxiliary fuel such as natural gas, propane, or fuel oil must be included to maintain heating during periods of low digester gas production or high heating demand, such as for digester startup. If a cogeneration system is used, waste heat from sludge gas–powered engines used to generate electricity or directly to drive pumps or blowers is sufficient to meet digester requirements.

5.5.5 Gas Usage

A well-operated aerobic digester system produces more gas than that required to supply the energy for heating the digester. In most small wastewater treatment plants, the excess gas produced is flared in a waste gas burner to avoid odor. In large treatment facilities, the excess gas may be (1) burned in a boiler to produce heat for buildings in the plant, (2) used to power an engine to generate electricity or directly to drive pumps or blowers, (3) used to fire incinerators, or (4) sold to a local utility for use in the domestic gas supply.

Digester gas can be used in cogeneration, which is a system for producing both electricity and thermal energy. In cogeneration, digester gas is used to power an engine-generator to generate electricity; and cooling water discharged from the engine at a temperature of 70 to 82°C (160 to 185°F) and the hot exhaust gas from the engine are used for digester and building heating. Figure 5.14 is a schematic of typical in-plant use of digester gas.

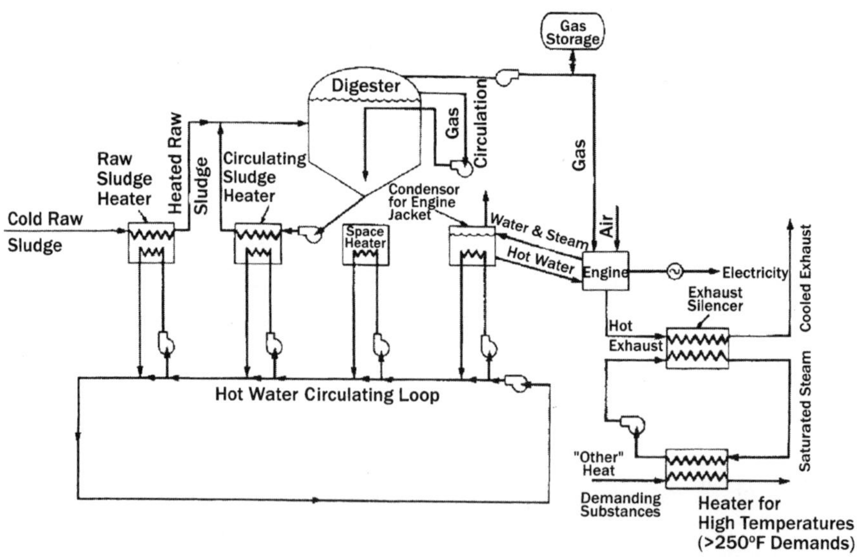

Figure 5.14 Process schematic of a typical digester gas use.

Digester gas, which is only about 65 to 75% methane, has a lower heating value of approximately 22,400 kJ/m^3 (600 Btu/ft^3). Lower heating value is the heating value of gas when none of the water vapor formed by combustion has been condensed. By comparison, methane has a lower heating value of 35,800 kJ/m^3 (960 Btu/ft^3), and natural gas has a heating value of 37,300 kJ/m^3 (1000 Btu/ft^3).

Design Example 5.1 The following example illustrates the basic sizing of a single-stage mesophilic anaerobic digester system for a 10-mgd (37,800-m^3/d) activated sludge wastewater treatment plant.

1. Design parameters:

Primary sludge:

solids produced:	12,000 lb/d (5443 kg/d)
solids concentration:	5%
specific gravity:	1.02
volatile solids (VS):	65%

Thickened waste activated sludge:

solids produced:	6000 lb/d (2722 kg/d)
solids concentration:	4%
specific gravity:	1.00
VS:	75%

2. Daily sludge volume:

$$\text{primary sludge volume} = \frac{12{,}000\ \text{lb/d}}{(1.02)(62.4\ \text{lb/ft}^3)(0.05\ \text{lb/lb})}$$
$$= 3770\ \text{ft}^3/\text{d}\ (107\ \text{m}^3/\text{d})$$

$$\text{WAS volume} = \frac{6000\ \text{lb/d}}{(1.00)(62.4\ \text{lb/ft}^3)(0.04\ \text{lb/lb})}$$
$$= 2400\ \text{ft}^3/\text{d}\ (68\ \text{m}^3/\text{d})$$

$$\text{total sludge volume} = (3770 + 2400)\ \text{ft}^3/\text{d}$$
$$= 6170\ \text{ft}^3/\text{d}\ (175\ \text{m}^3/\text{d})$$

3. Digester volume:

$$\text{sizing criteria: 15 days of SRT (see Table 6-3)}$$

$$\begin{aligned}\text{digester volume, V} &= Q(\text{SRT})\\ &= (6170\,\text{ft}^3/\text{d})(15\,\text{d})\\ &= 92,550\,\text{ft}^3\ (2621\,\text{m}^3)\end{aligned}$$

4. Solids loading rate:

$$\begin{aligned}\text{total VS loading} &= (12{,}000\,\text{lb/d})(0.65\,\text{lb/lb}) + (6000\,\text{lb/d})(0.75\,\text{lb/lb})\\ &= 12{,}300\,\text{lb/d}\ (5579\,\text{kg/d})\end{aligned}$$

$$\text{VS loading rate} = \frac{12{,}300\,\text{lb/d}}{92{,}550\,\text{ft}^3} = 0.13\,\text{lb/ft}^3\text{-d}\ (2.1\,\text{kg/m}^3\cdot\text{d})$$

5. Combined feed solids concentration: Assume that the specific gravity of the combined sludge is 1.01.

$$\begin{aligned}\text{total solids feed} &= (12{,}000 + 6000)\,\text{lb/d}\\ &= 18{,}000\,\text{lb/d}\ (8165\,\text{kg/d})\end{aligned}$$

$$\begin{aligned}\text{solids concentration} &= \frac{18{,}000\,\text{lb/d}}{(1.01)(62.4\,\text{lb/ft}^3)(6170\,\text{ft}^3/\text{d})} \times 100\\ &= 4.6\%\end{aligned}$$

6. Digester sizing: Provide two digesters.

$$\begin{aligned}\text{active volume of each digester} &= \frac{92{,}550\,\text{ft}^3}{2}\\ &= 46{,}275\,\text{ft}^3\ (1310\,\text{m}^3)\end{aligned}$$

Assume a 50-ft (15.2-m)-diameter tank.

$$\begin{aligned}\text{area of tank} &= \pi(25^2)\\ &= 1963\,\text{ft}^3\ (182\,\text{m}^2)\end{aligned}$$

$$\begin{aligned}\text{active depth} &= \frac{46,275\,\text{ft}^3}{1965\,\text{ft}^2}\\ &= 23.6\,\text{ft}\ (7.2\,\text{m})\end{aligned}$$

Add additional depths as follows:

$$\text{grit deposit (in addition to conical bottom)} = \underline{2\,\text{ft}}$$

$$\text{scum blanket} = \underline{2\,\text{ft}}$$

space below cover at maximum level $= \underline{2\,\text{ft}}$

total additional depth $= \underline{8\,\text{ft}}$

Note: Some designers provide an allowance for maximum sludge production conditions. However, in this example, an SRT of 15 days is selected (the minimum SRT should be 10 days) to account for the fluctuations in solids production.

$$\begin{aligned}\text{total sidewall height} &= (23.6 + 6)\,\text{ft}\\ &= 29.6\,\text{ft, say 30 ft (9.1m)}\end{aligned}$$

7. VS destruction and gas production: From Figure 6-1, VS destruction is 56% for an SRT of 15 days.

$$\begin{aligned}\text{VS destoryed} &= (12{,}300\,\text{lb/d})(0.56)\\ &= 6888\,\text{lb/d}\ (3120\,\text{kg/d})\end{aligned}$$

Assume that the gas produced is $16\,\text{ft}^3/\text{lb}$ ($1\,\text{m}^3/\text{kg}$) of VS destroyed.

$$\begin{aligned}\text{total gas produced} &= (6888\,\text{lb/d})(16\,\text{ft}^3/\text{d})\\ &= 110{,}208\,\text{ft}^3/\text{d}\ (3120\,\text{m}^3/\text{d})\end{aligned}$$

Because digester gas is about two-thirds methane,

$$\begin{aligned}\text{total methane produced} &= (110{,}208\,\text{ft}^3/\text{d})(0.67)\\ &= 73{,}840\,\text{ft}^3/\text{d}\ (2090\,\text{m}^3/\text{d})\end{aligned}$$

8. Digested sludge solids:

$$\begin{aligned}\text{fixed solids in feed sludge} &= (18{,}000 - 12{,}300)\,\text{lb/d}\\ &= 5700\,\text{lb/d}\\ \text{VS remaining after digestion} &= (12{,}300 - 6888)\,\text{lb/d}\\ &= 5412\,\text{lb/d}\\ \text{total solids in digested sludge} &= (5700 + 11{,}112)\,\text{lb/d}\\ &= 11{,}112\,\text{lb/d}\,(5040\,\text{kg/d})\end{aligned}$$

Single-stage digesters operate without supernatant withdrawal. Therefore, the volume of $6170\,\text{ft}^3/\text{d}$ fed is the same as the volume withdrawn. Assume that the specific gravity of digested sludge is 1.02.

$$\begin{aligned}\text{solids concentration in digested sludge} &= \frac{11{,}112\,\text{lb/d}}{(1.02)(6170\,\text{ft}^3/\text{d})(62.4\,\text{lb/ft}^3)} \times 100\\ &= 2.8\%\end{aligned}$$

9. Digester heating: Assume the following temperature conditions:

air:	25°F (–4°C)
average for earth around wall:	35°F (1.7°C)
earth below floor:	45°F (7.2°C)
raw sludge feed:	55°F (12.8°C)
digester contents:	95°F (35°C)

Heat transfer coefficients, U (see Table 6.5):

insulated wall exposed to air:	0.12 Btu/ft^2-°F-hr
wall exposed to dry earth:	0.11 Btu/ft^2-°F-hr
moist earth below floor:	0.15
insulated roof:	0.17

$$\text{sludge feed to each digester} = \frac{6170\,\text{ft}^3/\text{d}}{2} = 3085\,\text{ft}^3/\text{d}$$

Compute heat required for each digester for raw sludge using equation (5.2).

$$Q_1 = (3085\,\text{ft}^3/\text{d})(62.4\,\text{gal}/\text{ft}^3)(1\ \text{Btu/lb-°F})(95 - 55°\text{F})$$
$$= 7{,}700{,}168\,\text{Btu/d} = 320{,}840\,\text{Btu/d}\,(3.38 \times 10^8\ \text{J/h})$$

Compute the area of each component:

$$\text{wall area} = \pi(50)(30) = 4712\,\text{ft}^2$$

Assume that one-half of the wall is below grade.

$$\text{areas exposed to dry earth or air} = \frac{4.712\,\text{ft}^2}{2} = 2356\,\text{ft}^2$$

center depth of conical tank floor at a slope of 1:5 = 5 ft

floor area exposed to earth = $\pi(25)(25^2 + 5^2)^{1/2} = 2002\,\text{ft}^2$

roof area = $\pi(25) = 1963\,\text{ft}^2$

Compute the heat loss for each component using equation (6.3):

$$\text{wall above grade} = (0.12\,\text{Btu/ft}^2\text{-°F-hr})(2356\,\text{ft}^2)(95 - 25°\text{F})$$
$$= 19{,}790\,\text{Btu/hr}$$

$$\text{wall below grade} = (0.11\,\text{Btu/ft}^2\text{-°F-hr})(2356\,\text{ft}^2)(95-35\text{°F})$$
$$= 15{,}550\,\text{Btu/hr}$$

$$\text{floor} = (0.11\,\text{Btu/ft}^2\text{-°F-hr})(2002\,\text{ft}^2)(95-45\text{°F})$$
$$= 15{,}015\,\text{Btu/hr}$$

$$\text{roof} = (0.17\,\text{Btu/ft}^2\text{-°F-hr})(1963\,\text{ft}^2)(95-25\text{°F})$$
$$= 23{,}360\,\text{Btu/hr}$$

$$\text{total heart loss} = 73{,}715\,\text{Btu/hr}\ (7.78\times10^7\ \text{J/h})$$

If separate heat exchangers are used for heating raw sludge feed and the recirculating sludge, the heat exchanger capacities should be 641,680 Btu/h (3.38 7.78 × 10^7 J/h) and 73,715 Btu/h (7.78 × 10^7 J/h), respectively. For a single heat exchanger for the combined raw sludge and recirculating sludge, the heat exchanger capacity should be 394,555 Btu/h (320,840 + 73,715) (4.16 × 10^8 J/h). The values above are for each digester.

5.6 OPERATIONAL CONSIDERATIONS

It is important that stable operating conditions be maintained in an anaerobic digester because it is sensitive to changes. For example, sharp and frequent fluctuations in temperature affect the methane-producing bacteria. Therefore, changes in digester temperature greater then 0.6°C should be avoided. If uncontrolled, the changing conditions can result in digester upset and failure.

5.6.1 Reactor Performance

All process streams in an anaerobic digester system should be monitored for the volume of flow per day, and sampled and analyzed for various constituents and physical conditions (WEF, 1998):

- *Feed sludge*: TS, VS, pH, alkalinity, temperature
- *Digester content*: TS, VS, volatile acids, alkalinity, temperature
- *Digested sludge draw-off*: same as digester content
- *Digester gas*: percentage of methane, carbon dioxide, and hydrogen sulfide
- *Heating fluid* (hot water): TDS, pH
- *Supernatant*: pH, BOD, COD TS, total nitrogen, ammonia nitrogen, phosphorus

There is no single parameter that can be isolated as the best indicator of digester performance. A decrease in methane production and a corresponding increase in the CO_2 content of the gas produce an increase in volatile acid concentration, and decreases in the alkalinity and pH of the digester content indicate unbalanced anaerobic treatment operation. Anaerobic digestion may become unbalanced temporarily by the sudden change in temperature, organic loading, composition of feed sludge, or toxic loading, or a combination of these. This type of imbalance can be remedied by not feeding the digester temporarily, or by providing sufficient time for the microorganisms to adjust to the new environment.

The relationships among the CO_2 content of digester gas and the alkalinity and pH of the digesting sludge has been described in Section 6.2.3 and presented in Figure 5.4. As shown in the figure, at the normal anaerobic digestion condition of pH between 6.6 and 7.4 and CO_2 content of 30 to 40%, the bicarbonate alkalinity will be in the range 1000 to 5000 mg/L as $CaCO_3$. The concentration of bicarbonate alkalinity in the digester content should be about 3000 mg/L as $CaCO_3$.

Prolonged and relatively permanent imbalance of the anaerobic system may be the result of the presence of toxic materials or an extreme drop in pH. Imbalance due to an extreme drop in pH is encountered frequently in starting up a new digester. The doubling time of acid-forming bacteria is measured in hours, whereas it is measured in days for methane-forming bacteria. Therefore, the methane-forming bacteria require time to develop in sufficient numbers to convert the volatile acids to methane and CO_2. If the pH drops below the minimum value tolerated by the methane-forming bacteria, raising the pH by the addition of chemicals such as sodium bicarbonate is required until the digester returns to normal efficiency.

Figure 5.4 indicates that system pH is controlled by CO_2 concentration of the gas phase and bicarbonate alkalinity of the liquid phase. A digester with a given gas-phase CO_2 concentration and liquid-phase bicarbonate alkalinity can exist at only one pH. If bicarbonate alkalinity is added to the digester and the portion of the CO_2 in the gas phase remains the same, the digester pH must increase. For any fixed gas-phase CO_2 concentration, the amount of sodium bicarbonate required to achieve the desired pH change is given by the equation

$$D = 0.60\ (\text{BA at initial pH} - \text{BA at final pH}) \tag{5.4}$$

where D is the sodium bicarbonate dose, (mg/L) and BA is the bicarbonate alkalinity in digester, (mg/L $CaCO_3$).

Inorganic and organic materials that are toxic to anaerobic digestion process are described in Section 5.2.4 and are listed in Tables 5.1 and 5.2. The presence of toxic materials in wastewater treatment plants has been reduced as a result of the implementation of industrial pretreatment programs nationwide. However, if one or more toxic materials caused prolonged imbalance of a digester system, these materials must be removed.

5.6.2 Odor Control

Hydrogen sulfide gas and gaseous ammonia are the main odor-causing by-product of anaerobic digestion. Odor can also be caused by volatile acids in the incompletely digested sludge. Hydrogen sulfide can cause corrosion problems when the gas is used in boilers or engine generators. Therefore, the gas should be chemically scrubbed to remove hydrogen sulfide before use.

A leak-free gas-handling system will not cause an odor problem. Odor problems in an anaerobic digestion system are associated primarily with supernatant liquid and with storage and dewatering of digested sludge. Supernatant is not drawn off from single-stage digesters. In two-stage systems also, prethickening of sludge can eliminate, or at least reduce, supernatant discharge volume.

Dewatering operations of digested sludge create odors that are very difficult to control, especially when open dewatering equipment such as a belt filter press is used for dewatering. An option is to scrub the ventilation air around the dewatering equipment and from the digested sludge day tank using chemicals. Odorous air can also be treated by passing the air through an odor control biofilter. Because of their enclosed design, centrifuges contain odorous air, thereby facilitating efficient off-gas collection for treatment. Belt filter presses are presently available with factory-built enclosures for the same reason.

5.6.3 Supernatant

Supernatant from an anaerobic system generally contains high concentrations of dissolved and suspended solids, organic materials, nitrogen, phosphorus, and other materials. When returned to the treatment plant influent, these materials may impose an extra load on the liquid treatment processes. Table 5.6 presents reported characteristics of supernatant from two-stage digestion systems at different wastewater treatment plants. These characteristics represent a summary of the wide range of data observed at facilities in which the feed sludge has been (1) primary sludge alone, (2) primary and trickling filter sludge, or (3) primary and activated sludge.

As can be seen from the table, supernatant quality varies widely. Suspended solids, BOD, ammonia, and phosphorus can all cause problems in a treatment plant. The finely divided suspended solids in supernatant settle poorly and can build up in the plant, causing process overloading. High BOD imposes a high oxygen demand, requiring increased aeration requirement in activated sludge treatment systems. In plants that must achieve nitrogen limits in their effluents, the high ammonia loadings increase the cost of providing the oxygen required for treatment. In biological phosphorus removal, phosphorus is taken up by the growing cell mass and is removed from the wastewater stream in the waste biological sludge. This cell-bound phosphorus may be released during anaerobic digestion. Return of this phosphorus to the

TABLE 5.6 Characteristics of Supernatant from Two-Stage Anaerobic Digestion Systems

	Concentration[a] (mg/L)									
Parameter	Primary Sludge		Primary and Trickling Filter Sludge			Primary and Activated Sludge				
Total solids	9,400	—	4,545	—	—	1,475	2,160	—	—	—
Total volatile solids	4,900	—	2,930	—	—	814	983	—	—	—
Suspended solids										
Average	—	4,277[b]	2,205	1,518	7,772[c]	383	143	740	1,075	4,408[b]
Maximum	—	17,300	—	—	12,400	—	—	—	—	14,650
Minimum	—	660	—	—	100	—	—	—	—	100
Volatile suspended solids										
Average	—	2,645	1,660	—	4,403	299	118	—	750	3,176
Maximum	—	10,850	—	—	17,750	—	—	—	—	10,650
Minimum	—	420	—	—	60	—	—	—	—	75
BOD										
Average	—	713	—	—	1,238	—	—	—	515	667
Maximum	—	1,880	—	—	6,000	—	—	—	—	2,700
Minimum	—	200	—	—	135	—	—	—	—	100
COD	—	—	4,565	2,230	—	1,384	1,310	1,230	—	—
TOC	—	—	1,242	—	—	443	320	—	—	—
Total (PO_4)-P	—	—	143	85	—	63	87	100	—	—
NH_3N	—	—	853		—	253	559	—	480	—
Organic nitrogen	—	—	291	678	—	53	91	360	560	—
pH	8.0	—	7.3	7.2	—	7.0	7.8	7.0	7.3	—
Volatile acids	—	—	264	—	—	322	250	—	—	—
Alakalinity (as $CaCO_3$)	2,555	—	3,780	—	—	1,349	1,434	—	—	—
Phenols										
Average	—	0.23	—	—	0.23	—	—	—	—	0.35
Maximum	—	0.80	—	—	0.50	—	—	—	—	1.00
Minimum	—	0.06	—	—	0.06	—	—	—	—	0.08

Source: U.S. EPA, 1979.

[a] Unless noted, all values are average for the sampling period studied.
[b] Values in this column are a composite from seven treatment plants.
[c] Values in this column are a composite from six treatment plants.

TABLE 5.7 Alternatives for the Treatment of Anaerobic Digestion Supenatant

Constituent	Treatment Alternatives
Suspended solids	Coagulation with iron salts; filtration
BOD	Removal with suspended solids; stripping of volatile acids; aerobic biological treatment; activated carbon adsorption
Carbon dioxide	Precipitation with lime; stripping; ion exchange
Nitrogen	Removal of organic nitrogen with suspended solids; chemical precipitation; ion exchange; ammonia stripping may occur at $pH > 8.3$
Phosphorus	Removal with suspended solids; chemical precipitation; ion exchange

Source: Lue-Hing et al., 1998.

liquid stream can substantially reduce the net phosphorus removal efficiency of the plant and may necessitate removal by chemical treatment.

Problems associated with the recycling of supernatant from an anaerobic digestion system to the headworks of the plant include odor problems, possible sludge bulking, increased chlorine demand, and higher concentrations of nutrients in the effluent (Lue-Hing et al., 1998). High concentrations of some of the constituents may require removal by physical, biological, or chemical treatment of the supernatant before it is recycled to the liquid stream. Process alternatives for the treatment of supernatant are summarized in Table 5.7.

It is preferable to eliminate, rather than treat, a highly polluted digester supernatant. Modern digesters are designed as single-stage digesters without supernatant draw-off. In addition to eliminating supernatant discharge, prethickening the feed sludge reduces digester volume and heating requirements.

5.6.4 Struvite

Struvite is magnesium ammonium phosphate, a precipitate formed in digester systems. It is caused largely by environmental factors unique to anaerobic digestion. Ammonium and phosphate ions are released continuously by the digesting biomass, adding to those ions already in the sludge. Solvated magnesium ions also exist in the sludge. Thus, the digesting biomass can sometimes become supersaturated with magnesium ammonium phosphate. Precipitation of the compound occurs when the concentration exceeds the conditional solubility product of magnesium ammonium phosphate at the temperature and pH conditions in the digester. Conditional solubility refers to the specific set of actual conditions under which a given metal–salt compound may or may not be soluble.

Formation of struvite does not adversely affect the anaerobic digestion process. However, struvite scale deposits in piping and heat exchangers will cause maintenance problems. Scale deposits can also occur on parts of equipment that dewater digested sludge, especially the rollers of belt filter presses. Struvite particles in sludge are extremely hard and have been reported to damage pumps, especially progressive cavity and other positive-displacement pumps that pump anaerobically digested dewatered sludge.

Struvite scale deposits in pipelines can be controlled by frequent pigging of pipelines. Acid washing can remove the deposits, but it can be costly. A preventive solution to struvite formation is to precipitate the phosphorus in the digesting sludge by adding an iron compound such as ferrous chloride. An added benefit to adding ferrous chloride is that it controls hydrogen sulfide generation. Care should be taken to ensure that sufficient phosphorus remains to meet the nutritional needs of the anaerobic bacteria.

5.6.5 Digester Cleaning

Digesters can become partially filled with a bottom layer of settled grit and a top layer of floating scum. These accumulations reduce the active volume of the digesters and degrade their performance. When this happens, the digesters must be drained and deposits removed. The cleaning process is usually expensive and time consuming. Therefore, attention should be given to reducing the rate at which grit and scum accumulate, and in making it easy to clean the digesters when it becomes necessary.

Effective removal of grit from wastewater in the headworks of plants is the best preventive approach in reducing the amount of grit entering the digesters. Similarly, separate processing of scum collected from the clarifiers, such as hauling to a rendering plant, can reduce scum accumulation in the digesters. However, grit and scum entering a digester cannot be eliminated completely. Therefore, the design of an efficient mixing system that maintains a homogeneous mixture within the digester and keeps the grit and scum from separating out can practically eliminate the need for digester cleaning.

Providing multiple withdrawal points, as in a waffle bottom digester, can improve the grit removal process. Pipes extending into the upper levels of the digesting sludge can be used to remove floating materials periodically while the digester is in operation, before the materials form a thick mat. One of the advantages of an egg-shaped digester is that because of its small areas at the bottom and top, grit and scum accumulations are minimal, and any amount accumulated can easily be removed. When a digester needs emptying and cleaning, a dedicated and adequately sized drain pump for that purpose can make the cleaning activity faster and easier.

REFERENCES

Baker, H. A. (1956), *Bacterial Fermentations,* Wiley, New York.

Eastman, J. A., and Ferguson, J. F. (1981), Solubilization of Particulate Organic Carbon During the Acid Phase of Anaerobic Digestion, *Journal of the Water Pollution Control Federation,* Vol. 53, p. 352.

Estrada, A. A. (1960), Design and Cost Consideration in High Rate Digestion, *ASCE Proceedings of the Sanitary Engineering Division,* Vol. 86, No. 1, p. 111.

Fisher, A. J. (1934), Digester Overflow Liquor: Its Character and Effect on Plant Operation, *Sewage Works Journal,* Vol. 6, p. 30.

Ghosh, S., et al. (1975), Anaerobic Acidogenesis of Wastewater Sludge, *Journal of the Water Pollution Control Federation,* Vol. 47, p. 30.

——— (1995), Pilot and Full-Scale Studies on Two-Phase Digestion of Municipal Sludge, *Water Environment Research,* Vol. 67, p. 206.

Golueke, C. G. (1958), Temperature Effects on Anaerobic Digestion of Raw Sewage Sludge, *Sewage and Industrial Waste Journal,* Vol. 30, p. 1225.

Kappe, S. E. (1958), Digester Supernatant: Problems, Characteristics, and Treatment, *Sewage and Industrial Wastes,* Vol. 30, p. 937.

Lee, K. M., et al. (1989), Destruction of Enteric Bacteria and Virus During Two-Phase Digestion, *Journal of the Water Pollution Control Federation,* Vol. 61, p. 1421.

Liptak, B. G. (1974), *Environmental Engineering Handbook,* Chilton Book Co., Radnor, PA.

Lue-Hing, C., Zeng, D. R., and Kuchenither, R. (Eds.) (1998), *Water Quality Management Library,* Vol. 4, *Municipal Sewage Sludge Management: A Reference Text on Processing, Utilization and Disposal,* Technomic Publishing Co., Lancaster, PA.

Malina, J. F., Jr. (1962), Variables Affecting Anaerobic Digestion, *Public Works,* Vol. 93, No. 9, p. 113.

——— (1962), The Effect of Temperature on High-Rate Digestion of Activated Sludge, *Proceedings of the 16th Purdue Industrial Waste Conference,* Purdue University, Lafayette, IN, Vol. 46, No. 2, p. 232.

——— (1964), Thermal Effects on Completely Mixed Anaerobic Digestion, *Water and Sewage Works,* Vol. 95, p. 52.

———, and Pohland, F. G. (Eds.) (1992), *Design of Anaerobic Process for the Treatment of Industrial and Municipal Wastes,* Technomic Publishing Co., Lancaster, PA.

McCarty, P. L. (1964), Anaerobic Waste Treatment Fundamentals, *Public Works,* Vol. 95, pp. 9–12.

McFarland, M. J. (2000), *Biosolids Engineering,* McGraw-Hill, New York.

Metcalf & Eddy, Inc. (2003), *Wastewater Engineering: Treatment and Reuse,* 4th ed., Tchobanoglous, G., Burton, F. L., and Stensel, H. D. (Eds.), McGraw-Hill, New York.

Mignone, N. A. (1977), Elimination of Anaerobic Digester Supernatant, *Water and Sewage Works,* Vol. 108, p. 48.

Parkin, G. F., and Owen, W. F. (1986), Fundamentals of Anaerobic Digestion of Wastewater Sludge, *Journal of Environmental Engineering,* Vol. 112, p. 5.

Schafer, P. L., and Farrell, J. B. (2000a), Performance Comparisons for Staged and High-Temperature Anaerobic Digestion System, *Proceedings of WEFTEC 2000,* Water Environment Federation, Alexandria, VA.

——— (2000b), Turn Up the Heat, *Water Environment and Technology,* Vol. 12, p. 11.

Torpey, W. N. (1954), High-Rate Digestion of Concentrated Primary and Activated Sludge, *Sewage and Industrial Wastes,* Vol. 26, p. 479.

——— (1955), Loading to Failure of High Rate Digester, *Sewage and Industrial Wastes,* Vol. 27, p. 121.

Turovskiy, I. S. (1999), Beneficial Use of Wastewater Sludge in Russia, *Florida Water Resources Journal,* May, pp. 23–26.

U.S. EPA (1979), *Process Design Manual for Sludge Treatment and Disposal,* EPA 625/1-79/011.

——— (1987), *Stabilization of Sewage Sludge by Two-Phase Anaerobic Digestion,* Ghosh, S., et al., EPA-7600/2-87/040.

WEF (1987), EPA Design Information Report: Anaerobic Digester Mixing Systems, *Journal of the Water Pollution Control Federation*, Vol. 59, p. 152.

——— (1995), *Wastewater Residuals Stabilization,* Manual of Practice FD-9, Water Environment Federation, Alexandria, VA.

——— (1996), *Operation of Municipal Wastewater Treatment Plants,* 5th ed., Manual of Practice 11, Water Environment Federation, Alexandria, VA.

——— (1998), *Design of Municipal Wastewater Treatment Plants,* 4th ed., Manual of Practice 8 (ASCE 76), Water Environment Federation, Alexandria, VA.

6

ALKALINE STABILIZATION

6.1 INTRODUCTION

The treatment of wastewater sludge with an alkaline material can be an effective stabilization process. Lime is the most widely used alkaline material in the wastewater industry. Lime has traditionally been used in wastewater

Wastewater Sludge Processing, By Izrail S. Turovskiy and P. K. Mathai

treatment plants to raise the pH in stressed anaerobic digesters, to remove phosphorus in advanced wastewater treatment, and to condition sludge prior to vacuum filtration and pressure filter press dewatering. The original objective of lime conditioning of sludge was to improve its dewaterability, but in time it was observed that odors and pathogen levels were also reduced. Subsequently, lime addition became a major sludge stabilization alternative. Stabilization and disinfection of sludge with quicklime or hydrated lime are in wide use in Sweden, Finland, Germany, and Eastern European countries (Turovskiy, 1988). In the United States, lime stabilization or pasteurization is used to reduce odor, pathogens, and the putrecibility of sludge in PSRP (process to significantly reduce) and PFRP (process to further reduce pathogens). According to U.S. EPA's 1998 Need Survey (U.S. EPA, 1989), over 250 municipal wastewater treatment plants use lime to stabilize sludge. Larger treatment plants that use the process include those in Pittsburgh, Pennsylvania; Memphis, Tennessee (80 mgd); Toledo, Ohio (100 mgd); Washington, DC (300 mgd); and Las Vegas, Nevada (WEF, 1998).

The standard approach to lime stabilization is to add lime slurry to liquid sludge in sufficient quantity to achieve a pH of 12 or higher for two hours following addition. The high pH creates an environment that halts or substantially retards microbial reactions that could otherwise lead to odor production and vector attraction. This type of stabilization is classified by U.S. EPA as a PSRP that meets class B requirements as stated in the 40 CFR Part 503. Many of the advanced alkaline stabilization technologies in which dewatered sludge is treated with a dry alkaline material meet the class A requirements.

6.1.1 Advantages and Disadvantages

Both liquid lime and dry lime stabilization processes have several advantages over other sludge stabilization processes:

- The capital cost is low.
- The processes are reliable and easier to operate.
- They are capable of fast startup and shutdown.
- Pathogen reduction is reported to be as effective as or better than digestion processes.
- The processes greatly reduce odors if homogeneous lime feed and mixing occur.
- Stabilized sludge demonstrates improved dewaterability.
- Stabilized sludge may partially or fully replace liming agents used on acid soils for increasing alkalinity.
- The elevation of pH acts to fix or immobilize specific metal ions in sludge and soil and therefore restricts the possible uptake of metals by plants.

In addition to agricultural applications, beneficial end uses of lime-stabilized sludge include daily, intermediate, final, and vegetative cover in landfills; manufactured organic topsoil blends; bulk-fill applications such as slope stabilization and dike construction; and horticulture use in nurseries and sod farms. With respect to applying lime-stabilized biosolids to agricultural land, it may not be appropriate where soils are naturally alkaline, as in many parts of the western United States.

Lime stabilization has some disadvantages, including the following:

- Compared to digestion, there is no reduction in solid mass in lime stabilization.
- Increased mass from lime and chemical formation may result in higher transportation and ultimate disposal costs.
- Lime handling requires moderate operator attention and housekeeping because of the inherent dust in lime.
- Lime-stabilized biosolids have lower concentrations of nitrogen and phosphorus then do anaerobically digested biosolids.
- The lime stabilization process produces ammonia and possibly other odorous gases that may have to be treated before being exhausted.
- Dewatered sludge treated by lime can harden during storage.

6.1.2 Process Theory

Lime addition to sludge reduces levels of odor-causing microorganisms and pathogens by creating a high-pH environment that is hostile to biological activity and vector attraction. Gases containing nitrogen and sulfur that are evolved during anaerobic decomposition of organic matter are the principal sources of odor in sludge. When lime is added, the microorganisms involved in this decomposition are strongly inhibited or destroyed in the highly alkaline environment. Similarly, pathogens and other microorganisms are inactivated or destroyed by lime addition.

In addition to raising the pH, lime also affects the chemical and physical characteristics of sludge. Although the complex chemical reactions between lime and sludge are not well understood to date, it is likely that complex molecules of hydrolysis, saponification, and acid neutralization split in the high-pH environment. The following equations, simplified for illustrative purposes, show the types of reactions that may occur (WEF, 1998):

Reactions with inorganic constituents:

$$\text{Calcium: } Ca^{2+} + 2HCO_3^- + CaO \rightarrow 2CaCO_3 + H_2O \tag{6.1}$$

$$\text{Phosphorus: } 2PO_4^{3-} + 6H^+ + 3CaO \rightarrow Ca_3(PO_4)_2 + 3H_2O \tag{6.2}$$

$$\text{Carbon dioxide: } CO_2 + CaO \rightarrow CaCO_3 \tag{6.3}$$

Reactions with organic constituents:

$$\text{Acids: RCOOH} + \text{CaO} \rightarrow \text{RCOOCaOH} \tag{6.4}$$

$$\text{Fats: fat} + \text{Ca(OH)}_2 \rightarrow \text{glycerol} + \text{fatty acids} \tag{6.5}$$

Initially, lime raises the pH of sludge. Then reactions occur such as those in the equations above. If sufficient lime is not added, the pH decreases as these reactions take place. Subsequently, during sludge storage, biological activity resumes creating odor problems as well as possible regrowth of pathogenic organisms. Therefore, excess lime is required. The change in pH during storage of primary sludge using different lime dosages is illustrated in Figure 6.1. Excess lime in the range of 5 to 15% of the amount necessary for initial pH elevation is required to maintain the elevated pH because of slow reactions that continue to occur between lime and both atmospheric carbon dioxide and sludge solids.

In addition to the reactions above and raising the pH, hydration of calcium oxide generates heat. One mole (56 g) of calcium oxide generates 65 kJ of heat, as shown in the following equation:

$$\text{CaO} + \text{H}_2\text{O} \rightarrow \text{Ca(OH)}_2 + 64.5\,\text{kJ heat} \tag{6.6}$$

One kilogram of chemically pure lime (100% CaO) produces 1152 kJ of heat and requires 320 g of water. The heat in Kilojoules produced by the hydration of lime, taking its activity into consideration, is

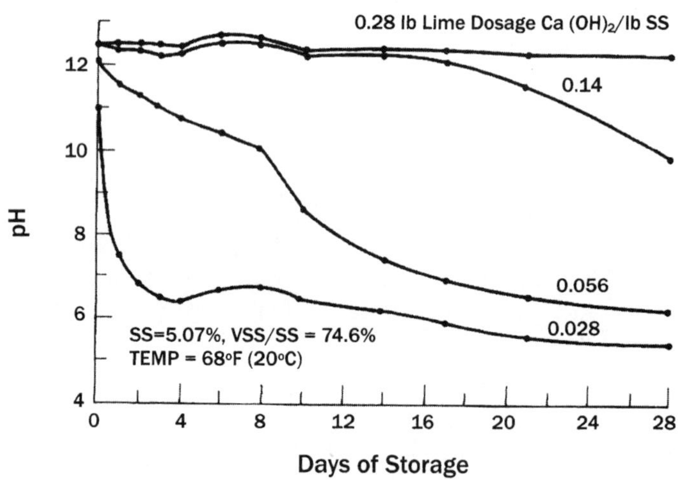

Figure 6.1 Changes in pH during storage of primary sludge using different lime dosages.

$$Q_R = 1152AM_L \tag{6.7}$$

where

Q_R = heat produced, kJ
A = lime activity, %
M_L = mass lime, kg

The amount of heat in Kilojoules that is required to heat sludge by lime may be determined by the formula

$$Q = (M_{SL}C_{SL} + M_LC_L)\Delta T \tag{6.8}$$

where

Q = amount of heat, kJ
M_{SL} = mass of sludge, kg
C_{SL} = specific heat of sludge, kJ/kg·°C
M_L = mass of lime, kg
C_L = specific heat of lime, kJ/kg·°C = 0.93
ΔT = difference between initial temperature of sludge and final raised temperature, °C

6.2 PROCESS APPLICATION

Lime stabilization is appropriate at small treatment plants where land is available for agricultural utilization of biosolids, although large treatment plants have used lime stabilization as the primary stabilization process. It is also practical at small plants that store stabilized sludge for later transportation to larger facilities for further treatment and disposal. Lime stabilization can be used to supplement an existing sludge stabilization process when the treatment plant is expanded. For example, when it is not practical to expand an existing aerobic digestion system, lime stabilization can be used to treat the excess sludge or the entire quantity of partially digested sludge.

Several different alkaline stabilization technologies are available. These can be grouped into three commonly used methods: (1) liquid lime stabilization, (2) dry lime stabilization, and (3) advanced alkaline stabilization technologies. Each system has advantages and disadvantages; therefore, it is important to evaluate and select a process that will produce the desired product on a site-specific basis.

6.2.1 Liquid Lime Stabilization

In liquid lime stabilization (also known as *prelime stabilization*), lime slurry is mixed with liquid sludge to produce class B biosolids. Figure 6.2 is a schematic of a liquid lime stabilization system. This is usually a batch treatment system in which the contact tank for mixing lime slurry with liquid sludge is

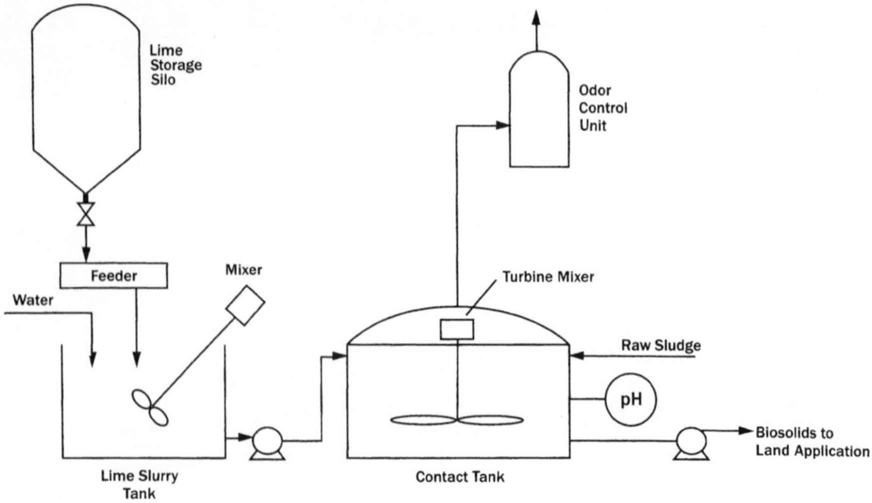

Figure 6.2 Typical liquid lime stabilization process schematic.

sized, based on the quantity of sludge and the number of batches to be run, for a minimum of two hours of detention time following treatment. The gases (predominantly ammonia) evolved from the process may have to be treated before being discharged to the atmosphere. Because of the large quantities of liquid biosolids that need to be transported to land disposal sites, liquid lime stabilization has been limited to smaller wastewater treatment plants. A storage tank is required for long-term storage of biosolids when land application is not practical throughout the year.

Prelime stabilization can be part of a sludge conditioning process prior to dewatering. Lime addition alone, without the addition of other conditioners, such as polymers or aluminum or iron salts, may produce a sludge that does not dewater well. Dewatering of lime-stabilized sludge is typically accomplished by vacuum filters or pressure filter presses. Centrifuge or belt filter press dewatering is seldom practiced because of the abrasive wear and scaling problems associated with lime-conditioned sludge.

Liquid lime stabilization of sludge requires more lime per unit weight of sludge solids processed than that necessary for dewatering alone. Excess lime is required to attain and maintain the pH above 12 for two hours so that bacterial regrowth is prevented. The dosages required for different types of sludge and solids concentrations are described later in the chapter.

6.2.2 Dry Lime Stabilization

In dry lime stabilization (also known as *postlime stabilization*), dry lime (hydrated lime or quicklime) is mixed with dewatered sludge cake to raise the pH of the mixture. The process requires adequate mixing to avoid pockets of putrescible material and to produce a homogeneous mixture. An effective mixer is a pug mill in which two screw conveyors or paddle mixers rotate in opposite directions. Other types of mixers include plow blender, paddle mixer, and screw conveyor. Figure 6.3 shows the schematic of a dry lime stabilization system. The system consists of a dry lime storage silo, a volumetric feeder and conveyor to transfer the lime, a dewatered sludge cake conveyor, and the mixer described above. The stabilized sludge is granular or crumbly in texture, depending on the moisture content of the dewatered sludge cake, and can be stored for long periods or distributed on land using a conventional manure spreader.

Quicklime, hydrated lime, or other dry alkaline materials, such as lime kiln dust or cement kiln dust, can be used for dry lime stabilization. Hydrated lime is typically limited to small installations. Quicklime is less expensive than hydrated lime and is easier to handle in large-scale facilities using lime in excess of 3 to 4 tons/day. Additionally, the heat generated from the exothermic reaction of quicklime and the water in sludge cake can enhance pathogen destruction.

Significant advantages of dry lime stabilization include no additional water added to sludge in the form of lime slurry, no special requirements for sludge dewatering equipment, and no lime-related abrasion and scaling problems to the dewatering equipment.

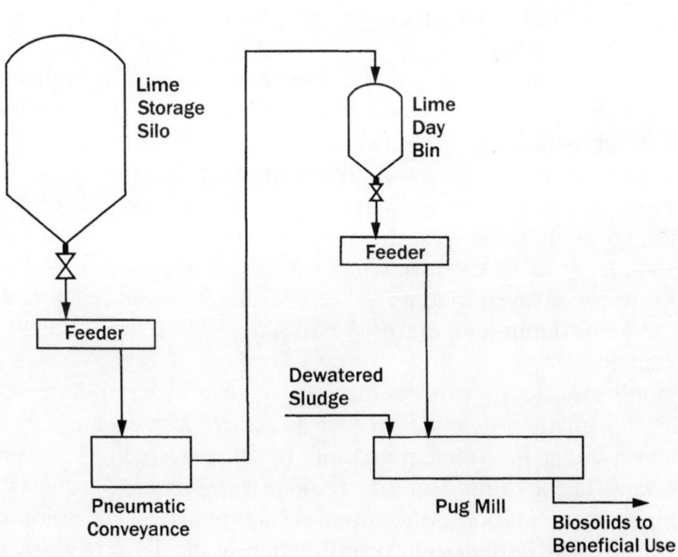

Figure 6.3 Typical dry lime stabilization process schematic.

6.2.3 Advanced Alkaline Stabilization Technologies

Technologies using materials other than lime for alkaline stabilization have been used by a number of municipalities. These technologies are modifications of traditional dry lime stabilization, and most of them use materials such as cement kiln dust, lime kiln dust, portland cement, or fly ash. The most common modifications are the supplemental drying and addition of other chemicals or bulking agents.

The principal advantages of advanced alkaline stabilization processes include: (1) the product meets class A stabilization requirements and is easy to handle; (2) the product has decreased odor potential and has value as a liming agent; (3) the capital cost is low compared with other class A stabilization processes; and (4) the processes are easy to operate, start up, and shut down. The disadvantages include high operating costs, the extensive odor control system that is required to treat ammonia and other off-gases, the increase in total solids/chemical mass to transport, and the product not being appropriate for alkaline soils.

Many of these advanced alkaline stabilization technologies are proprietary and are available through private companies. In one of the systems, class A pasteurization criteria are met by utilizing the exothermic reaction of quicklime with water in the sludge. Each kilogram of 100% quicklime produces approximately 15,300 cal/g·mol (27,500 Btu/lb-mol) of heat. The reaction between quicklime and carbon dioxide, also exothermic, releases approximately 43,300 cal/g·mol (78,000 Btu/lb-mol) of heat. Depending on the quantity of lime used, this reaction achieves a process temperature in excess of 70°C for more than 30 minutes as required to meet class A criteria. Odor control reagents and nutrient enhancements may also be incorporated as required.

In another pasteurization process, dewatered sludge is preheated in an insulated electrically heated screw conveyor prior to being transferred to a pug mill mixer, where the heated sludge and quicklime are mixed. Because supplemental heat is used to elevate the temperature of the sludge, quicklime must only be added in sufficient quantities to elevate the pH. This results in a lower lime dose than if the exothermic reaction of lime and water is the only source available to elevate sludge temperature. From the pug mill, the sludge–lime mixture is conveyed to a heated and insulated vessel reactor, where it is retained at a minimum temperature of 70°C for 30 minutes to meet the class A requirements.

In another proprietary process that satisfies the PFRP requirements of 40 CFR Part 503, an alkaline material such as cement kiln dust or lime kiln dust is mixed with sludge in sufficient quantity to achieve a pH of 12 or greater for at least 7 days. The alkaline-stabilized sludge is then dried for at least 30 days, until a minimum solids concentration of 65% is achieved. The biosolids need to be kept above 60% solids concentration before the pH drops below 12, and the mean ambient temperature must be above 5°C for the first 7 days. In a variation of this, the pH is increased to 12 or greater for only 72 hours, and the mixture is then heated to a temperature of at least 52°C and maintained for a minimum of 12 hours. The stabilized sludge is then dried to achieve a solids concentration of minimum 50%.

In a chemical stabilization/fixation technology process, pozzolanic materials are added to dewatered sludge. The addition of these materials causes cementitious reactions and after drying, produces a soil-like material of 35 to 50% solids content that is not subject to liquefaction under mechanical stress. The principal use of this material is as landfill cover.

6.3 PROCESS DESIGN

A number of design considerations should be evaluated before implementing an alkaline stabilization process. The design considerations for liquid and dry lime stabilization include the following:

- Sludge characteristics
- Contact time and pH
- Lime dosage
- Alkaline material storage
- Lime feeding
- Liquid lime mixing
- Dry lime mixing

6.3.1 Sludge Characteristics

The amount, source, and composition of the sludge to be processed determine the overall size of the alkaline stabilization facility. Equipment capacities must be sized to accommodate raw sludge storage and feed, chemical storage and feed, and stabilized product storage. Dosages of lime vary considerably for different types of sludge; for example, compared to primary sludge, waste activated sludge typically requires double the dosage of lime per unit dry weight of sludge solids. Compared to raw sludge, anaerobically digested sludge contains five to eight times the concentration of ammonia nitrogen, which is released from sludge in gaseous form at the elevated pH required for the alkaline stabilization process. Consequently, the potential for odors from ammonia and other nitrogen compounds such as amines is increased, and odor control facilities may be required, depending on the intensity of the odors and proximity to residential and commercial areas. The solids concentration of sludge to be stabilized has an effect on the process because it affects the chemical dosage and facility sizing. The nutrient content of the sludge affects the characteristics of the final product. The agronomic benefits of alkaline-stabilized biosolids depend on the amount of plant nutrients contained in the biosolids.

6.3.2 Contact Time and pH

The primary objective of lime stabilization is to inhibit bacterial decomposition and inactivate pathogenic organisms. The effective factor in achieving this objective is the pH level; however, as with most disinfection processes, the time of exposure is equally important. This is confirmed by the fact that the addition of lime for conditioning of sludge just before dewatering does not produce sludge cake with adequate bacteriological destruction. The alkaline material added must provide enough residual alkalinity to maintain a high pH until the product is used or discarded. The high pH prevents growth or reactivation of odor-producing and pathogenic organisms.

The drop in pH (referred to as *pH decay*) of stabilized sludge occurs because of two reactions that occur in succession. Atmospheric carbon dioxide is absorbed and then gradually consumes the residual alkalinity of the stabilized sludge, and the pH gradually decreases. Eventually, a pH is reached (below 11.0) where bacterial action resumes, continuing the drop in pH from the production of organic acids. Therefore, the design objective is to maintain pH above 12 for two hours or more to ensure pathogen destruction and to provide enough residual alkalinity so that the pH does not drop below 11 for several days, allowing sufficient time for disposal or use without the possibility of renewed putrefaction. The procedure recommended to accomplish these objectives is to bring the sludge to pH 12.5 by lime addition and maintain it at that level for 30 minutes, which keeps the pH above 12 for two hours.

6.3.3 Lime Dosage

The amount of lime required to stabilize sludge is determined by the type of sludge, its chemical composition, and the solids concentration. In liquid lime stabilization, dosages of lime [$Ca(OH)_2$] required to raise the pH to 12.5 or above and to maintain it at minimum pH 12 for 30 minutes for various types of sludge vary from 16 to 40% of sludge dry solids.

Chemical composition also determines the lime dosage required. This composition is a combination of the type of sludge as well as the type of treatment process from which it is produced, such as whether chemical coagulation treatment is used. The third factor affecting lime dosage is solids concentration. For liquid lime stabilization, lime dose per unit mass of sludge solids required to attain a particular pH level is relatively constant. That is, the lime requirement, rather than the sludge volume, is more closely related to the total mass of sludge solids. However, the lime dose required per unit mass of solids tends to be somewhat higher for very dilute sludge because more lime is required to raise the pH of water.

For dry lime stabilization, the dosage of lime required depends greatly on the type of final product required. Figure 6.4 shows the theoretical dry lime stabilization dose for both class B and class A stabilization (WEF, 1998). The

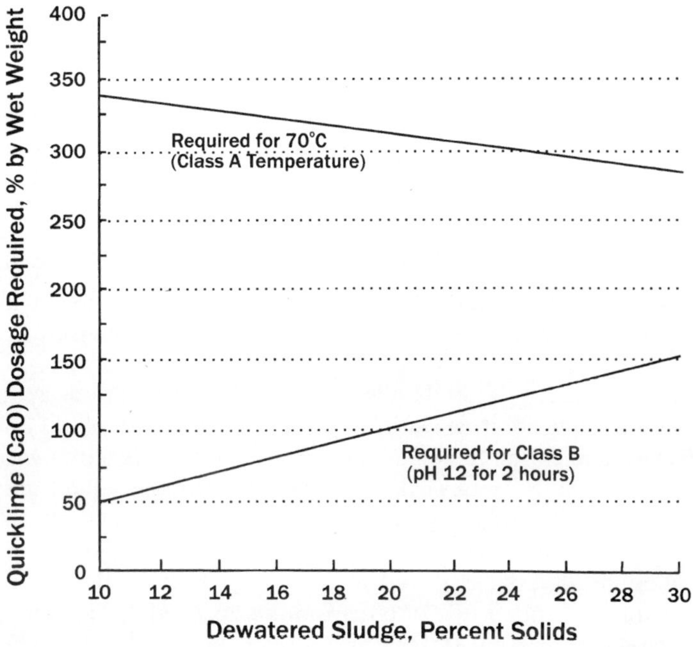

Figure 6.4 Theoretical quicklime dosages for class A and class B dry lime stabilization.

lower line shows the dose required for class B pH requirements, and the upper line shows the dose for class A temperature requirements. The figure shows that the quicklime requirement for class B stabilization theoretically increases with increased solids, whereas the quicklime requirement for class A stabilization decreases with increased solids. This is because of the predominance of the heating requirements of the water for class A versus the pH effects of class B. In practice, the amount of quicklime required to achieve class A can be 50% greater the theoretical class A temperature line in Figure 6.4. A drier, more easily crumbled product can be obtained by increasing the quicklime addition up to as much as twice the theoretical value shown.

Chemical dosages for advanced alkaline stabilization technologies vary by process, type of chemical and final product requirements. The exact dosage for any particular technology can be estimated through bench-scale testing.

6.3.4 Alkaline Material Storage

Several types of alkaline chemicals are available: quicklime (CaO), hydrated lime [$Ca(OH)_2$], lime kiln dust, cement kiln dust, and portland cement. In dry lime and advanced alkaline stabilization technologies, different types of additives result in different product texture and granulation. Quicklime is available in pebble, granular, and pulverized (virtually all passes a No. 20 sieve) forms. Hydrated lime is available in powder form (at least 75% passes a No. 200 sieve). Lime kiln dust and cement kiln dust are by-products from the respective industries, and it is important to ensure that these materials do not introduce contaminants or additional pollutants, jeopardizing the quality of the product.

Lime can be delivered in bags or in bulk. Bagged lime is preferred only where daily requirements are less than 300 kg (660 lb) of lime per day. Bagged lime must be stored under cover to prevent it from getting wet. Proper handling is especially important when quicklime is used, because it is highly reactive with water, producing heat and swelling that can cause the bags to burst.

Hydrated lime may be stored under dry conditions for periods up to a year without serious deterioration with atmospheric carbon dioxide, known as *recarbonation*. Quicklime deteriorates more rapidly by reacting with moisture from the air, which causes caking. Under good storage conditions, quicklime may be held as long as six months, but in general should not be stored for more than three months.

Bulk alkaline material is stored in conventional steel silos with hopper bottoms that have a side slope of 60° from the horizontal. The storage silos must be airtight to prevent slaking and recarbonation. Pebble quicklime is free flowing and will discharge readily from storage bins. Pulverized quicklime and especially hydrated lime have a tendency to arch and therefore require some type of mechanical or pneumatic agitation to ensure uniform discharge from silos.

Storage silos should be sized on the basis of the daily lime demand, the type and reliability of delivery, and an allowance for flexibility and expansion. As a minimum, storage should be provided to supply a 7-day lime demand, with a two- to four-week supply preferred. In any case, the total storage volume should be at least 50% greater than the capacity of the delivery truck to endure adequate lime supply between shipments.

6.3.5 Lime Feeding

In a liquid sludge stabilization process, lime is always delivered to the liquid sludge mixing vessel as a calcium hydroxide slurry (milk of lime). Dry lime should not be added to liquid sludge because caking will occur. After being mixed into slurry, hydrated lime and quicklime are chemically the same.

Feeding of Hydrated Lime In small treatment plants where bagged hydrated lime is used, the dry chemical is simply mixed with water in a batch tank. Lime solution is not corrosive, so that an unlined steel tank is sufficient for mixing and storing the slurry. An impeller type of mixer is generally used for mixing. Lime slurry is fed as a 10 to 30% solution by weight, the percentage depending on application and operator preference. The slurry can be discharged to the sludge mixing tank in one batch or metered continuously through a solution feeder.

In larger operations, where hydrated lime is stored in a silo, a more automated mixing and feeding scheme is appropriate. A dry chemical feeder that is positioned at the base of the bulk storage silo is used for continuous delivery of a measured amount of dry lime to a dilution tank. Two general types of automated dry feeders are available: volumetric feeders and gravimetric feeders. Volumetric feeders deliver a constant, preset volume of chemical in a unit time, regardless of changes in material density. Gravimetric feeders discharge a constant mass of chemical. Gravimetric feeders cost approximately twice as much as volumetric feeders, but they are more accurate. Gravimetric feeders are preferred because of their accuracy and dependability, but less expensive volumetric feeders may be sufficient when greater chemical feeding accuracy is not required. The feeder should be isolated from the storage silo with a slide gate to allow easy removal of the metering equipment if it becomes clogged.

Dust is a major problem associated with most dry chemical feed systems. Poorly fitting slide gates and leaking feeders are obvious sources of dust. The vertical drop between the feeder and slurry tank should be reduced or enclosed to reduce dust problems.

Dry hydrated lime is delivered by the feeder to a dilution tank that is normally located directly under the feeder. Level sensors in the tank and an automated waterline valve and feeder allow a preset amount of lime and water to be mixed in batches. The tank content is agitated by compressed air, water jets, or impeller-type mixers. The lime slurry is then transferred to the sludge mixing tanks. This transfer operation is the most troublesome single operation in the lime-handling process. Milk of lime reacts with atmospheric carbon dioxide or carbonates in the dilution water to form hard calcium carbonate scales, which, with time, can plug the slurry transfer line. Because the magnitude of this problem is directly proportional to the distance over which the slurry must be transferred, lime feed facilities should be located as close as possible to the lime–sludge mixing tanks. The slurry tank and all appurtenances should be cleaned and the line flushed thoroughly at the end of each day's operation.

Slaking and Feeding of Quicklime Feeding of quicklime is similar to that for hydrated lime, except that there is the additional step of slaking, in which the quicklime reacts spontaneously with water to form hydrated lime. Bagged quicklime can be slaked in batches simply by mixing 1 part quicklime with 2 to 3 parts water in a steel trough while blending with a hoe. Proportions should be adjusted so that the heat of hydration maintains the temperature of the reacting mass near 93°C (200°F). The resulting paste should be held for 15 to 30 minutes after mixing to complete hydration. Manually operated batch slaking is a potentially hazardous operation and should be avoided if possible. Uneven distribution of water can produce explosive boiling and splattering of lime slurry.

Continuous slaking is accomplished in automated machines that also degrit the lime slurry. Several types of continuous slakers are available. They vary mainly in the proportion of lime to water mixing initially. A volumetric or gravimetric dry chemical feeder is used to measure and deliver quicklime to the slaker. Since quicklime is available in a wide range of particle sizes, it is important to match the dry feeder with the type of quicklime to be used in the particular application. A paste is created in the slaker by a water-to-lime ratio of 2:1. The paste is then diluted further, as required, in a slurry tank. The paste should be held for approximately 5 minutes to allow complete hydration in the slaking chamber before discharg into the slurry tank.

6.3.6 Liquid Lime Mixing

A tank must be provided for mixing raw sludge with lime slurry and then holding the mixture for a minimum contact time. The contact time recommended is a minimum of 30 minutes after the pH reaches 12.5. The tank can be constructed of steel or concrete, although circular steel tanks with a coating that can withstand the high pH and the abrasion from the particles in sludge and lime, and with a concrete floor, are the most common. The size of the tank depends on whether the mixing is done on a batch or a continuous basis.

In the batch mode, the tank is filled with sludge and sufficient lime is added and mixed to maintain the pH of the sludge–lime mixture above 12.5 for the next 30 minutes. After this minimum contact time, the mixture is held for an additional 2 hours before the stabilized sludge is transferred either to the dewatering facilities or tank trucks for land application. Once the tank is emptied, the cycle begins again. In land application of liquid sludge, a storage tank may be required to store the biosolids during periods of the year when biosolids cannot be land applied.

In the continuous-flow mode, the pH and volume of the sludge in the mixing tank are held constant. Feed sludge displaces an equal volume of treated sludge. Lime is added continuously in proportion to the flow of the incoming raw sludge. The lime dose must be sufficient to keep the contents of the tank at a pH of 12.5 for at least 2 hours.

It is most common to operate lime stabilization systems in batch mode. Batch operations are very simple and are well suited for small-scale and manually operated systems. In very small treatment plants, the tank should be sized to treat the daily sludge production in one batch, because small plants have operating staff present during only one shift. In such instances, the mixing tank can also be used to gravity-thicken the lime-treated sludge before disposal. Continuous-flow systems require automated control of lime feeding and therefore are not usually cost-effective for small treatment plants. The primary advantage of continuous-flow systems over batch systems is that a smaller tank size may be possible.

Thickening of raw sludge before lime stabilization will reduce the mixing tank capacity requirement in direct proportion to the reduction in sludge volume. However, the lime requirement will be reduced only slightly because most of the lime demand is associated with the quantity of sludge solids.

Sludge–lime mixing can be accomplished with either diffused air or mechanical mixers. The agitation should be great enough to keep the sludge solids suspended and to distribute the lime slurry evenly and rapidly. Diffused air systems have the added advantage of keeping feed sludge fresh preceding lime addition in batch mode operations. However, ammonia will be stripped from sludge, producing odors and reducing the fertilizer value of the biosolids. Although mechanical mixers are subject to fouling with rags and other debris in the sludge, newer nonclog mechanical systems are available.

With air mixing, coarse bubble diffusers should be used. An air supply of 1.2 to 2.4 $m^3/m^3 \cdot h$ (20 to 40 cfm per 1000 ft^3) is required for adequate mixing. Mechanical mixing design criteria is based on (1) maintaining the bulk fluid velocity (defined as the turbine agitator pumping capacity divided by the cross-sectional area of the mixing tank) above 8.5 m/min (26 ft/min), and (2) using an impeller Reynolds number greater than 1000. For mechanical mixing in circular tanks, baffles that are one-twelfth the diameter of the tank should be placed 90° apart at the tank periphery. Consultation with agitation equipment manufacturers is recommended to select the most efficient and nonclog mechanical mixers.

6.3.7 Dry Lime Mixing

Mixing of dewatered sludge cake and lime (or other alkaline materials) is the most critical component of dry lime stabilization. The goal is to provide intimate contact between the sludge cake and lime and to produce a homogenized product. The mixing step is typically accomplished using a mechanical mixer such as a pug mill, plow blender, paddle mixer, or screw conveyor. Dewatered sludge and lime are introduced at the head end of the mixer. Sludge cake with a low moisture content and in large chunks, such as pressure filter press dewatered sludge, may require a "cake-breaker" conveyor to feed the sludge cake to the mixer. The proportioning of lime dosage to the rate at which cake is introduced is important. Dry lime can be supplied to the mixer with a screw conveyor if the distance from the bulk storage silo to the mixer is short, or with a pneumatic device. Large facilities with multiple mixing units convey lime pneumatically from the storage silo to a day tank (usually, a small storage silo) at each unit. Lime is then conveyed to the mixer using a volumetric or gravimetric feeder and a short screw conveyor.

A large number of variables affect the mixing process and, consequently, the characteristics of the product. The mixing characteristics of dewatered sludge cake vary with moisture content, conditioning chemicals used before dewatering, alkaline chemical type and dosage, temperature, mixing intensity, and mixing retention time. To adjust mixing intensity and retention time, mixers can be equipped with a variable-speed drive, adjustable mixer paddle configuration, weir plates, and other options. Many mixer manufacturers have mobile pilot testing units available. Whenever possible, this equipment should be used in the evaluation and selection of the proper and most efficient mixing equipment.

Design Example 6.1 Design a liquid lime stabilization system to treat a mixture of primary and thickened waste activated sludge for a wastewater treatment plant with the average flow of 3 mgd. The stabilized sludge is to be land applied. Stabilization and land application operate one shift a day (7 hours effective), five days a week. Assume the following design criteria:

Primary sludge:

average day sludge produced:	3000 lb/d (1361 kg/d)
peak day sludge produced:	4500 lb/d (2041 kg/d)
dry solids concentration:	4%
specific gravity of sludge:	1.03

Thickened waste activated sludge:

average day sludge produced:	2000 lb/d (907 kg/d)
peak day sludge produced:	3000 lb/d (1361 kg/d)
dry solids concentration:	3%
specific gravity of sludge:	1.01

Note: Critical components should be sized to meet the critical condition of peak day sludge production. Two mixing tanks should be provided, each with the capacity to treat the peak day sludge production. While one tank is filling, sludge in the other is treated with lime, mixed for 30 minutes, held for about 2 hours, and then loaded in tank trucks for land application. Alternatively, stabilized sludge can be dewatered for disposal. Hydrated lime is used in this facility.

1. Mixing tank:

volume of primary sludge at peak day production

$$= \frac{4500\,\text{lb/d}}{(62.4\,\text{lb/ft}^3)(0.04)(1.03)}$$

$$= 1750\,\text{ft}^3/\text{d}\ (50\,\text{m}^3/\text{d})$$

volume of WAS at peak day production

$$= \frac{3000\,\text{lb/d}}{(62.4\,\text{lb/ft}^3)(0.03)(1.01)}$$

$$= 1587\,\text{ft}^3/\text{d}\ (45\,\text{m}^3/\text{d})$$

volume of concrete tank required

$$= [(1750 + 1587)\,\text{ft}^3/\text{d}]\left(\frac{7\,\text{d/wk}}{5\,\text{d/wk}}\right)\left(\frac{7\,\text{hr}}{24\,\text{hr/d}}\right)$$

$$= 1363\,\text{ft}^3\ (39\,\text{m}^3)$$

tank surface area with 8-ft liquid depth

$$= \frac{1363\,\text{ft}^3}{8\,\text{ft}} = 170\,\text{ft}^2\ (16\,\text{m}^2)$$

With a 2-ft (0.6-m) freeboard, the tank size is 15 ft in diameter × 10 ft high (4.5 m × 3.0 m).

Note: Circular tanks are usually made of steel. Square concrete tanks may also be used, in which case each tank is 13 ft × 13 ft × 10 ft (4 m × 4 m × 3 m).

2. Mixing system: Air or mechanical mixing can be provided. Use an air-mixing criterion of 30 ft^3 per 1000 ft^3. Provide one blower per tank.

$$\text{blower capacity} = \frac{(1363\,\text{ft}^3)(30\,\text{cfm})}{1000\,\text{ft}^3}$$

$$= 41\,\text{cfm}\ (1.2\,\text{m}^3/\text{min})$$

If mechanical mixing is to be provided, mixer manufacturers should be consulted to design the proper mixer. A turbine mixer with a 5-ft-diameter impeller at 37 rpm and with a 7.5-hp motor may be provided.

3. Lime storage: Use a lime storage criterion of 30-day storage for average loading conditions.

Hydrated lime characteristics:

purity:	90% $Ca(OH)_2$
bulk density:	30 lb/ft^3

Lime dosage (see Table 6.1):

primary sludge:	12% by dry weight
WAS:	30% by dry weight

$$\text{average daily lime required} = \frac{(3000\,\text{lb/d})(0.12) + (2000\,\text{lb/d})(0.30)}{0.9}$$

$$= 1067\,\text{lb/d}\ (484\,\text{kg/d})$$

TABLE 6.1 Typical Lime Dosages for Liquid Lime Stabilization

	Dry Solids Concentration, %		Lime [$Ca(OH)_2$] % by Dry Weight		Average pH	
Type of Sludge	Range	Average	Range	Average	Initial	Final
Primay	3–6	4.3	6–17	12	6.7	12.7
Waste activated	1.0–3.0	1.3	21–43	30	7.1	12.6
Anaerobically digested Combined	6–7	5.5	14–25	19	7.2	12.4
Septage	1.0–4.5	2.7	9–51	20	7.3	12.7

$$\text{storage volume} = \frac{(1067\,\text{lb/d})(30\,\text{d})}{30\,\text{lb/ft}^3} = 1067\,\text{ft}^3\ (30\,\text{m}^3)$$

Note: A full truckload of hydrated lime is about 20 tons, which is about 1333 ft^3. It is usually economical to deliver a full truckload of lime. Storage volume should be 150% of the lime delivery volume so that a delivery can be ordered when two-thirds of the silo volume is expended. Therefore, the storage silo should be sized for 2000 ft^3 (57 m^3), which is about 56 days of lime requirement.

4. Lime slurry: Assume that 20% hydrated lime slurry to be used in two batches at 10 minutes per batch.

$$\text{lime feeder capacity} = \frac{(1067\,\text{lb/d})(7\,\text{d/wk})(1\text{ batch})}{(2\text{ batches/d})(5\,\text{d/wk})(10\,\text{min})} = 75\,\text{lb/min}\ (34\,\text{kg/min})$$

Assume a 25% lime slurry with a specific gravity of 1.1.

$$\text{lime slurry volume per batch} = \frac{(75\,\text{lb/min})(10\,\text{min})}{(62.4\,\text{lb/ft}^3)(0.25)(1.1)} = 44\,\text{ft}^3\ (1.2\,\text{m}^3)$$

Provide a slurry tank 4 ft in diameter × 5 ft high (1.5 ft of freeboard) (1.2 m × 1.5 m).

5. Long-term storage: Stabilized sludge generally needs to be stored during winter when sludge cannot be land-applied. Assume that 90 days of storage is required.

$$\text{average-day volume of sludge} = \frac{3000\,\text{lb/d}}{(62.4\,\text{lb/ft}^3)(0.04)(1.03)} + \frac{2000\,\text{lb/d}}{(62.4\,\text{lb/ft}^3)(0.03)(1.01)} = 2225\,\text{ft}^3/\text{d}\ (63\,\text{m}^3/\text{d})$$

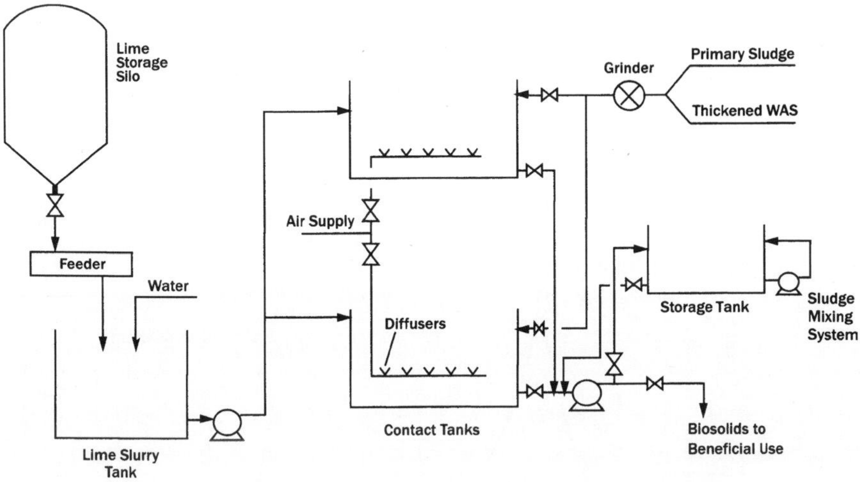

Figure 6.5 Schematic of lime stabilization in design example.

$$\text{volume of storage required} = 2225\,\text{ft}^3/\text{d} \times 90\,\text{d}\;(36{,}421\,\text{m}^3)$$
$$= 200{,}250\,\text{ft}^3\;(5671\,\text{m}^3)$$

Provide a storage tank 110ft in diameter × 22ft high (1ft of freeboard) (33.5m × 6.7m).

Figure 6.5 is a schematic of the lime stabilization system in Design Example 6.1. Prior to stabilization, sludge feed should be passed through an in-line grinder. This improves sludge mixing and flow characteristics, protects downstream pumping, and eliminates unsightly conditions such as rags, sticks, and plastics at the disposal site. An odor control unit may be needed if air mixing is used to treat the ammonia stripped from the sludge.

6.4 PROCESS PERFORMANCE

Alkaline stabilization reduces odors and odor production potential in sludge, reduces pathogen levels in sludge, and improves dewatering characteristics of sludge. The nature and extend of the effects are described in the following paragraphs.

6.4.1 Odor Reduction

Lime treatment creates a high-pH environment in sludge, thereby eliminating or suppressing the growth of microorganisms that produce odorous gases. Sufficient lime must be added to retard pH decay because odor generation will generally resume once the pH of the sludge falls below 10.

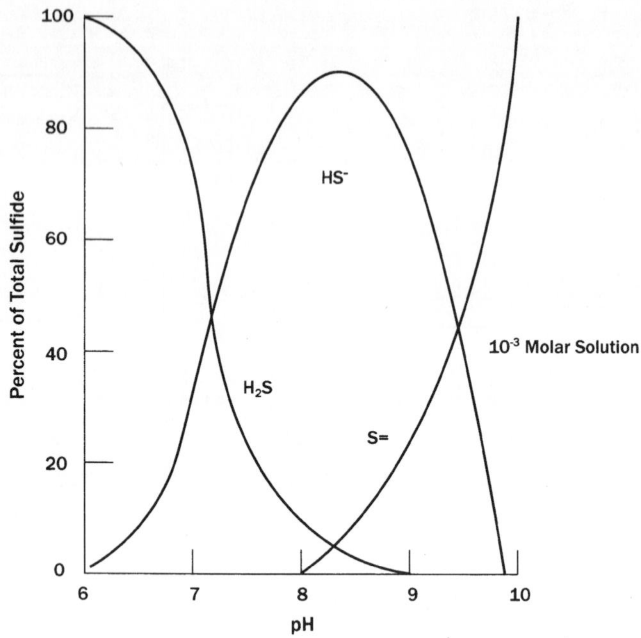

Figure 6.6 Effects of pH on the equilibrium between hydrogen sulfide and ionized sulfides.

Hydrogen sulfide gas present in dissolved form in sludge is a major cause of sludge odors. Figure 6.6 shows that as the pH of the sludge is raised, the fraction of total sulfide in the hydrogen sulfide form decreases from about 50% at pH 7 to essentially zero at pH 9 because it is converted to nonvolatile ionized forms. Consequently, above pH 9, there is no longer any hydrogen sulfide odor.

If an air mixing system is used for sludge–lime mixing, initial odors increase as a result of ammonia stripping. After the initial odors are dispersed or treated, odors will be reduced by a factor of 10 (Westphal and Christensen, 1983).

6.4.2 Pathogen Reduction

Several studies have demonstrated that liquid lime and dry lime stabilization achieve significant pathogen reduction, provided that a sufficiently high pH is maintained for an adequate period of time. Bacteria levels measured during full-scale studies performed at Lebanon, Ohio indicated that liquid lime stabilization of raw sludge reduced total coliform, fecal coliform, and fecal streptococci concentrations reduced by more than 99.9% (U.S. EPA, 1979). The number of *Salmonella* and *Pseudomonas eruiginosa* were reduced below the level of detection. The same studies showed that pathogen concentrations in different types of lime-stabilized sludge ranged from 10 to 1000 times less than those in anaerobically digested sludge from the same plant.

Research by Christensen on the pathogen reduction performance of dry lime stabilization using quicklime at doses of 13 and 40% as $Ca(OH)_2$ indicated that dry lime stabilization can achieve a reduction in fecal coliform and fecal streptococcus pathogens of at least two orders of magnitude (Christensen, 1982). This was as good as and in some cases better than standard liquid lime stabilization and then liquid lime stabilization followed by vacuum filtration. Neither organism had grown by day 7 (Westphal and Christensen, 1983). Westphal and Christensen also reported that liquid lime and dry lime stabilization processes performed as well as or better than mesophilic aerobic digestion, anaerobic digestion, and mesophilic composting in reducing densities of fecal coliform and fecal streptococcus.

There is little information about the amount of virus reduction during lime stabilization, but a few studies suggest rapid destruction at a pH of 12 (U.S. EPA, 1982). Qualitative analysis has shown substantial survival of higher organisms, such as hookworms, and amebic cysts, after 24 hours of contact time at high pH. It is not known whether long-term contact would eventually destroy these organisms. However, helminthes do not survive for a long time without a host. What is more important is that helminth ova can be destroyed in the alkaline stabilization process by heating the sludge to 50°C for two hours, to 60°C for several minutes, or to 70°C for several seconds.

6.4.3 Dewatering Characteristics

Lime has been used extensively in the past as a conditioning agent to improve the dewaterability of sludge (although lime conditioning and lime stabilization are different processes). Lime-conditioned sludge usually needs additional conditioners, such as ferric chloride, to improve its dewaterability. Additional lime in excess of the dose required for conditioning is typically necessary for stabilization, and this also results in improved dewaterability over that of untreated sludge. One potential drawback to mechanical dewatering of lime-stabilized sludge is the likelihood of scaling problems from the high lime doses.

REFERENCES

AWWA (1983), *Standard for Quicklime and Hydrated Lime*, AWWA B202-83, American Water Works Association, Denver, CO.

Bitton, G., et al. (1980), *Sludge: Health Risks of Land Application*, Ann Arbor Science Publishers, Ann Arbor, MI.

Christensen, G. L. (1982), Dealing with the Never-Ending Sludge Output, *Water Engineering and Management*, Vol. 129, p. 25.

——— (1987), Lime Stabilization of Wastewater Sludge, in *Lime for Environmental Uses*, Gutschick, K. A. (Ed.), ASTM, Philadelphia.

Farrell, J. B., et al. (1974), Lime Stabilization of Primary Sludges, *Journal of the Water Pollution Control Federation*, Vol. 46, No. 1, pp. 113–122.

Federal Register (1993), 40 CFR Part 503, *Standards for the Use or Disposal of Sewage Sludge*.

Lue-Hing, C., Zeng, D. R., and Kuchenither, R. (Eds.) (1998), *Water Quality Management Library*, Vol. 4, *Municipal Sewage Sludge Management: A Reference Text on Processing, Utilization and Disposal*, Technomic Publishing Co., Lancaster, PA.

Metcalf & Eddy, Inc. (2003), *Wastewater Engineering: Treatment and Reuse*, 4th ed., Tchobanoglous, G., Burton, F. L., and Stensel, H. D. (Eds.), McGraw-Hill, New York.

NLA (1995), *Lime: Application and Treatment in Treatment Processes*, Bulletin 213, National Lime Association, Arlington, VA.

Otoski, R. M. (1981), *Lime Stabilization and Ultimate Disposal of Wastewater Sludge*, EPA 600/S2-81/076.

Ramirez, A., and Shuckrow, A. J. (1980), Chemicals Disinfect Sludge, *Water and Sewage Works*, Vol. 127, No. 4, p. 52.

Reimers, R. S., et al. (1981), *Parasite in Southern Sludge and Disinfection by Standard Sludge Treatment*, EPA 600/2-81/166.

Turovskiy, I. S. (1988), *Wastewater Sludge Treatment,* Stroyizdat, Moscow.

U.S. EPA (1975), *Lime Stabilized Sludge: Its Stability and Effects on Agricultural Land*, EPA 670/2-75/012.

——— (1977), *Full Scale Demonstration of Lime Stabilization*, EPA 600/2-77/214.

——— (1979), *Process Design Manual for Sludge Treatment and Disposal*, EPA 625/1-79/011.

——— (1982), *Guide to the Disposal of Chemically Stabilized and Solidified Waste*, SW-872, Office Solid Waste Emergency Response, Washington, DC.

——— (1989a), *Environmental Regulations and Technology: Control of Pathogens in Municipal Wastewater Sludge*, EPA 625/10-89/096.

——— (1989b), *1988 Needs Survey of Municipal Wastewater Treatment Facilities*, EPA 430/09-89/001.

WEF (1998), *Design of Municipal Wastewater Treatment Plants*, 4th ed., Manual of Practice 8 (ASCE 76), Water Environment Federation, Alexandria, VA.

Westphal, A., and Christensen, G. L. (1983), Lime Stabilization: Effectiveness of Two Process Modifications, *Journal of the Water Pollution Control Federation*, Vol. 55, p. 1381.

7

COMPOSTING

Wastewater Sludge Processing, By Izrail S. Turovskiy and P. K. Mathai

7.1 INTRODUCTION

Composting of domestic solid waste and manure has been practiced for thousands of years and wastewater sludge has been used as a constituent in composting for a hundred years. However, producing compost from wastewater sludge on a large scale began only in the 1960s. Early methods in Europe used static piles without aeration (anaerobic decomposition) and later involved filling open trenches with periodic mixing.

Composting wastewater sludge began in France, Germany, Hungary, and Japan using turf and manure as bulking materials. Later, the northern European countries of Finland, Sweden, and the Netherlands followed the practice, despite the colder climate. Russia started composting using municipal solid waste and later a mixture of solid waste and dewatered sludge. Composting of wastewater sludge in the United States started only in the early 1970s. Examples of composting facilities in Europe and North America are described later in this chapter.

7.1.1 Composting Process

Composting of wastewater sludge is an aerobic biothermal process that decomposes the organic constituents. It can be described by the formula

$$C_6H_{12}O_6 + 6O_2 = 6CO_2 + 6H_2O + 674\,\text{kcal}$$

It can also be performed as an anaerobic process described by the formula

$$C_6H_{12}O_6 = 2C_2H_5OH + 2CO_2 + 27\,\text{kcal}$$

The aerobic process provides a higher caloric content. It goes much faster than the anaerobic process, and decomposition of the organic constituents produces a stable humuslike material.

Microbiology Composting represents the combined activity of a succession of mixed populations of bacteria, actinomycetes, and fungi. Although the interrelationship of these microbial populations is not fully understood, it is known that bacteria are responsible for the decomposition of a major portion of the organic matter.

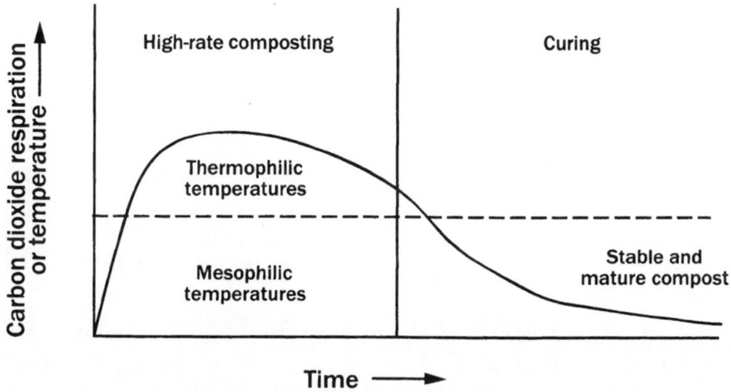

Figure 7.1 Phases during composting.

Composting occurs in three successive phases: the mesophilic, thermophilic, and curing phases. In the initial mesophilic phase, the temperature of the composting pile increases from ambient to 40°C (104°F). In the thermophilic phase, the temperature increases from 40°C to 70°C (104°F to 160°F). In the final curing phase (also known as the *cooling phase*), the microbial activity is reduced, and the composting process is completed. Figure 7.1 shows the three phases as related to carbon dioxide respiration and temperature.

All three categories of microorganisms are active during all three phases. In the mesophilic phase, acid-producing bacteria metabolize carbohydrates, sugars, and proteins. In the thermophilic phase, thermophilic bacteria become active and metabolize proteins, lipids, and fats. They are also responsible for much of the heat energy produced. Actinomycetes and fungi are present at both the mesophilic and thermophilic stages and are responsible for the destruction of a wide variety of complex organic compounds and cellulose. In the final curing phase, the compost matures through further microbial activity into a stable product. Compost that is insufficiently mature will reheat in storage and upon rewetting.

7.1.2 Composting Methods

There are three classifications of sludge composting:

- Windrow process
- Aerated static pile process
- In-vessel processes

In the windrow process, the dewatered sludge is agitated periodically for aeration. Agitation also controls the temperature and produces an end product that is uniform. In the aerated static pile process, air is blown through the materials using a blower. In-vessel composting systems are proprietary systems in which composting is performed in a reactor with air forced through the reactor, but with or without periodic agitation. In-vessel systems are enclosed processes. Windrow and aerated static pile processes are not enclosed, but a roof may be provided to protect the compost from precipitation.

The fundamental process of each composting method is similar (see Figure 7.2). The basic steps are as follows:

- A bulking agent is added to the dewatered sludge for porosity and moisture control. Amendments, if needed, are also added in this step to supplement limiting nutrients such as carbon. The bulking agents and amendments can be a wide variety of materials, such as wood chips, ground bark, yard waste, sawdust, wood ash, peat, agricultural residues such as rice hulls, and composted sludge. The mixture must be porous, structurally stable, and capable of self-sustaining the biothermal decomposition reaction.
- Next, a high-rate decomposition step must take place in which the mixture is aerated by mechanical turning, by providing air with a blower, or by both. A temperature in the thermophilic range 40 to 70°C should be attained. This also ensures destruction of pathogenic organisms and reduces the moisture content by evaporation.
- A storage and curing step that allows the compost to further stabilize and cool to lower temperatures.
- If a bulking agent is used, it can be recovered at the end of the high-rate decompositions step or the curing phase. Some compost may be recycled as a bulking agent. Additional air drying may be required between the high rate decomposition and curing step if the compost is too wet for screening.

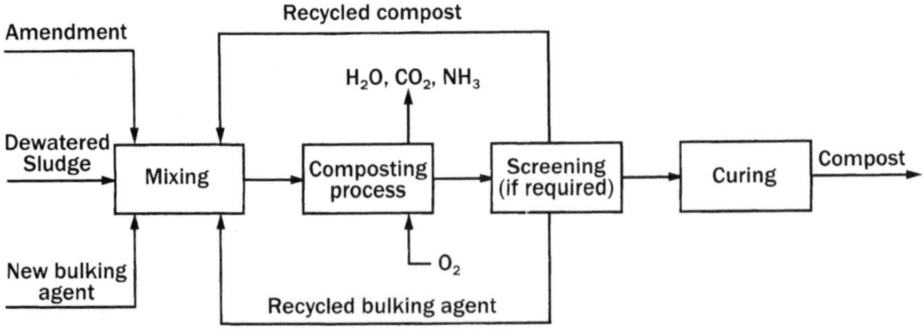

Figure 7.2 Generalized schematic of composting.

7.1.3 Advantages and Disadvantages of Composting

The main advantages of sludge composting are the following:

- Compost has an abundance of nutrients and is suitable for a wide variety of end uses, such as landscaping, topsoil blending, and growth media.
- Compost has less nitrogen than biosolids from other stabilization processes, due to the loss of ammonia during composting. However, nitrogen in compost is released more slowly and is available to plants over a long period of time, which is more consistent with plant uptake needs.
- Well-composted sludge can meet the requirements for class A biosolids and can be sold to distributors and the public.
- Compost increases the water content and retention of sandy soils.
- Compost increases aeration and water infiltration of clay soils.
- Windrow and aerated static pile processes have the flexibility to handle changing feed characteristics and peak loads, require relatively simple mechanical equipment, and are simple to operate.
- In-vessel processes require relatively small areas and have the ability to control odors.

The main disadvantages of composting are the following:

- Windrow and aerated static pile composting require relatively large areas, and odor control is a common problem.
- Ambient temperatures and weather conditions influence windrow and aerated static pile composting.
- In-vessel reactors have limited flexibility to handle changing conditions and are maintenance intensive.

7.1.4 Zoological Characteristics of Compost

Compost has the ability to enrich soil with beneficial invertebrates that stimulate the growth of soil organisms. The zoological population influences the physical, chemical, and microbial factors in soil. Some zoological populations in the mixture of sludge, bulking agent, and amendment survive the composting process. They penetrate the compost pile when temperature decreases to 30 to 35°C during the cooling stage. When the compost is stored, the zoological populations increase. The number of species in the cured compost can increase to more than 200. Some scientists are of the opinion that by determining the zoological population in the compost, one can predict how much biological activity occurs in the soil when the compost is applied to the soil.

7.2 PROCESS DESCRIPTION

7.2.1 Factors Influencing Composting

Composting represents the combined activity of a succession of mixed populations of bacteria, actinomycetes, and fungi at different stages of the process. The principal factors that affect the biology of composting are moisture, temperature, pH, nutrient concentration, and oxygen supply.

Moisture Decomposition of organic matter depends on moisture. Less than 40% moisture may limit the rate of decomposition. The optimum moisture content is 50 to 60%. Moisture content is also important for the structural integrity and sufficient porosity of the composting pile. If the initial compost mixture has more than 60% moisture, proper structural integrity will not be achieved and the mixture will not decompose well.

Dewatered municipal sludges are usually 18 to 35% solids (65 to 82% moisture), depending on the type of dewatering equipment used. Such sludge cakes are too wet for composting. Mixing the cake with a dry bulking material can reduce the moisture content of the sludge cake. Table 7.1 lists some of the typically used bulking agents and their characteristics. Figure 7.3 shows the effect the solids content of dewatered sludge cake has on the required mixing ratio of wood chips to sludge by volume. This illustration is site specific and the curve will shift depending on the relative volatility and solids contents of the sludge and bulking material. In this illustration, the amount of wood chips needed for sludge with 18% solids is about three times the amount required for a 35% sludge cake. High-solids sludge cake will reduce the amount of bulking materials required and the material management costs for mixing and screening the bulking materials. However, sludge cake of 30 to 35% solids may not break into small clumps uniformly when mixed with bulking agents. This necessitates more sophisticated mixing equipment, which can add cost and maintenance requirements. If the mixture is not homogeneous, airflow through the composting materials will not be uniform, allowing some zones to become anaerobic and cause odor. A site-specific economic analysis or pilot testing on a given feed stock with various bulking agents and blending strategies should be made to determine the best approach for a particular composting facility.

Temperature For the most efficient operation, composting process depends on temperatures of 50 to 65°C, but not above 70°C. For best results, temperatures should be maintained between 50 and 55°C for the first few days and between 55 and 60°C for the remainder of the composting process. Temperatures above 65 to 70°C for a significant period of time are detrimental to microbial activity. However, thermophlic microflora consists of bacteria whose temperature activity can oscillate between 60 and 75°C. Moisture content, aeration rates, size and shape of pile, atmospheric conditions, and nutrients

TABLE 7.1 Characteristics of Bulking Agents

Bulking Agent	Characteristics
Wood chips	Typically, will be purchased. High recovery rate in screening (60 to 80%). Good source of supplemental carbon.
Shredded bark	Typically, will be purchased. Medium recovery rate in screening (50 to 70%). Good source of supplemental carbon.
Chipped brush	Possibly available as a waste material. Low recovery rate in screening (40 to 60%) because of high percentage of fines. Good source of supplemental carbon. Long curing time for compost because of continued decomposition of unrecovered fines.
Sawdust and ground waste lumber	Possibly available as a waste material. Good source of supplemental carbon if relatively fresh. Poor source of supplemental carbon if old and extremely dry because more volatile forms of carbon are missing. Very low recovery rate, resulting in a large volume of compost.
Leaves and yard waste	Available as waste materials. Must be shredded. Wide range of moisture content requires close attention to material balance. Readily available source of supplemental carbon. Relatively low porosity. Not recoverable, resulting in a large volume of compost.
Shredded tires	Often mixed with other bulking agents. Provide no supplemental carbon. Nearly 100% recoverable. May contain metals.
Agricultural residues such as rice hulls	Availability is regional. Possibly available as a waste material. Good source of supplemental carbon. Very low recovery rate.

Source: Adapted in part from WEF, 1998.

influence the distribution of temperature in a composting pile. For example, if excessive moisture is present, heat will be carried off by evaporation, and thus the temperature rise will be less. On the other hand, low moisture content will decrease the rate of microbial activity, thereby reducing the rate of heat evolution and retention.

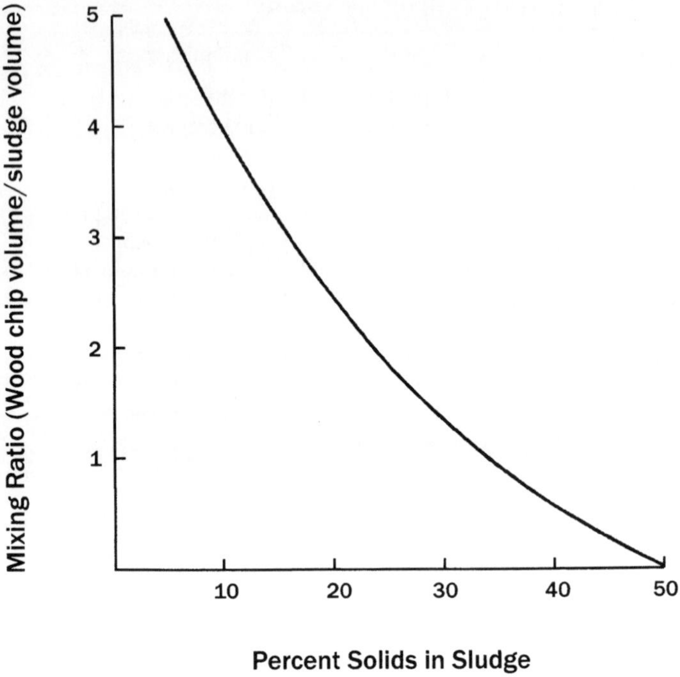

Figure 7.3 Effects of solids content on the ratio of wood chips to sludge by volume.

pH The pH of composting mixture should generally be in the range 6 to 9. Optimum pH range for the growth of most bacteria is between 6 and 7.5 and between 5.5 and 8 for fungi. Although the pH in the pile varies throughout the composting period, it is essentially self-regulating. It is difficult to alter the pH for optimum biological growth. Therefore, it has not been found to be an effective operational control. The effect of pH is described further later in the chapter.

Nutrient Concentration Both carbon and nitrogen are required as energy sources for the growth of microorganisms. Approximately 30 parts by weight of biodegradable carbon are used by microorganisms for each part of nitrogen. Therefore, the most desirable carbon-to-nitrogen ratio in the composting mix is in the range 25:1 to 35:1 by weight. Lower ratios increase the loss of nitrogen by volatilization as ammonia, resulting in the loss of nutrient value of the compost and the emission of ammonia odor. Higher ratios lead to progressively longer composting time, and the organic material remains active well into the curing stage (Poincelot, 1977).

The carbon-to-nitrogen ratio of wastewater sludge is generally in the range 20:1 to 40:1. Therefore, sludges with lower than a 25:1 ratio require additional biodegradable carbon for active microbial growth. Bulking agents and amendments provide the supplemental carbon and improve both the energy balance and the carbon-to-nitrogen ratio. In wood chips, only a thin layer of wood is available as carbon. Carbon in amendments such as sawdust is more readily available.

Oxygen Supply Oxygen concentration in the composting mass should be maintained between 5 and 15% by volume of gas mass. Although in the windrow process oxygen concentrations as low as 0.5% have been observed without anaerobic symptoms, a minimum of 5% is generally required for aerobic conditions. Oxygen concentrations higher than 15% will result in a temperature decrease because of the higher airflow. Oxygen should reach all parts of the composting materials for optimum results, especially in in-vessel systems.

7.2.2 Windrow Process

In the windrow process, dewatered sludge mixed with a bulking agent is formed in long parallel rows or windrows. The width of a typical windrow is 2 to 4.5 m (6 to 14 ft) at the base and the height is 1 to 2 m (3 to 6.5 ft). Depending on the characteristics of the equipment used for mixing and turning of the windrows, the cross section of the pile may be triangular or trapezoidal. Windrow composting is commonly performed at open outdoor sites. However, in areas of significant precipitation, it may be desirable to provide a roofed structure to cover the windrows.

The windrows are turned periodically to expose materials to the air and loosen the materials for ease of air movement through the materials. It also helps to reduce moisture. A frond-end loader or a dedicated turning machine can be used for turning the windrows (see Figure 7.4).

Bulking agents may include the recycled composted sludge, and external agents such as wood chips, sawdust, ground bark, straw, yard waste, or rice

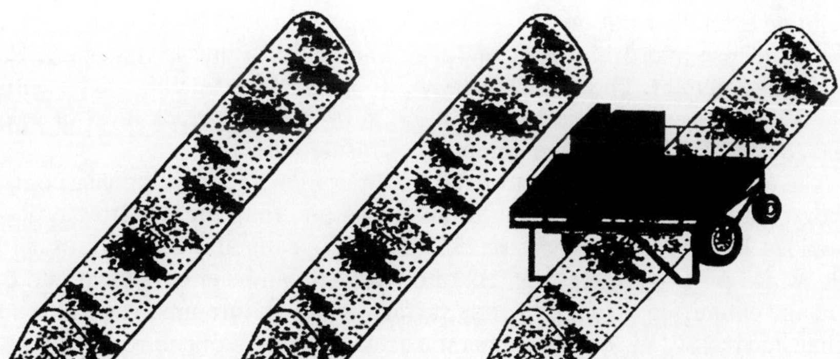

Figure 7.4 Windrow composting. (Reprinted with permission from WEF, 1998.)

hulls. Enough bulking agent is added to the dewatered sludge to obtain the mixture solids content of 40 to 50%. Higher solids content may impede bacterial activity and limit the rate of decomposition. Bulking agent greatly improves the porosity of the mixture, which in turn improves the aeration characteristics. External bulking agent is a good source of carbon for the biological decomposition. The ideal carbon-to-nitrogen ratio is in the range 25:1 to 35:1. Bulking agent also increases the structural integrity of the mixture and thus its ability to maintain a properly shaped windrow.

The composting period for a windrow process is about 21 to 28 days, although this is climate and feed stock dependent. Typically, windrows are turned every 4 to 5 days. Temperature in the central portion of the windrow reaches as high as 65°C as a result of the decay process. The temperature should be maintained at or above 60°C for optimum biological activity. Temperature in the outer layer is considerably lower and may reach ambient conditions. During winter conditions, temperature in the center portion may reach only 50 to 60°C, which would prolong the composting period.

Aerobic conditions are difficult to maintain throughout the windrows. If anaerobic conditions exist, offensive odor will be released when the windrows are turned. Some facilities provide a trench with a perforated pipe under each windrow for aeration with a fan to maintain aerobic conditions within the materials.

7.2.3 Aerated Static Pile Process

The identifying feature of the aerated static pile system is a grid of aeration piping for forced aeration. A blower or fan aerates the pile. The aerated static pile process consists of mixing of dewatered sludge with a bulking agent (usually, wood chips), construction of the composting pile over the grid of aeration piping, composting, screening of the compost, and curing and storage. Figure 7.5 shows an aerated static pile process of composting.

The aeration grid is usually made of 100- to 150-mm (4- to 6-in.) perforated plastic pipe laid inside a 0.3-m (1-ft) plenum of wood chips. The wood chips facilitate even distribution of air during composting and absorb moisture that may condense and drain from the pile. The compost pile is constructed on top of the plenum. The pile is usually 2 to 2.5 m (6 to 8 ft) high. The entire pile is then covered with a 150- to 200-mm (6- to 8-in.)-thick layer of wood chips or unscreened finished compost for insulation.

The forced air provides a more flexible operation and more precise control of oxygen and temperature conditions in the pile than the windrow process provides. It also reduces the chances of developing anaerobic conditions in the pile, which in turn reduces the risk of odors. Air can be either blown into the grid and exhausted through the pile surface or drawn through the surface and exhausted by the fan. The latter method affords better odor control by exhausting the air through a biofilter made from the cured compost (see Figure 7.5). Figure 7.6 illustrates examples of aeration plenum used in Europe.

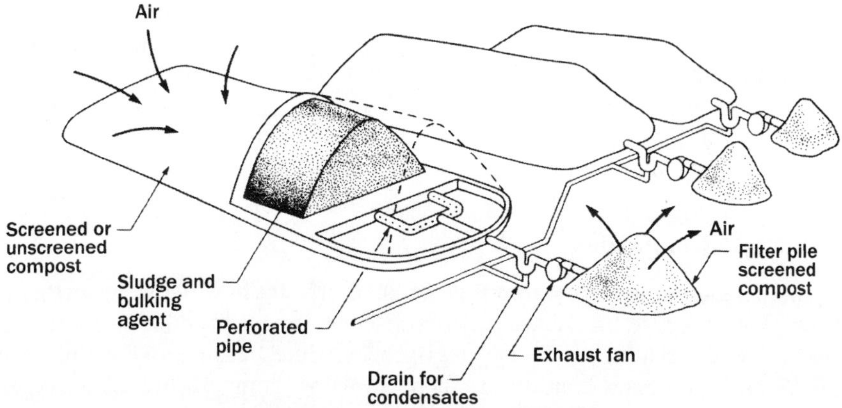

Figure 7.5 Aerated static pile composting.

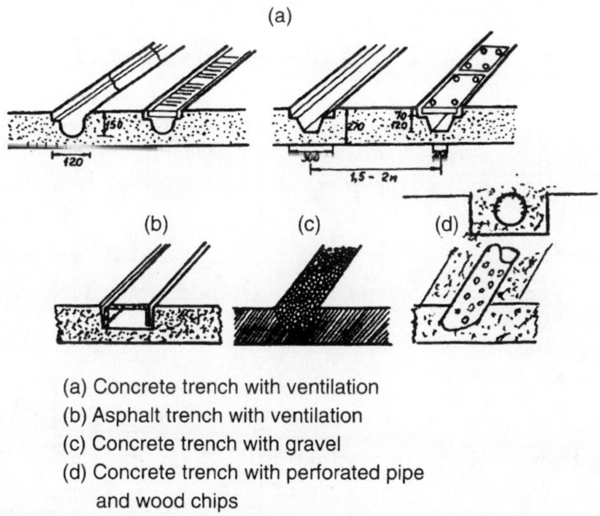

Figure 7.6 Examples of aeration plenum in static pile composting.

An active composting period is typically 21 to 28 days followed by a curing period of 30 days minimum. Curing further stabilizes the material. The compost is screened before or after the curing step for recovering of the bulking agent. For ease of screening, it is critical that the compost moisture content not be higher than 45%. Moisture content can be reduced, if required, by an intensive drying step with a higher-than-normal aeration rate before screening.

In small facilities, individual piles may be constructed with a dedicated fan for each pile. In large facilities, to make more efficient use of available space, an extended pile is constructed by adding each day's contribution to the side of the preceding day's pile. In a typical operation of the extended aerated pile system, the pile is extended daily for 28 days. The first section is removed after 21 days. After seven sections are removed in sequence, there is adequate space for operating the equipment, so that a new extended pile can be started from the beginning (see Figure 7.7).

7.2.4 In-Vessel Process

An in-vessel composting system is a confined process in which composting is accomplished inside an enclosed container or basin. It produces a more stabilized and consistent product in less process time because of its ability to control environmental conditions, such as airflow, temperature, and oxygen concentration. It also affords better containment and control of odors. Detention time in a reactor varies from 10 to 21 days depending on the system supplier's recommendations, regulatory requirements, and desired product characteristics. A schematic of a typical in-vessel process is shown in Figure 7.8.

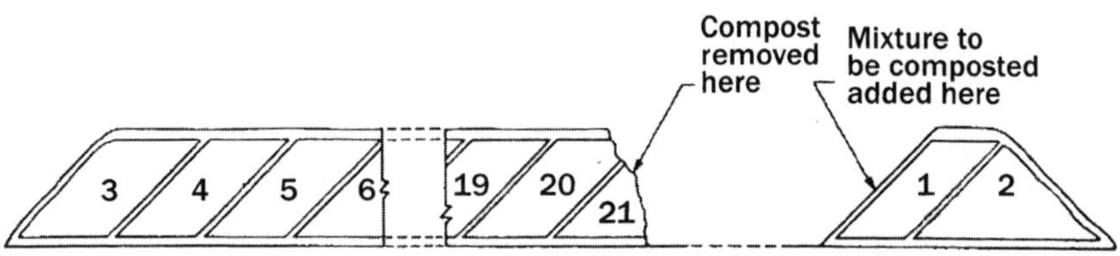

Figure 7.7 Configuration of extended aerated pile.

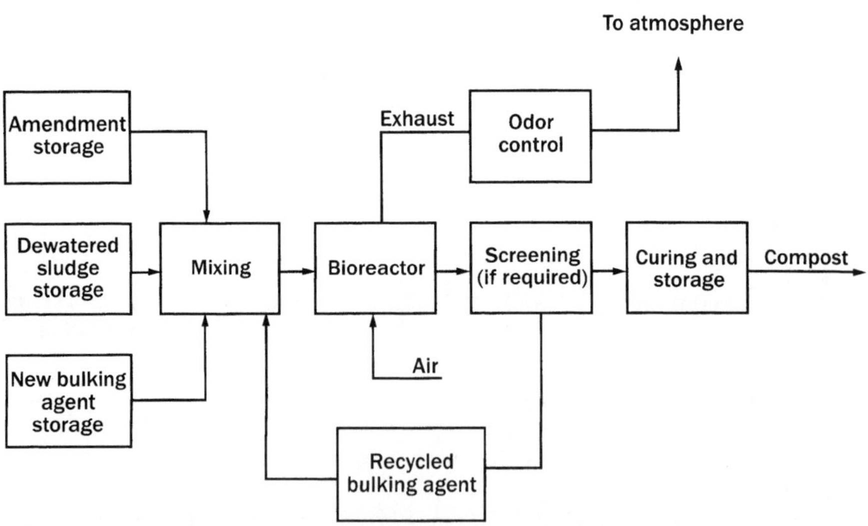

Figure 7.8 Schematic of in-vessel composting process.

Reactors are loaded with the composting mixture for only a portion of a day (in smaller plants sometimes only five days per week). However, the sludge dewatering operation is normally for a longer period of time than the loading period of the reactors. Therefore, a storage facility is required for storing the dewatered sludge in addition to the storage facilities required for storing bulking material, amendment, and compost. The storage facilities can be covered piles, live-bottom bins, or silos.

In-vessel composting system is divided into two general classes: plug flow reactors and agitated (also known as dynamic) reactors. The plug flow reactors are divided further into vertical and horizontal plug flow reactors. Figures 7.9 and 7.10 show the various types of reactors.

Vertical plug flow reactors are constructed of steel with a corrosion-resistant coating, reinforced fiberglass, or concrete. Dewatered sludge cake is mixed with recycled compost and amendment or bulking agent and is added to the top of the preceding day's fill. The material in the vessel is aerated but not agitated or mixed. When the bottom compost layer is removed by a live bottom auger discharge or by a sliding frame and conveyor, the entire content of the vessel moves as a plug to the bottom.

There are three major vertical plug flow systems in the United States. The primary differences among them are shape and materials of construction (cylinder made of steel, rectangular concrete boxes, and square or rectangular

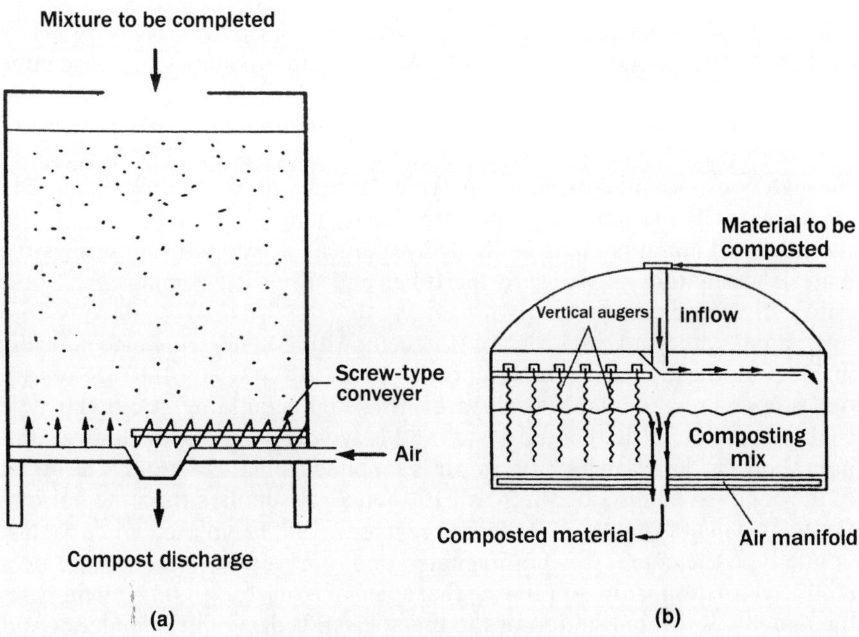

Figure 7.9 Vertical reactors.

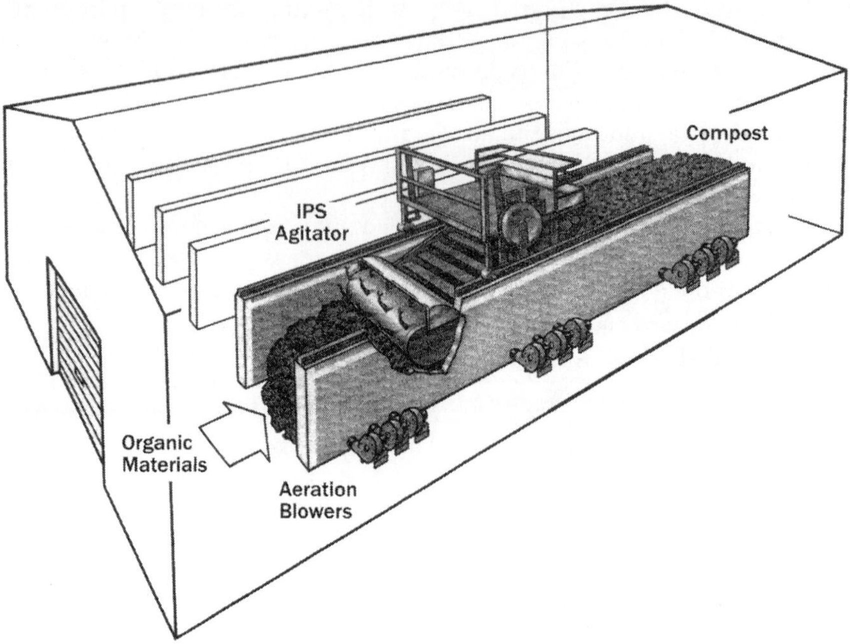

Figure 7.10 Horizontal agitated reactor. (From UF Filter, a Siemens business, Sturbridge, MA.)

fiberglass boxes), configuration of the aeration systems, and discharge devices.

Horizontal plug flow reactors (also known as *tunnel reactors* because of their shape) are typically constructed of concrete. Air is supplied by a blower through floor-mounted diffusers. A typical reactor unit is approximately 5 m (18 ft) wide, 3 m (10 ft) deep, and 20 m (65 ft) long. Fresh mix is loaded at the inlet end and a pneumatically or hydraulically operated ram pushes the materials longitudinally toward the other end. Finished compost then falls onto a discharge conveyor.

In an agitated in-vessel reactor, the composting material is mechanically mixed periodically during the composting process. Physically, the reactors are open-topped rectangular bins and are enclosed in a building. A circular type is also available in the United States and is typically covered with an aluminum geodesic dome. In both types, air is supplied from the bottom. The depth of the compost bed is 2 to 3 m (6 to 10 ft) and air is supplied from the bottom. Of the two major rectangular types of reactors available in the United States, in one type the compost is both agitated and removed from the reactor by a mobile extraction–conveyor device that spans the reactor. In the second type, the agitator is located on top of the reactor and it digs out the material and redeposits it behind the machine. In the circular reactor type, fresh material is loaded on the perimeter, and vertical augers agitate and move the mix to the center of the reactor.

7.2.5 Design Considerations

In this section we describe the factors that need to be considered in the design and operation of composting facilities. These factors, the factors influencing composting described earlier, and the theoretical aspects of composting described in another section should be considered to meet the requirements of each composting system.

Energy Balance Haug (1993) has shown that the organic decomposition in a composting operation is self-sustaining when the ratio W is less than 10, where

$$W = \frac{\text{mass of water in initial compost mixture}}{\text{mass of degradable organics in the mixture}}$$

The degradability of the mixture can be adjusted by the addition of a bulking agent or amendment that contain high concentrations of degradable organic material. The bulking agents and amendments are usually dry, and they increase the volatile fraction and decrease the moisture fraction of the mixture, thereby reducing the ratio, W.

The ratio W can also be defined as an energy balance equation:

$$W = \frac{\text{weight of water evaporated}}{\text{weight of volatile solids lost}}$$

Organic decomposition produces water and generates heat. To keep this ratio below 10, it is important that sufficient moisture is removed from the mixture by evaporation. However, the composting process temperature should be maintained for proper decomposition. The temperature in the mixture will not rise if rate of heat loss exceeds the rate of heat generation.

Detention Time When the aerated static pile composting system was developed by the U.S. Department of Agriculture in the 1970s, it established the active composting period to be 21 days followed by 30 days of curing period without aeration for the composting mix of dewatered wastewater sludge and wood chips. Several states have established this detention time in their regulations. However, in-vessel suppliers have established their own active composting periods. Some systems recommend as little as 14 days of active composting, although most horizontal agitated systems use 21 days.

Even after a long curing period, with the compost considered stable, the organics in stored compost can continue to decompose, especially if it not screened because of the biodegradable organics that remain in the bulking agent. The stability of the compost can be measured either by the carbon dioxide production or oxygen consumption. Compost with a respiration rate of 3 mg CO_2 per gram of organic carbon per day or a consumption rate of 1 mg O_2 per gram of organic carbon per day is considered stable. From the aerobic decomposition equation shown earlier in this chapter, this oxygen consumption rate is equivalent to 1.4 CO_2 per gram of organic carbon.

Temperature Control and Aeration Temperature in a composting pile may exceed 70°C if not sufficiently controlled. A temperature in excess of 70°C is detrimental to microbial activity. Sufficient aeration should be provided to control the temperature rise. As the airflow rate is increased in a system, the temperature in the pile decreases. This also increases the rate of water vapor removal. In an agitated composting system, agitation also releases heat and water vapor.

An aeration rate of $34\,m^3/Mg \cdot h$ ($1100\,ft^3$/dry ton-hr) provides adequate drying, oxygen for organic decomposition, and high-enough temperatures for pathogen destruction. During the mesophilic stage of composting, a higher aeration rate is required to prevent excessive temperature buildup. Therefore, aeration capacities in the range 30 to $160\,m^3/Mg \cdot h$ (1000 to $5000\,ft^3$/dry ton-hr) should be provided.

Centrifugal blowers can provide the necessary pressure for aeration of the compost mixture and also push the air through the odor biofilter. Aeration blowers can be controlled by a timer to turn them on and off. The percentage of time the blowers are on is set based on temperature readings. The bowers can also be turned on and off automatically based on temperature readings and temperature feedback control. In an aerated static pile system, air is either exhausted or pulled through the pile. When air is exhausted through the pile, water vapor is driven to the surface and promotes drying and avoids accumulation of condensation in the pile and aeration piping. When air is pulled thorough the pile, moist air drawn through the pile condenses in cooler areas of the pile. When enough condensate accumulates, it will leach materials from the sludge. Condensate also collects in the aeration piping, which should be removed by water traps.

This condensate, the leachate, and any contaminated rainfall runoff should be collected and discharged to the treatment plant influent stream for treatment because the contaminated water can become a source of odors if allowed to accumulate in puddles around the piles. The main advantage of aeration by drawing air through the pile instead of exhausting is that the odorous drawn air can be pushed directly through a biofilter for odor control.

Screening Bulking agents such as wood chips should be recovered from the compost by screening. Fine materials such as sawdust cannot be recovered effectively. Recycling of the bulking agent can save 50 to 80% of the cost of new bulking agent. Drying of the compost is critical to screening because screens do perform well on compost with more than 50% moisture. Compost is dried by providing adequate aeration or agitation to drive off moisture. Two types of screen are generally available: vibrating deck screens and rotating screens. Depending on the number of screens in a deck, vibrating deck screens can separate materials into several classifications of sizes. However, rotating screens with cleaning brushes can screen compost with higher moisture content.

Site Considerations Windrows and aerated static piles are normally constructed on concrete or asphalt paved surfaces. Leachate and stormwater runoff from the composting surface are collected and discharged to the influent facilities of the treatment plant. Therefore, proximity to a treatment plant should be one consideration in selecting a site. Other factors in selecting a site include availability of land area and availability of buffer zone because of potential odor problems. Climatic considerations rarely create a problem, as composting is gaining popularity even in the colder countries of northern Europe. Increasingly newer aerated static pile composting facilities, including bulking agent and cured compost storage, are covered with a roof to keep precipitation out, thereby eliminating stormwater runoff from the facilities.

Odor Control Odor control is the primary environmental consideration in the operation of a composting facility because organic decomposition produces odorous volatile compounds. Odor sources include dewatered sludge storage facilities, mixing and turning of compost mixture, buildings, surface emission from active piles, leachate puddles around piles, and blower exhaust. Process and operational improvements can reduce odors considerably. Good odor control starts with prompt mixing of sludge and bulking agents. In addition, lumps of materials or puddles of liquid should not be allowed to remain in the mixing area. Condensate, leachate, and runoff from the piles should be collected and treated as quickly as possible. The compost should be adequately cured before it is removed, and any unstable material should be recycled back into the composting process for further treatment.

Composting odor is the result of a mixture of compounds that include mercaptans, organic sulfides, ammonia, organic nitrogen, fatty acids, and ketones. Containment and treatment of exhaust are trends in the newer composting facilities. Biofilters and wet scrubbers are two odor control technologies that are typically used. Well-stabilized and cured compost with bulking agents is used in building the biofilter beds or piles (see Figure 7.5). In a biofilter, odor compounds are oxidized by microorganisms. Biofilters are relatively inexpensive because compost is readily available. They are also easy and inexpensive to operate. Fabricated biofilters are also available commercially and rely on a fixed film developed on a structural packing medium for treatment. Wet scrubbers with as many as three stages, although more expensive to build and operate, are very effective in removing odorous compounds from the exhaust.

Safety and Health Issues Composting facilities use vehicles and conveyers for material handling, which are potential causes for injuries to operators. Another cause for danger to operators is poorly ventilated buildings where composting process is performed and materials are stored. Although *Salmonella*, fecal coliforms, total coliforms, and viruses increase in numbers during the initial stage of composting process, they are essentially destroyed within two weeks. However, exposure to pathogens can occur during the composting process by body contact of materials and inhalation of aerosols containing microorganisms. Airborne microorganisms include bacteria, fungi, and actinomycetes and microbial toxins. Fungus is also present in high concentrations in wood chips. Buildings that enclose composting process and stored materials should be ventilated a minimum of six air changes per hour for the comfort and safety of operators. When handling materials and when operating conveyors and screens, operators should wear dust masks. Good housekeeping, including dust control, should also be practiced for health reasons.

7.3 THEORETICAL ASPECTS OF COMPOSTING

For the best results in a composting process, it is important to have appropriate mixing of sludge cake with bulking agents and recycled materials. For the process to operate in good condition, it needs to have the optimum mass balance, moisture, temperature, pH, nutrients, and air. Figure 7.11 shows a mass balance diagram that can be used for all three composting systems: windrow, aerated static pile, and in-vessel. In the diagram:

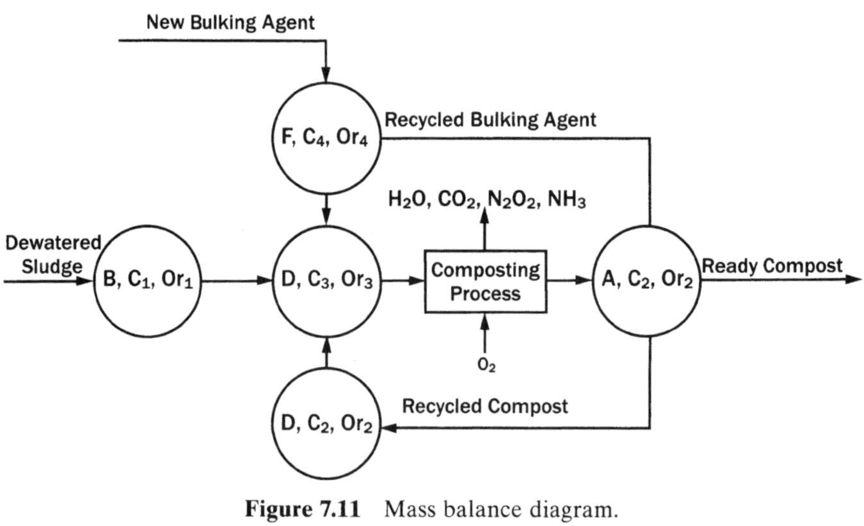

Figure 7.11 Mass balance diagram.

A' = weight of ready compost in one day

A = weight of recycled compost in one day

B = weight of dewatered sludge in one day

D = weight of a mixture of dewatered sludge, bulking agent, recycled bulking agent, and recycled compost in one day

F = weight of bulking agent in one day

C_1, C_2, C_3, C_4 = concentration (%) of dry solids in dewatered sludge, recycled compost, composting mixture, and bulking agents, respectively

Or_1, Or_2, Or_3, Or_4 = concentration (%) of organics in dewatered sludge, recycled compost, composting mixuture, and bulking agents, respectively

The quantity of mixture being composted is described by the equation

$$D = A + B + F \tag{7.1}$$

The quantity of mixture without bulking agents is

$$D = A + B \tag{7.2}$$

The mass of dry solids being composted without bulking agents is

$$AC_2 + BC_1 = DC_3 \tag{7.3}$$

Incorporating equation (7.2) yields

$$AC_2 + BC_1 = C_3(A + B) \tag{7.4}$$

That is,

$$AC_2 - AC_3 = BC_3 - BC_1 \tag{7.5}$$

so

$$A = \frac{BC_3 - BC_1}{C_2 - C_3} \tag{7.6}$$

$$= \frac{B(C_3 - C_1)}{C_2 - C_3} \tag{7.7}$$

Therefore,

$$\frac{A}{B}=\frac{C_3-C_1}{C_2-C_3} \tag{7.8}$$

The mass of recycled compost divided by the mass of dewatered sludge is

$$\frac{A}{B}=R_1 \tag{7.9}$$

where R_1 is the coefficient of compost recycling. Therefore,

$$R_1=\frac{C_3-C_1}{C_2-C_3} \tag{7.10}$$

From equation (7.5),

$$AC_2=AC_3+BC_3BC_1 \tag{7.11}$$

Dividing equation (7.11) by BC_1 gives

$$\frac{AC_2}{BC_1}=\frac{AC_3}{BC_1}+\frac{BC_3}{BC_1}-1 \tag{7.12}$$

But

$$\frac{AC_2}{BC_1}=R_2 \tag{7.13}$$

where R_2 is the coefficient of compost recycling on a dry weight basis. By incorporating equation (7.7), we obtain

$$R_2=\frac{B(C_3-C_1)C_3}{(C_2-C_3)BC_1}+\frac{C_3}{C_1}-1 \tag{7.14}$$

$$=\frac{C_3/C_1-1}{1-C_3/C_2} \tag{7.15}$$

Equations (7.10) and (7.15) can be used to determine the quantity of recycled compost.

The quantity of added bulking agent can be determined by the equation

$$F=\frac{DC_3-(AC_2+BC_1)}{C_4} \tag{7.16}$$

Adding the mass of bulking agents and the mass of recycling compost to dewatered sludge, we have

$$f_1 = A + \frac{F}{B} \tag{7.17}$$

Part of the bulking material has been lost by organic decomposition in the composting process:

$$f_2 = \frac{F}{F + A} \tag{7.18}$$

From equations (7.10) and (7.15), it can be seen that one of the most important factors is the dry solids concentration in dewatered sludge. Reducing the concentration of dry solids in dewatered sludge increases the coefficient R_1. Increasing the concentration of dry solids in recycled compost decreases the quantity of recycled compost to be used (coefficient R_2 decreased). The concentrations of dry solids in the mixture of dewatered sludge and bulking agents exert a big influence on the duration of composting process and the quantity of air required for aeration. The concentration of dry solids in the mixture has to be not less than 35% and not more than 50%. Seventy percent moisture in the composting mixture will increase the quantity of air required by several times.

Decomposition of organics by microorganisms releases carbon dioxide and water, resulting in decreased mass and volume. Assuming that Or_3 is fully decomposed as carbon dioxide and water, from the mass balance diagram,

$$BC_1\mathrm{Or}_1 + AC_2 2\mathrm{Or}_2 = DC_3\mathrm{Or}_3 \tag{7.19}$$

Using equation (7.2) yields

$$BC_1\mathrm{Or}_1 + AC_2 2\mathrm{Or}_2 = (A + B)C_3\mathrm{Or}_3 \tag{7.20}$$

Therefore,

$$\mathrm{Or}_3 = \frac{BC_1\mathrm{Or}_1 + AC_2 2\mathrm{Or}_2}{(A + B)C_3} \tag{7.21}$$

Dividing numerator and denominator by B and using equation (7.9) gives us

$$\mathrm{Or}_3 = \frac{\mathrm{Or}_1 C_1 + \mathrm{Or}_2 C_2 R_1}{C_3(1 + R_1)} \tag{7.22}$$

Or by using the coefficient of recycling in dry solids, we obtain

$$\mathrm{Or}_3 = \frac{\mathrm{Or}_1 + \mathrm{Or}_2 R_2}{1 + R_2} \tag{7.23}$$

The organics contained in the composting mixture depend on the initial organics contained in sludge. If the quantity of dry solids in sludge is decreased, the quantity of recycled compost has to be increased, but that will decrease the concentration of organics in the composting mixture. Decomposition of organic material in composting process decreases organics by 25 to 30%. The heat generated by decomposition of organics is approximately 20 to 21 MJ/kg. Because raw sludge contains more organics than digested sludge, composting of raw sludge is more effective than composting digested sludge. One of the effective ways to increase organics in composting process is to use bulking agents. Fresh sawdust and wood chips contain up to 98% organics, and when recycled from ready compost, up to 80% organics. However, stored sawdust and wood chips including recycled wood chips can raise temperature in a composting pile much faster than fresh sawdust and wood chips because they contain more living microorganisms. That is one of the reasons for recycling wood chips from the compost. Also, this recycling of bulking agent puts the carbon-to-nitrogen ratio in the composting mixture within the optimum range of 25:1 to 35:1.

The quantity of air in the composting process in the absence of a bulking agent, per the mass balance, can be determined indirectly from the quantity of water evaporated in one day by the equation

$$\mathrm{H_2O} = (B - C_1 B) - (A' - C_2 A') \tag{7.24}$$

The mass balance for the inorganic substance is

$$(1 - \mathrm{Or}_1) C_1 B = (1 - \mathrm{Or}_2) C_2 A' \tag{7.25}$$

or by using equation (7.24),

$$\frac{\mathrm{H_2O}}{BC_1} = \left(\frac{1 - C_1}{C_1} - \frac{1 - \mathrm{Or}_1}{1 - \mathrm{Or}_2} \right) \frac{1 - C_2}{C_2} \tag{7.26}$$

The airflow rate has to be monitored and controlled based on the temperature, oxygen, or carbon dioxide levels. The amount of air needed for the composting process is approximately 15 to 20 m^3/h for each metric ton of organic material. More air is required as the temperature rises to keep it below 70°C. If the temperature reaches 70°C, the quantity of microorganisms decreases and the process of composting stops. The carbon dioxide level should not exceed 8%, and oxygen should not be less than 5 to 15% in the pile. The stoichiometric demand of oxygen can be determined by the equation

$$C_{10}H_{19}O_3N + 12.5O_2 = 10CO_2 + 9H_2O + NH_3 \tag{7.27}$$

Ammonia (NH_3) generated during the decomposition of organics is volatilized. One gram of organics needs 2 g of oxygen for decomposition. However, during the first few days of the process, as the temperature increases, the demand for oxygen can be higher. This can increase the demand for oxygen by 3 to 6 g/g of organics. When moisture in the pile increases due to the production of water in the composting reaction, the demand for air also increases, but increasing the air can cause a decrease in temperature, and that in turn can decrease the decomposition rate.

During the decomposition of organics, heat will be lost through the surface of the pile such that

$$Q_l = k(T_c - T_a) \tag{7.28}$$

where

Q_l = heat lost
K = coefficient of heat exchange
T_c = temperature of composting mass
T_a = ambient temperature outside the pile

Several scientists have developed equations (Ettlich and Lewis, 1977; Sicora et al., 1981; Wilson, 1983; Epstein, 1997) for heat loss and heat exchange.

Design Example 7.1 A wastewater treatment plant composts combined primary and waste activated sludge dewatered on belt filter press. Given the following data, calculate the quantity of the bulking material, wood chips (F), and the dimensions of an aerated static pile composting system.

$$\text{wet weight of dewater sludge } (B) = 100 \text{ tons/day}$$

$$\text{wet wight of recycled compost } (A) = 20 \text{ tons/day}$$

$$\text{concentration of dry solids in dewatered sludge } (C_1) = 25\%$$

$$\text{concentration of dry solids in recycled compost } (C_2) = 50\%$$

$$\text{concentration of dry solids in composting mixture } (C_3) = 40\%$$

$$\text{mass of dry solid in wood chips} = 0.25 \text{ ton/m}^3 \text{ or } 25\%$$

It is also assumed that the process variables have the following values:

organics in sludge cake:	75%
organics in sludge cake decomposed:	45%
solids in wood chips:	70%
organics in wood chips:	90%
organics in wood chips decomposed:	10%
organics in recycled compost:	75%
organics in compost decomposed:	10%

From equation (7.2), the weight of the composting mixture without a bulking agent

$$\begin{aligned} D &= A + B \\ &= 20 + 100 = 120 \text{ tons/day} \end{aligned}$$

From equation (7.16), the quantity of wood chips

$$\begin{aligned} F &= \frac{(120)(0.40) - (20)(0.50) + (100)(0.25)}{0.25} \\ &= 52 \text{ tons/day, or} \\ &= 52 \text{ tons} \cdot \text{day}^{-1}/0.25 \text{ ton} \cdot \text{m}^{-3} \\ &= 208 \text{ m}^3/\text{day} \end{aligned}$$

$$\begin{aligned} \text{ratio of dewatered sludge to recycled compost} &= B : F \\ &= 100 \text{ tons/day to } 52 \text{ tons/day} \\ &= 2:1 \end{aligned}$$

$$\text{quantity of new wood chips} = 52 - 20 = 32 \text{ tons/day}$$

The ratio

$$\begin{aligned} W &= \frac{100(1-0.25) + 20(1-0.75) + 52(1-0.7)}{(100)(0.25)(0.7)(0.45) + (20)(0.5)(0.75)(0.1) + (52)(0.7)(0.9)(0.1)} \\ &= 8.0 \end{aligned}$$

Since W is less than 10, no additional recycling of compost or addition of external amendment is required.

The following bulk densities are used in this example:

dewatered sludge cake (25% solids): 890 kg/m^3
wood chips: 250 kg/m^3
screened (recycled) compost: 520 kg/m^3

$$\text{daily volume of sludge cake} = (100\ \text{tons/day})\left(\frac{1000\,\text{kg/ton}}{890\,\text{kg/m}^3}\right)$$
$$= 112.4\,\text{m}^3/\text{day}$$

$$\text{daily volume of wood chips} = (52\ \text{tons/day})\left(\frac{1000\,\text{kg/ton}}{250\,\text{kg/m}^3}\right)$$
$$= 208.0\,\text{m}^3/\text{day}$$

$$\text{daily volume of recycled compost} = (20\ \text{tons/day})\left(\frac{1000\,\text{kg/ton}}{520\,\text{kg/m}^3}\right)$$
$$= 38.5\,\text{m}^3/\text{day}$$

$$\text{total volume} = 358.9\,\text{m}^3/\text{day}$$

For a 2.5-m-high and 20-m-long pile, the width of the pile extended each day is

$$\left(\frac{358.9\,\text{m}^3}{2.5\,\text{m}}\right)(20\,\text{m}) = 7.2\,\text{m}$$

The amount of wood chips to construct the 0.3-m-thick pad for the pile is

$$(20\,\text{m})(7.2\,\text{m})(0.3\,\text{m}) = 43.2\,\text{m}^3/\text{day}$$

Unscreened compost required to cover the pile with a thickness of 0.2 m is

$$(20\,\text{m})(7.2\,\text{m})(0.2\,\text{m}) = 28.8\,\text{m}^3$$

7.4 NEW TECHNOLOGY IN COMPOSTING

The new technology presented in this section is based on the studies performed at several wastewater treatment plants in Russia, Bulgaria, and Hungary (Turovskiy and Westbrook, 2002). The finished product from this composting process provides class A biosolids that can be used as fertilizer and soil conditioners.

7.4.1 Organic Content

In general, the higher the sludge's organic content, the greater the quantity of heat released during composting. This greater quantity of heat results in the thermophilic phase (55 to 65°C) being reached earlier in the composting process. Raw sludge typically contains 60 to 80% organic material while digested sludge contains only 30 to 50% organic material. Since raw sludge (from primary clarifiers and secondary clarifiers) contains more organic material than digested sludge, it is reasonable to prepare compost from dewatered raw sludge. By utilizing raw sludge, the digesters, pipe, pumps, electrical power, personnel, and so on, normally utilized in the digestion process can be reduced or eliminated.

The composting process not only reduces the organic content in sludge by approximately 25% but also reduces the moisture content of the sludge. The reduction in moisture content is the result of the heat generated during the organic decomposition. The heat generated by composting 1.0 kg of organic material averages 21 MJ. Approximately 4.0 MJ of heat will evaporate 1.0 kg of moisture (taking into account heat losses and heating of the compost material). Thus, the composting of 1.0 kg of organic material facilitates the removal of approximately 5.0 kg of moisture from the residual (21 MJ/4 MJ/kg of water). Figure 7.12 illustrates the change of indexes during composting of

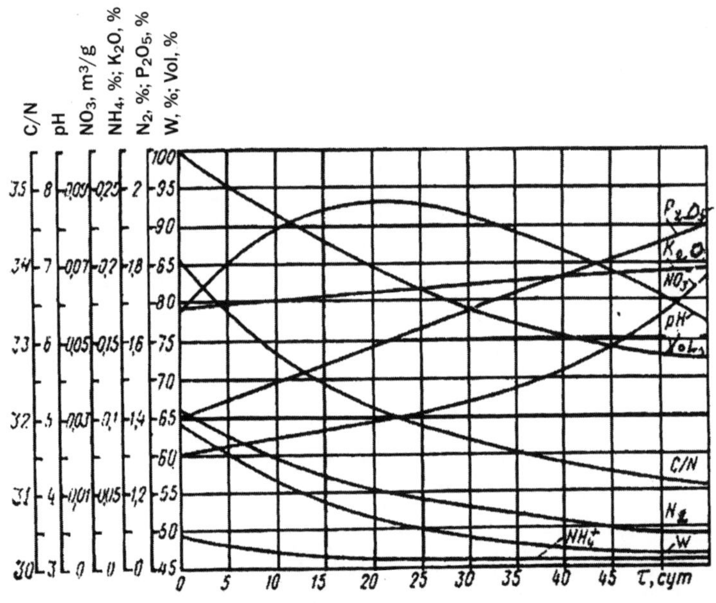

C/N Carbon-to-nitrogen ratio
NO_3 Nitrate
NH_4 Ammonia
K_2O Potassium
N_2 Nitrogen
P_2O_5 Phosphate
W Moisture
VOL Volatile solids

Figure 7.12 Changes of indexes in raw sludge composting.

dewatered raw sludge with the bulking agent sawdust at the ratio of 1:2 by volume (Chertes et al., 1988).

7.4.2 Odor

A problem with composting raw sludge as compared to digested sludge is the higher-intensity odor that can be released due to the higher percentage of organic material in raw sludge. Various methods can be used to control the odor, but the method favored is the addition of quicklime (CaO) to change the pH of the sludge. Experiments show that organic material loses its odor when the pH is raised from the typical 5.5 to 6.5 to a pH of 10.0 to 10.5.

The hydration of the quicklime (absorbing moisture from the sludge) causes the quicklime to release heat to the composting mixture. During the process of hydrating 1 kg of chemically pure quicklime (100% CaO), 320 g of moisture is absorbed from the sludge and 1152 kJ of heat is produced. This release of heat shortens the time span of the mesophilic phase (25 to 40°C) and drives the process to the thermophilic phase (55 to 65°C) quicker, resulting in an overall reduction in the composting time.

In addition to raising the pH, absorbing moisture, and releasing heat, the quicklime also binds with heavy metals that might be present in the residual. This binding of the heavy metals provides a better environment for the growth of the microorganisms needed for the composting process.

7.4.3 Temperature and Moisture

The temperature increase caused by a predetermined dose of quicklime may be calculated by the formula

$$\Delta T = \frac{1152 A M_L}{M_{SL} C_{SL} + M_L C_L} \tag{7.29}$$

where

ΔT = temperature increase of the residual, °C
A = quicklime activity factor, typically 0.9
M_L = mass of quicklime, kg
M_{SL} = mass of residual, kg
C_L = specific heat of quicklime = 0.92 kJ/kg · °C
C_{SL} = specific heat of residual in kJ/kg · °C may be calculated as

$$C_{SL} = 1.8(1 + 0.85 W_{SL}^3) \tag{7.30}$$

where W_{SL} is the moisture content (in %) of the dewatered residual.

Raising the residual temperature by 10°C doubles the speed of the microbiological activity that accomplishes the composting process.

The addition of quicklime absorbs moisture from the sludge, thereby reducing the moisture content of the compost mixture. The sludge moisture content after the addition of quicklime can be calculated using the formula

$$W_K = \frac{M_{\mathrm{SL}}W_{\mathrm{SL}} - 0.32AM_{\mathrm{L}}}{M_{\mathrm{SL}} + M_{\mathrm{L}}} \tag{7.31}$$

where W_K is the moisture content (in decimals) of the sludge after the addition of quicklime. The lowering of the moisture content of the sludge decreases the volume of the sludge. This decreases the amount of quicklime required to raise the pH to 10.5. Therefore, it is reasonable to dewater the residual prior to the addition of quicklime.

7.4.4 Composting Mixture

The new technology requires quicklime to be mixed with the dewatered sludge just prior to adding a bulking agent (sawdust, wood chips, bark, etc.) and recycled compost. Refer to Figure 7.13 for the schematic of the proposed

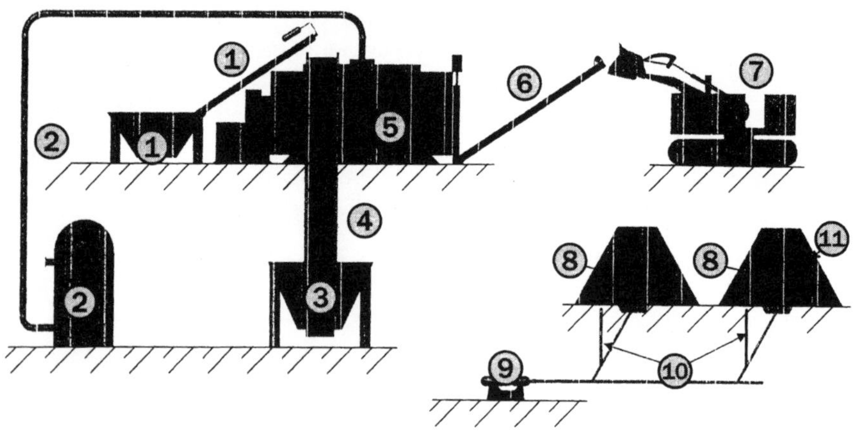

1. Hopper with conveyor for dewatered sludge
2. Silos for quicklime with processing unit and pneumatic pumping of lime
3. Hopper for bulking agents and recycled compost
4. Conveyor
5. Mixture device
6. Discharge conveyor for mixture to be composted
7. Loader
8. Composting piles
9. Air blowers
10. Air flowmeters
11. Cover over composting piles

Figure 7.13 Schematic of raw sludge composting.

composting process. Once the quicklime and the dewatered sludge are thoroughly mixed, the bulking agent and a portion of recycled compost are added and mixed. This mixture is then formed into piles and allowed to compost until a temperature of 55 to 65°C has been maintained for 3 to 11 days. The piles are often covered with a layer of bulking agent or recycled compost to protect the pile from heat loss as well as to avoid attracting flies, mosquitoes, and other undesirable insects.

If odor becomes a problem during the composting process, the simple procedure of drawing air through the compost piles and discharging the air to a biofilter can help reduce the associated odor. Experiments indicate the following recommendations:

moisture content of the dewatered raw sludge: 80% or lower

organic material in the composting mixture: 55% or higher

quantity of quicklime added to dewatered raw sludge: 2.0 to 2.2% of sludge mass

quantity of bulking agent added to dewatered raw sludge: 100 to 120% by volume

quantity of recycled compost added to dewatered raw residual: 20% of sludge mass

7.4.5 Composting Process Control

Experiments show that the type and population of microorganisms varies during the composting process. It is therefore critical to control the composting environment so that the microorganisms can flourish. The composting environment parameters include the compost pile temperature, moisture content of the compost, oxygen and carbon dioxide levels in the compost pile, and the availability of nutrients, including carbon, nitrogen, phosphorus, and potassium for the microorganisms. These parameters must be monitored, as they affect the vitality of the microorganisms.

The temperature in the compost pile affects most directly the types of microorganisms and their functions. The type of microorganism changes as the compost pile temperature increases from its initial temperature to the mesophilic (25 to 40°C) phase, to the thermophilic (55 to 65°C) phase, and to the slow decrease in temperature following completion of the composting process. Experiments show that the thermophilic phase must be maintained for 3 to 11 days to produce a class A biosolid. It is during the thermophilic phase that most pathogens are destroyed. As the type of microorganism changes in relation to the compost pile temperature, so does its requirements for moisture and oxygen. The moisture content of the compost, oxygen and carbon dioxide levels in the compost pile, and compost pile temperatures are closely related; a change in one affects the others directly.

Oxygen is supplied to the compost pile by the introduction of air, and the rate of air supplied depends on the moisture content of the compost pile: The higher the moisture content, the higher the rate of air is required. A minimum oxygen level must be maintained, while carbon dioxide levels must not be allowed to exceed a maximum level. As air is supplied, the porosity of the pile increases, which leads to increased evaporation and a resulting decrease in the moisture content of the pile. Supplying air can also lead to heat losses, which result in a temperature reduction within the compost pile. This temperature reduction results in a lower rate of decomposition functions. Therefore, the oxygen and carbon dioxide levels and the amount of air supplied must be monitored and controlled. Experiments indicate that the rate of air supplied is approximately 15 to 20 m^3/h for each ton of organic material being composted. Monitors and controllers should be employed to supply air to the pile automatically when the carbon dioxide level within the pile reaches 8%.

Data obtained from raw sludge composting experiments indicate that the following recommendations should be followed:

moisture content of compost material: 60 to 65% by weight

oxygen level during composting: no less than 10% of gas mass

carbon dioxide level during composting: no more than 8 to 9% of gas mass

20 to 35 parts of carbon is used for every part of nitrogen

75 to 100 parts of nitrogen is required for every part of phosphorus

temperature: 55 to 65 °C for 3 to 11 days including curing period

duration of composting process: 25 to 40 days including curing period, depending on climate conditions

Design Example 7.2 A wastewater treatment plant (WWTP) with a design capacity of 40 million gallons per day (mgd) generates approximately 700 wet tons per day of thickened mixture of primary/waste activated sludge with moisture content of 97% (3% solids) or 21 tons/day of dry solids. The WWTP's dewatering system of centrifuges with polymer feed dewaters the sludge to a moisture content of 80% (20% solids). This reduces the mass of residual to 105 wet tons/day [(700 tons/day)/(20%/3%) = 105 tons/day].

The composting process utilizes the addition of quicklime and the addition of sawdust as a bulking agent. (*Note*: The utilization of polymer by dewatered raw residual is almost half of that utilized in dewatering anaerobic digested residual.) The addition of 2.1 tons of quicklime per day (105 tons/day × 2%) increases the pH to 10.5, removes the odor from the residual, and increases the temperature of residual. Temperature rise can be calculated using equations (7.29) and (7.30), which are repeated below.

$$\Delta T = \frac{1152 A M_L}{M_{SL} C_{SL} + M_L C_L}$$

$$C_{SL} = 1.8(1 + 0.85 W_{SL}^3)$$
$$= 1.8[1 + (0.85)(0.8^3)] = 2.58$$

$$\Delta T = \frac{(1152)(0.9)(2100)}{(105{,}000)(2.85) + (2100)(0.92)}$$
$$= 8.0° \text{ C}$$

The duration of the composting period and thermophilic phase depend on process performance, quantity, and composition of the compost mass (i.e., moisture content, organic and chemical content of the sludge, type of bulking agent, viability of the recycled compost, etc.) and can last from several days to several weeks. For example, by using biodegraded wood chips and/or recycled compost, the temperature in the compost pile increases at a faster rate since these materials are already in a state of biodegradation. Quicklime added to the residual shortens the composting process by increasing the starting temperature through a chemical reaction. A comparison of temperature versus time during the composting of raw sludge with and without quicklime, and digested sludge without quicklime, is shown in Figure 7.14.

The residual moisture content after the addition of quicklime can be calculated using equation (7.31), which is repeated below.

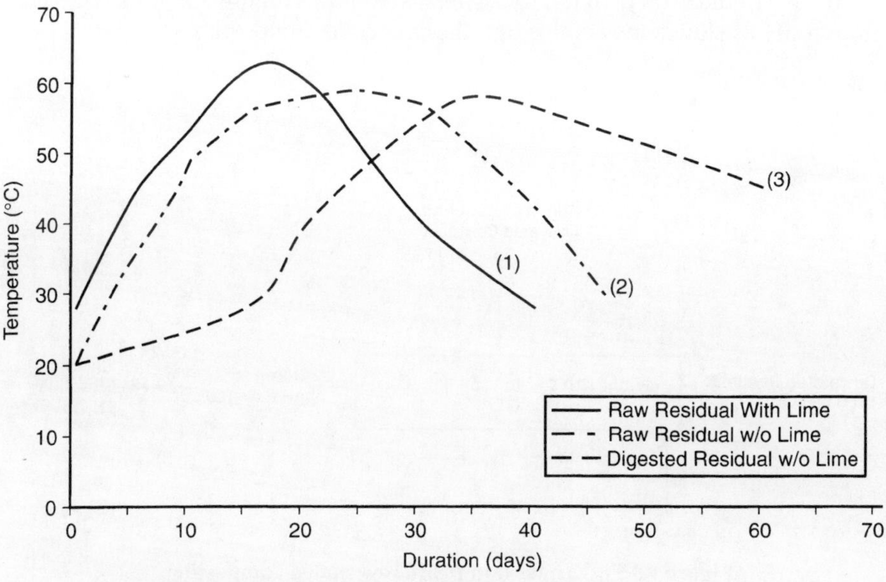

Figure 7.14 Comparison of composting with and without lime.

$$W_K = \frac{M_{SL} \times W_{SL} - 0.32AM_L}{M_{SL} + M_L}$$

$$= \frac{(105{,}000)(0.8) - (0.32)(0.9)(2100)}{105{,}000 + 2100} = 0.78 \text{ or } 78\%$$

The quantity of sawdust added is 105 tons/day (105 tons/day × 1.0) and recycled compost added is 21 tons/day (105 tons/day × 0.2).

It takes several days to reach the thermophilic temperature (55°C) with the addition of quicklime. Maintaining that temperature for 10 to 11 days (Figure 7.14) provides the highest level of pathogen reduction/vector control and produces a compost meeting class A requirements.

By comparison, an aerated static pile system requires longer composting time and more operational processes, resulting in a correspondingly larger area to store the composting materials. Also, compared with lime stabilization, the proposed composting process allows a decrease in the quantity of added lime to 25 to 33% of that required for lime stabilization.

7.4.6 pH

The use of quicklime to raise the pH of the sludge from 5.5–6.5 to 10.0–10.5 is known to inhibit the growth of microorganisms. To reduce the likelihood of this occurring, bulking agents and recycled compost are added to the sludge–lime mixture soon after the introduction of the quicklime. The resulting mixture's pH is reduced to 8.5 to 9.0, thus providing an environment conducive to the growth of microorganisms. Refer to Figure 7.15 for the variation in pH as sludge moves through the proposed composting process.

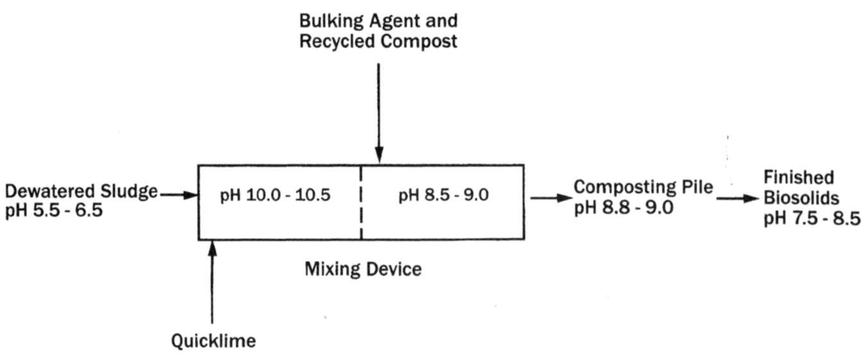

Figure 7.15 Variation in pH in raw sludge composting.

7.5 EXAMPLES OF COMPOSTING IN EUROPE

Several examples of different types of composting systems that have been in use in Europe are described below. All these systems are open-air processes, mostly on asphalt pavements.

In Finland, two wastewater treatment plants (city of Lappeenranta and city of Loensiy) compost sludge mixed with ground bark as the bulking agent at a ratio of 1:1 by volume in 3-m-high windrow piles. Every three weeks, piles are shoveled over by scoop loader. After two weeks of processing the temperature rises to 50°C, and after three to four weeks it rises to 60°C. Duration of composting during summer is four to six months. Cured compost has a moisture content of 40 to 60% (Paatero and Lehtokori, 1984).

The wastewater treatment plant in Blua, France, has composted digested dewatered sludge with sawdust at the ratio of 1:3 by volume. Windrow piles 1.5 to 2 m high and 4 to 5 m wide were formed and turned frequently by special machines two to five times a week during the first three weeks. The machines facilitate mixing, aeration, and homogenization of the composting material. After three weeks, mobile equipment removed the compost and reformed it to 2.5- to 3-m-high piles for curing for about two months. In Germany, a mixture of dewatered sludge and wood chips or municipal solid waste has been composted in windrows of 3.5-m-high piles. A view of the machine used for mixing and the formed piles in one of the composting facilities is shown in Figure 7.16.

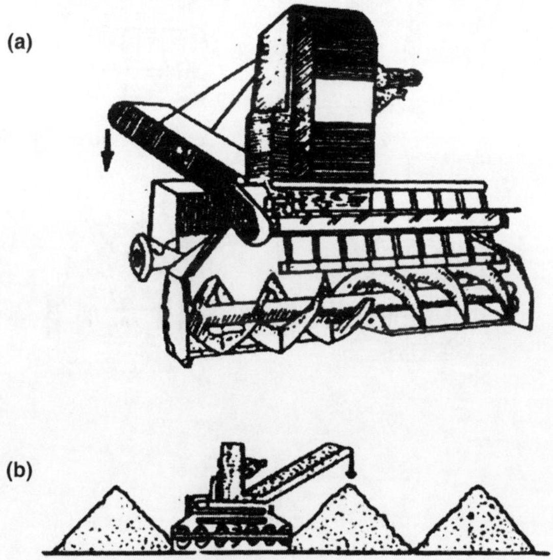

Figure 7.16 Machine for mixing and forming piles.

Several countries in Europe, including Russia, have built composting systems in trenches either in open air or with roofs to keep away precipitation. Figure 7.17 shows trenches with different types of mixing equipment. These installations can be categorized as in-vessel systems. In these installations, sludge mechanically dewatered or dried on drying beds is composted for several weeks to several months. The finished compost has been found to be well-stabilized, disinfected, and odorless.

Experiments on aerated static pile composting were performed in Petrosavodsk, Russia, in the 1980s (Chertes, 1988). Mechanically dewatered combined primary and waste activated sludge was mixed with wood chips, bark chips, or sawdust at the ratios 1:0.5, 1:1, 1:1.5, 1:2, and 1:2.5 by volume. Table 7.2 shows the characteristics of the bulking agents used. The moisture

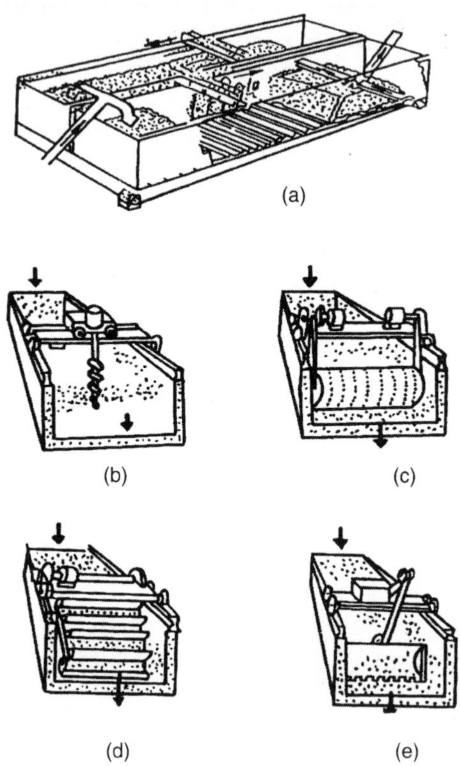

(a) Conveyor mixer
(b) Auger mixer
(c) Drum mixer
(d) Paddle mixer
(e) Bucket mixer

Figure 7.17 Composting in trenches and mixing systems.

TABLE 7.2 Characteristics of Bulking Agent

Type of Bulking Agent	Size of Material (mm)	Specific Surface Area (cm^2/g)	Dry Mass Density (ton/m^3)	Porosity (%)	Potential Water Absorption (%)
Wood chips	10–50	47–140	0.11–0.18	86–93	116–123
Ground bark	10–100	26–170	0.08–0.30	82–95	112–125
Sawdust	2.5–5.0	162–220	0.07–0.08	95–96	125–126

content of the mixture was 60 to 67%. Ten piles were made, each with a 0.4-m-thick base of bulking material, a 1.6-m-high composting mixture, and covered with a 0.2-m-thick layer of recycled compost. The 2.2-m-high piles were 3 to 4m wide. Aeration was provided using fans at the rate of 10 to $25\,m^3/h$ per ton of volatile solids. Aeration began after 8 to 10 days and continued for 8 hours every three days for a total of 30 days. Carbon dioxide generation was monitored. When the concentration of carbon dioxide was increased to more than 7% of the gas volume, microbial activity decreased, resulting in a decrease in temperature. The temperature variation in some of the piles is shown in Figure 7.18. The best result was obtained in piles with the sludge-to-bulking agent ratio of 1:2. Piles with the ratio of 1:0.5 did not have enough bulking agent to give adequate porosity for an even distribution of air. Piles with the ratio 1:2.5 resulted in the same well-stabilized compost as piles with the ratio 1:2, showing that excess bulking material did not improve the composting process. Characteristics of the well-stabilized compost are shown in Table 7.3.

In-vessel composting is also widely practiced in Europe. The firms Gersi in England, Carel-Fouchet in France, Hasemag in Germany, and Laxsa and Ingre-Lufgren in Sweden developed in-vessel composting in vertical towers. Examples of such towers are shown in Figure 7.19. The multistage towers have one to as many as 10 rotating platforms 0.4 to 0.8m apart. Dewatered sludge and bulking agent are transferred to the top of the tower. The rotating platforms mix the materials and transfer the mixture to the lower level. Air is supplied at the bottom of the tower by compressors or ventilating fans and released at the top. Depending on each manufacturer's equipment, the composting period lasts 4 to 30 days.

Some facilities in Europe compost wastewater sludge with municipal solid waste in rotating reactors called *biodrums*. Figure 7.20 shows two types of biodrums developed by the firm Dano in Denmark. The biodrums provide good environmental conditions for the bioreaction because of the effective mixing from the rotating motion, different sections in the vessel for the different stages, and controlled aeration. The process lasts from one to six days.

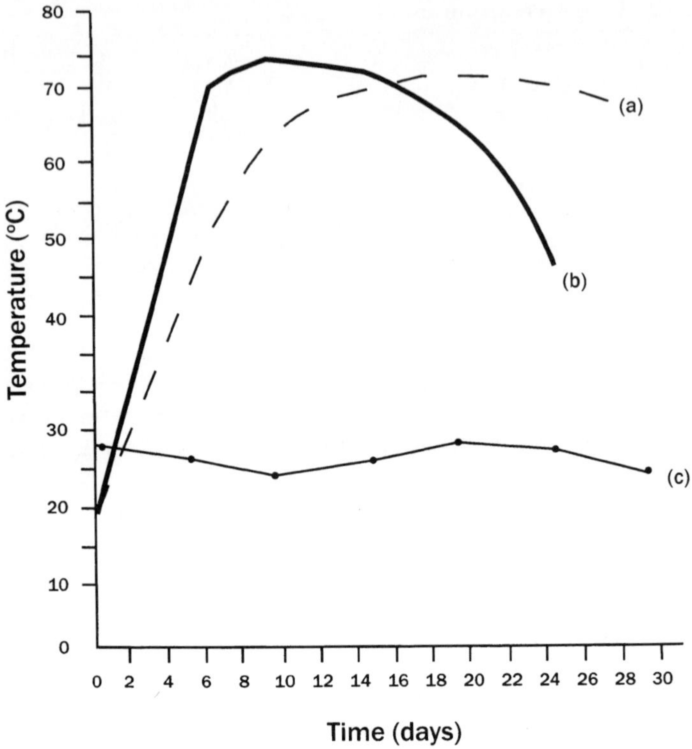

Sludge-to-bulking agent ratio: (a)1:2.5, (b)1:2, and (c)1:0.5

Figure 7.18 Temperature changes during composting.

TABLE 7.3 Characteristics of Compost

Characteristic	Value
Size fraction	
>30 mm (%)	6
3–30 mm (%)	48
<3 mm (%)	46
Moisture (%)	49–55
pH	5.4–7.1
Volatile solids (%)	65–75
N_2 (mg/kg)	1.5–2.0
P_2O_5 (mg · kg)	1.5–2.4
K_2O (mg/kg)	0.2–1.0
Heavy metals	
Chromium (mg/kg)	96–220
Nickel (mg/kg)	95–114
Zinc (mg/kg)	339–1240
Lead (mg/kg)	17–37
Cadmium	8–6

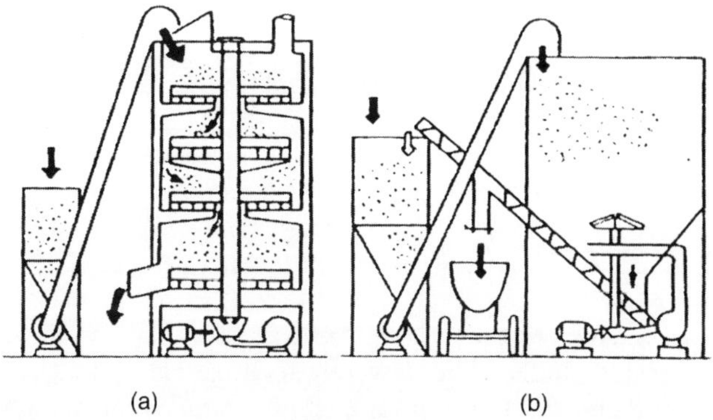

a) Multiple stage
b) Single stage

Figure 7.19 Vertical composting reactors.

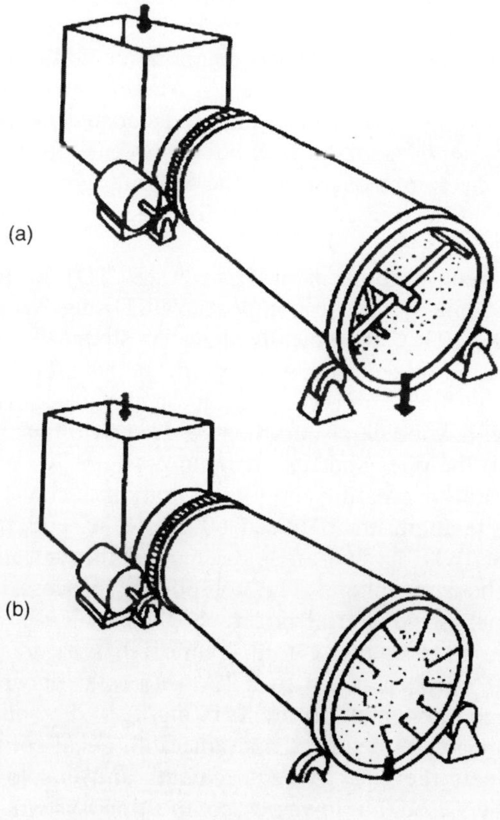

(a) With mixers inside drum
(b) With scrapers inside drum

Figure 7.20 Rotating biodrum reactors.

7.6 EXAMPLES OF COMPOSTING IN THE UNITED STATES

In the United States, studies of windrow and aerated static pile composting were conducted in the mid-1970s in Beltsville, Maryland, and in Carson, California. As of 2003, there were more than 200 wastewater sludge composting facilities in the United States. Most of the composting facilities are aerated static pile systems, and some are in-vessel systems. Windrow composting is rarely used because of the potential odor generation and the large area requirement.

The city of Eustis, Florida, uses the windrow process for composting aerobically digested sludge. The dewatered sludge cake of 12 to 14% solids mixed with wood chips at the ratio of 1:2 by volume is formed into windrow piles by a front-end loader on paved surface. The pile is mixed using a windrow machine every day for three to five days until the temperature reaches 55°C. The temperature is maintained between 55 and 65°C for 15 days, during which time the pile is turned about five times. The composting process takes about three weeks. The compost is then moved to a curing area and cured for three months. The compost is turned once every month during the curing process. The cured compost meets the class A requirements and is sold to distributors. The city of Austin, Texas, uses the windrow process for composting anaerobically digested sludge. The dewatered sludge cake of 15 to 25% solids is mixed with bulking agents, including yard waste, for composting. The compost is sold to the general public through registered vendors for use as a soil conditioner for residential lawns and flower gardens.

The Reedy Creek Improvement District (RCID) in Florida selected the aerated static pile system to compost Walt Disney World's wastewater sludge (Harkness, 1994). Aerobically digested sludge is dewatered to 20 to 22% solids and is mixed with wood chips at the ratio of 1:1 by weight (1:2 by volume). The mixture is then placed on a base of perforated aeration piping covered with wood chips and then covered with unscreened compost. Figure 7.21 shows the piles and the air piping system. Air is drawn through the pile 24 hours after creating the pile. Aeration is controlled by a timer to maintain the temperature between 60 and 65°C and the oxygen level at a minimum of 13%. Drawn air is discharged through a biofilter before discharging to the atmosphere. The composting process lasts about four weeks, after which the material is screened (see Figure 7.22) and cured for an additional four weeks. Cured compost has a moisture content of 35 to 40%, total nitrogen of 1.4 to 2.2%, and total phosphorus of 0.3 to 0.4%. The average heavy metal contents in mg/kg of dry solids are cadmium 15, copper 290, nickel 18, lead 15, and zinc 200. Fecal coliform is 5 MPN. The compost meets the class A requirements and is sold to distributors. RCID had earlier used an in-vessel composting system in two vertical plug flow reactors. Although the system produced stable compost, the facility

Figure 7.21 View of compost pile and aeration system.

experienced several problems during the operation. RCID conducted a cost analysis of three in-vessel systems and the aerated static pile system. Based on the analysis, the aerated static pile system was found to be the most cost-effective.

The wastewater treatment plants in Bristol, Tennessee, and Bristol, Virginia, have a shared in-vessel composting facility (Clifton et al., 1997). The combined raw primary and waste activated sludge at an average 4% solids is mixed with sawdust at the ratio 1:1 by volume and is dewatered with a cationic polymer in belt filter presses to 35% dry solids. There are two enclosed in-vessel units of 1400 m^3 each, one for the bioreaction process and the other for the curing. The dewatered mixture of sludge and sawdust from the storage silo is transferred by a conveyor to the top of the bioreactor, where a rotating finned disk layers the mixture over the previous loadings. Air is introduced at the bottom of the reactor. The temperature is maintained at 55°C for a minimum of three consecutive days. The bottom layer is removed from the reactor after 15 day of detention time and transferred to the top of the second reactor for curing. After 15 days of curing, the compost is stored in rows on a paved concrete surface for two to three months. The compost from the facility meets the class A requirements and is distributed to the public.

Figure 7.22 View of screened compost.

REFERENCES

Chernova, N. M. (1966), *Zoological Characteristics of Compost*, Academy of Science, Nauka, Moscow.

Chertes, K., Aukaev, R., Turovskiy, I., Lubavsky, V., Dokudovska, S., Borisov, V., and Kuricov, A. (1988), Instruction for Composting Wastewater Sludge in Petrozavodsk, *Water Supply and Sanitary Technology*, Vol. 5, p. 4.

Citton, F. W., Jr., Adams, T. E., and Dohoney, R. W. (1991), Managing Sludge Through In-Vessel Composting, *Water Engineering and Management*, December, p. 21.

Epstein, E. (1997), *The Science of Composting*, Technomic Publishing Co., Lancaster, PA.

Ettlich, W., and Lewis, A. (1977), *A Study of Forced Aeration Composting of Wastewater Sludge*, No. 11, U.S. EPA, Cincinnati, OH.

Foess, G. M., and Singer, R. B. (1993), Pathogen/Vector Attraction Reduction Requirement of the Sludge Rules," *Water Engineering and Management*, June, p. 25.

Garvey, D., Guario, C., and Davis, R. (1993), Sludge Disposal Trends Around the Globe, *Water Engineering and Management*, December, p. 17.

Goldfarb, L., Turovskiy, I., and Belaeva, S. (1983), *The Practice of Sludge Utilization*, Stroyizdat, Moscow.

Golueke, C. G. (1983), Epidemiological Aspects of Sludge Handling and Management, Part 2, *BioCycle*, Vol. 24, No. 3, p. 52.

Harkness, G. E., Reed, C. C., Voss, C. J., and Kunihiro, C. I. (1994), Composting in the Magic Kingdom, *Water Environment and Technology*, Vol. 6, No. 8.

Haug, R. T. (1993), *The Practical Handbook of Compost Engineering*, Lewis Publishers, Boca Raton, FL.

Kulik, A. (1996), Europe Cultivates Organics Treatment, *World Wastes*, Vol. 39, No. 2, p. 37.

Lue-Hing, C., Zenz, D. R., and Kuchenither, R. (Eds.) (1992), *Water Quality Management Library,* Vol. 4, *Municipal Sewage Sludge Management: Processing, Utilization and Disposal*, Technomic Publishing Co., Lancaster, PA.

McDonald, G. J. (1995), Applying Sludge to Agricultural Land: Within the Rules, *Water Engineering and Management*, February, p. 28.

Metcalf & Eddy, Inc. (2003), *Wastewater Engineering: Treatment and Reuse*, 4th ed., Tchobanoglous, G., Burton, F. L., and Stensel, H. D. (Eds.), McGraw-Hill, New York.

Murray, C. M., and Thompson, J. L. (1986), Strategies for Aerated Pile Systems, *BioCycle*, Vol. 6.

Outwater, A. B. (1994), *Reuse of Sludge and Minor Wastewater Residuals*, CRC Press/Lewis Publishers, Boca Raton, FL.

Poincelot, R. P. (1977), The Biochemistry of Composting, *Proceedings of the National Conference on Composting on Municipal Residues and Sludges*, Information Transfer, Inc., Rockville, MD.

Sicora, L. G., Wilson, G. B., Colacicco, D., and Parr, G. E. (1981), Material Balance in Aerated Static Pile Composting, *Journal of the Water Pollution Control Federation*, Vol. 53, No. 12.

Sommers, L. E. (1977), Chemical Composition of Sewage Sludges and Analysis of Their Potential Use as Fertilizers, *Journal of Environmental Quality*, No. 6, p. 225.

Spellman, F. R. (1996), *Wastewater Biosolids to Compost*, Technomic Publishing Co., Lancaster, PA.

Turovskiy, I. S. (1988), *Wastewater Sludge Treatment*, Stroyizdat, Moscow.

———, and Chertes, K. A. (1991), *Technology of Wastewater Sludge Composting*, Union Science Institute of the Wood Industry, Moscow.

———, and Westbrook, J. D. (2002), Recent Advances in Wastewater Sludge Composting, *Water Engineering and Management*, October, pp. 29–32.

———, Bucreeva, T. E., and Astachova, A. V. (1989), *Biothermal Treatment of Wastewater Sludge*, ZBTI Ministry of Water Management, Moscow.

U.S. EPA (1979), *Process Design Manual for Sludge Treatment and Disposal*, EPA 625/1-79/011.

——— (1989), *Summary Report: In-Vessel Composting of Municipal Wastewater Sludge*, EPA 625/8-89-016.

——— (1993), *Standards for the Use or Disposal of Sewage Sludge*, 40 CFR Part 503, Federal Register 58 FR9248 to 9404.

——— (1995), *A Guide to Biosolids Risk Assessment for EPA Part 503 Rule*, EPA 832/B-93/005.

——— (1999), *Biosolids Generation, Use, and Disposal in the United States*, EPA 530/R-99/009.

Ward, R. L., McFeters, G. A., and Yeager, J. G. (1984), *Pathogens in Sludge*, Sandia Report 83-0557, TTC-0428, Sandia National Laboratory, Albuquerque, NM.

WEF (1995), *Biosolids Composting*, Water Environment Federation, Alexandria, VA.

——— (1998), *Design of Municipal Wastewater Treatment Plants*, 4th ed., Manual of Practice 8 (ASCE 76), Water Environment Federation, Alexandria, VA.

Wilson, G. (1983), Forced Aeration Composting, *Water Science and Technology*, Vol. 15, No. 1.

8

THERMAL DRYING AND INCINERATION

Wastewater Sludge Processing, By Izrail S. Turovskiy and P. K. Mathai

8.1 INTRODUCTION

Thermal processing of wastewater sludge includes thermal conditioning, thermal drying, and incineration. *Thermal conditioning* is the process of applying heat and pressure to sludge to release bound water from solids, thereby enhancing its dewaterability without the addition of conditioning chemicals. This process is described in Chapter 3. Dry lime stabilization of sludge, which is a pasteurization process by virtue of the exothermic reaction of quicklime and sludge, is described in Chapter 6. In this chapter we describe both *thermal drying*, which is the process of evaporating water from sludge by thermal means, and *incineration*, which is total destruction of the organic solids in sludge by thermal means.

Increasingly strict regulations governing disposal and decreasing availability of disposal sites have renewed interest in thermal drying and incineration as a means of producing a marketable product (thermally dried biosolids) and reducing sludge volume. Economic and environmental analyses provide the best basis for deciding whether to use these methods of sludge processing. Although no treatment plant has produced a net profit from thermal drying and the sale of biosolids, several of these operations have been considered successful because the revenues from the sale of biosolids reduce the net processing costs.

8.2 THERMAL DRYING

Thermal drying of sludge reduces the moisture content below that achievable by conventional dewatering methods. The minimum moisture content practically attainable with thermal drying depends on the design and operation of the dryer, moisture content of the sludge feed, and the chemical composition of the sludge. For ordinary municipal wastewater sludge, moisture content as low as 5% may be achievable. The advantages of thermal drying include reduced product transportation costs, further pathogen reduction, improved storage capability, and marketability. Disadvantages include high capital and operating costs and requirements for skilled operating personnel.

8.2.1 Methods of Thermal Drying

Dryers are commonly classified on the basis of the predominant method of transferring heat to the wet solids being dried. These methods are convection (direct drying), conduction (indirect drying), radiation (infrared drying), or a combination of these.

In *convection* (direct drying), heat transfer is accomplished by direct contact between the wet sludge and hot gases. The heat of the inlet gas provides the latent heat required for evaporating the liquid from the sludge. The vaporized liquid is carried by the hot gases. Under equilibrium conditions of constant-rate drying, mass transfer is proportional to (1) the area of wetted surface exposed, (2) the difference between the water content of the drying air and saturation humidity at the wet-bulb temperature of the sludge–air interface, and (3) other factors, such as velocity and turbulence of drying air expressed as a mass transfer coefficient (Metcalf & Eddy, 2003). Direct dryers are the most common type used in thermal drying of sludge. Flash dryers, direct rotary dryers, and fluidized-bed dryers employ this method.

In *conduction* (indirect drying), heat transfer is accomplished by contact of the wet sludge solids with hot surfaces. A metal wall separates the sludge and the heating medium (usually, steam or oil). The vaporized liquid is removed independent of the heating medium. Indirect dryers for drying municipal sludge include horizontal paddle, hollow-flight or disk dryers, and vertical indirect dryers.

In *radiation* (infrared or radiant heat drying), heat transfer is accomplished by radiant energy supplied by electric resistance elements, gas-fired incandescent refractories, or infrared lamps. An example of a radiant heat dryer is the multiple-hearth furnace, commonly used for the incineration of sludge.

Thermal drying is preceded by dewatering of the sludge, usually by mechanical means. It is an important pretreatment step because it reduces the volume of water that must be removed in the dryer. Water in the dewatered sludge is evaporated in the dryer without destroying the organic matter in the sludge solids. This means that the solids temperature must be kept between 60 and 93°C (140 and 200°F). A portion of the dried sludge is often blended with the raw dewatered sludge feed to the dryer. This makes the drying operation more efficient by reducing agglomeration and thus exposing a greater solids surface area to the drying medium. Dried sludge and exhaust gases are separated in the dryer itself or in a cyclone, or both. The gas stream is exhausted through a pollution control system for removal of odors and particulates.

Flash Dryer Flash drying is the rapid removal of moisture by spraying or injecting the sludge into a hot gas stream. In a flash drying system (see Figure 8.1) the wet sludge cake is blended with previously dried sludge in a mixture to improve pneumatic conveyance. The blended sludge and hot gases from the furnace at 704°C (1300°F) are mixed ahead of a cage mill, and flashing of the water vapor begins. Gas velocities on the order of 20 to 30 m/s (65 to 100 ft/sec) are used. The cage mill mechanically agitates the sludge–gas mixture, and drying is virtually complete by the time the sludge leaves the cage mill, with a mean residence time of a few seconds. The dried sludge is conveyed to a cyclone pneumatically. The sludge at this stage has moisture content of only 8 to 10%. The sludge is then separated from the spent drying gases in the cyclone. The temperature of the dried sludge is about 71°C (160°F) and that of the exhaust gas is 104 to 149°C (220 to 300°F).

The exhaust gas treatment facility consists of a deodorizing preheater, a combustion air heater, an induced-draft fan, and a gas scrubber. Odors are

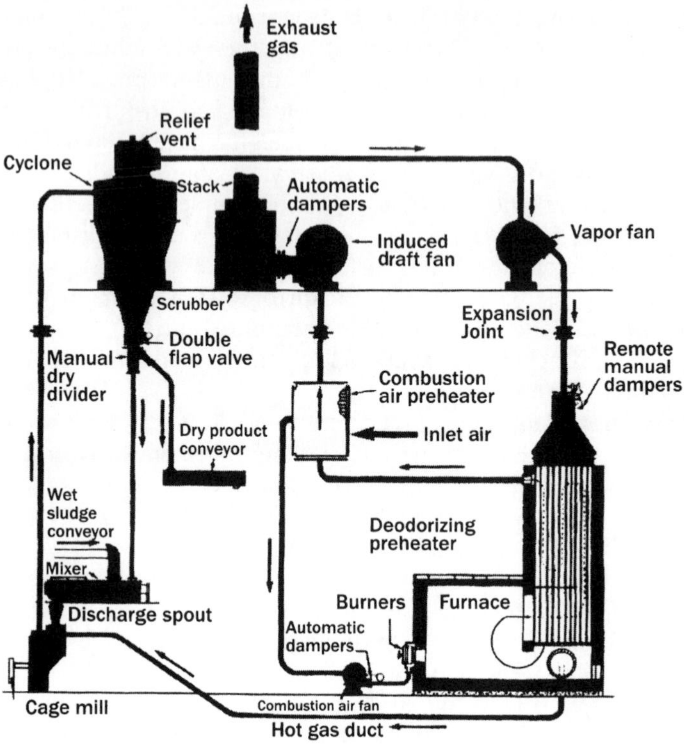

Figure 8.1 Schematic of a flash drying system.

destroyed when the temperature of the gas from the cyclone is elevated in the deodorizing preheater. Part of the heat absorbed is recovered in the combustion air preheater. The gas then passes through a scrubber to remove dust and discharges to the atmosphere.

Although approximately 50 municipal flash dryer facilities have been installed in the United States since the 1940s, only five or six are still in operation (WEF, 1998). Most of these flash dryers were shut down because of high energy and operation and maintenance costs. Today, other types of dryers are preferred over flash dryers.

Rotary Dryer The main component of a rotary drying system is a rotary drum dryer installed with an inclination of 3 to 4° to the horizontal. Figure 8.2 shows a rotary dryer in which the sludge moves under gravity along the drum from its raised (charging) end to the lower (discharge) end. The cocurrent movement of the gases and the rotation of the drum also aid the movement of the sludge. The rotational speed of the drum is 5 to 8 rpm. For uniform distribution of sludge along the drum cross section, the drum is

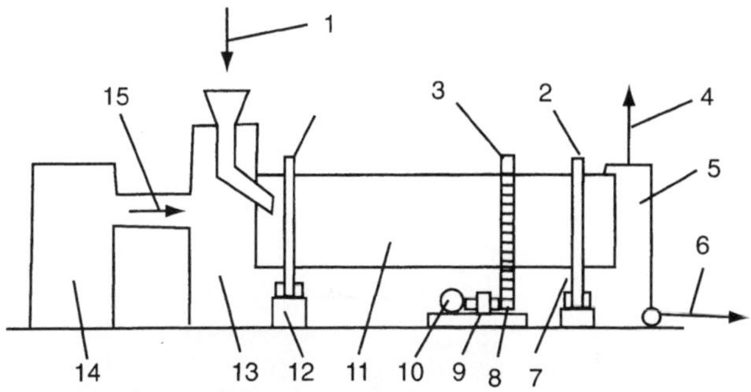

1 - Cake charge, 2 - Band, 3 - Drum drive, 4 - Spent gases
5 - Discharge chamber, 6 - Dry sludge discharge, 7 - support rollers, 8 - Drive gear, 9 - Reducer, 10 - Variable speed electric motor, 11 - Rotating drum, 12 - Rotating rollers, 13 - Charging chmber, 14 - Furnace, 15 - Furnace gases

Figure 8.2 Rotating dryer system.

equipped with fins of various types. To break up and mix the sludge, there are also chains freely suspended at the beginning and at the end of the drum. The chains prevent the sludge from sticking to the walls at the beginning of the dryer, intensify the sludge drying process, and avoid the necessity of having a dry sludge crusher in the system.

Auxiliary components of drum dryer system include a mixer for blending raw dewatered sludge and recycled dry sludge, a drum feeding screw, a furnace with a fuel burner for heating the air, a cyclone and scrubber for separating particulates from the hot exhaust gases, a heat exchanger, a dried sludge extraction screw, and a storage silo.

Drying systems with either direct or indirect rotary dryers are available. Various models from one manufacturer are available with evaporation capacities from 1000 to 10,000 kg/h (2200 to 22,000 lb/hr) for direct dryers and with evaporation capacities from 350 to 2000 kg/h (770 to 4400 lb/hr) for indirect dryers.

A rotary drying system is in operation at the Morris Forman wastewater treatment plant in Louisville, Kentucky. Primary sludge is anaerobically digested first and then blended with thickened waste activated sludge. The sludge mixture is then dewatered in centrifuges to about 26% solids and fed into four drying trains, each comprising a rotary drum dryer. Each dryer is sized to evaporate water from dewatered sludge at a rate of 8500 kg/h (18,700 lb/hr). Total installed evaporative capacity is 34 metric tons/h (75,000 lb/hr). The methane generated from the digesters provides half of the energy required by the dryers. Heat recovered from the dryers is used to heat the anaerobic digesters (Shimp and Childress, 2002).

Fluidized-Bed Dryer The fluidized-bed dryer consists of a stationary vertical chamber with a perforated bottom through which hot gases (usually air or steam) are forced. The flow schematic of a system for drying sludge in a fluidized-bed dryer is shown in Figure 8.3. The dewatered sludge from the hopper enters the cylindrical dryer by means of a feeder. The air and the furnace gases, under pressure created by the high-pressure ventilator, pass through the gas distribution grate and create a fluidized bed of drying sludge and inert materials. The dried sludge is discharged in the forms of granules over a baffle of adjustable height into the dry sludge hopper. The dust fractions present in the spent gases are recovered in the cyclone and go to the sludge feed hopper. Then the gases go to the wet scrubber, where they are purified, partially cooled, and vented to the atmosphere through the stack by the ventilator.

Table 8.1 shows the operating parameters of fluidized-bed dryers. The inert material in a fluidized bed can be quartz sand or slag. The exact design parameters of the process are determined for each particular case experimentally. The range of heat carrier temperature is 500 to 600°C (932 to 1112°F) and the drying time is 10 to 15 minutes. The sludge drying temperature is

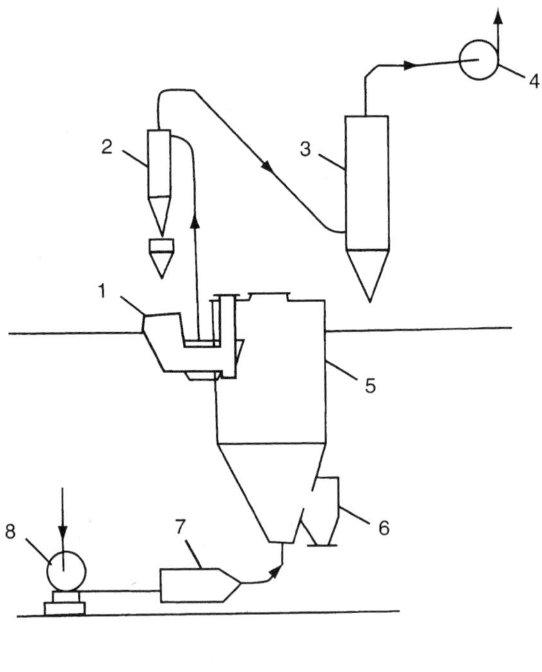

1 - Dewatered sludge hopper, 2 - Cyclone, 3 - Wet scrubber, 4 - Discharge ventilator (exhaust), 5 - Cylindrical dryer, 6 - Dry sludge discharge hopper, 7 - Furnace, 8 - High pressure ventilator (air blower).

Figure 8.3 Schematic of a fluidized-bed drying system.

TABLE 8.1 Operating Parameters of Fluidized-Bed Dryers

Parameter	Range
Temperature	
Heat carrier (°C)	500–600
Sludge drying (°C)	300–900
Gas in combustion zone (°C)	800–1100
Spent gases (°C)	120–250
Drying time (min)	10–15
Capacity of moisture evaporation (kg/h per unit volume)	80–140
Specific consumption per kilogram of evaporated moisture	
Air (kg)	10–25
Heat (J)	3770–5870
Velocity of heat carrier at dryer intake (m/s)	15–25

between 300 and 900°C (572 and 1652°F) and gas temperature in the combustion chamber is 800 to 1100°C (1472 to 2012°F). Heating to such high temperatures requires a large amount of energy; therefore, thermal drying needs an energy recovery device.

The principal advantages of fluidized-bed dryers are the ability of controlling the sludge drying time and the intensity of the heat transfer, the absence of moving parts, and the simplicity of design. A disadvantage is that the dust content of the exhaust gases is high, which is approximately 0.6 to 0.7 g/m^3.

Opposing Jet Dryer The opposing jet dryer is a dryer developed in Russia (Turovsky, 1998). It is a two-stage device (see Figure 8.4), the lower stage of which has the sludge drying chamber with the opposing jet element and the upper stage has the product/air handling device. After dewatering, the sludge cake is conveyed by a belt conveyor and then by double shaft screw feeders to the opposing jet element. The element is designed in the form of two horizontal jet pipes inserted coaxially in a standpipe. The drying scheme includes the recycling of fine dried particles to the sludge feed and discharge of the dried product from the air spout device. The sludge cake is mixed with a portion of the dried sludge in the double-shaft auger feeders, making the dryer feed homogeneous in composition and moisture content and intensifying the drying process. The airflow separator in the second stage increases the time of contact of the drying agent with the sludge. In addition, the dried sludge is classified in it by fractions.

The drying method in opposing jet dryers involves the formation of gaseous suspension of the sludge particles in the hot gas flow. The process is very effective, due to the vibratory motion generated by the collision of the jets. This leads to an increase in the sludge concentration in the drying zone. At sufficiently high velocities of the drying agent, the sludge is pulverized so that the total heat and mass transfer surface area are increased. Operating parameters of opposing jet dryers are listed in Table 8.2.

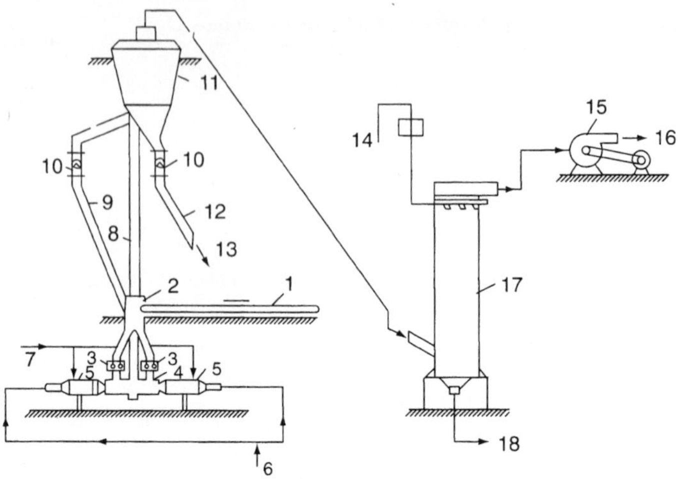

1 - Belt conveyor to feed the sludge, 2 - Receiving chamber, 3 - Double-axle screw feeders, 4 - Dryer chamber with jet pipes, 5 - Combustion chamber, 6 - Air inflow, 7 - Fuel (gas), 8 - Standpipe, 9 - Return pipe, 10 - Rotating gates, 11 - Airflow separator, 12 - Dry sludge pipe, 13 - Dry sludge feed to final product hopper, 14 - Water supply, 15 - Ventilator, 16 - Purified gases to atmosphere, 17 - Water scrubber, 18 - Sludge discharge.

Figure 8.4 Schematic of opposing jet dryer.

TABLE 8.2 Operating Parameters of Opposing Jet Dryers

Parameter	Range
Drying agent temperature	
Initial (°C)	600–800
Final (°C)	100–150
Air pressure (gage) upstream of nozzle (MPa)	0.01–0.03
Evaporated moisture load on dryer volume ($kg/m^3 \cdot h$)	600–1200
Specific consumption per kilogram of evaported moisture	
Heat (MJ)	3.4–3.8
Standard fuel (kg)	0.114–0.128
Electrical energy (kWh)	0.05–0.08
Sludge moisture (%)	
Before drying	60–80
After drying	20–50

Horizontal Indirect Dryer Horizontal indirect dryers for drying municipal wastewater sludge include the paddle dryer, hollow-flight dryer, and disk dryer. Figure 8.5 is a schematic diagram of a horizontal indirect dryer. The dryer consists of a horizontal jacked vessel with one or two rotating shafts fitted with paddles, flights, or disks which agitate and transport the sludge through the dryer. The heat transfer medium (usually, steam) circulates through the jacketed shell and through the hollow-core shafts and hollow agitators (paddles, flights, or disks). A weir at the discharge end of the dryer ensures complete submergence of the heat transfer surface in the material being dried. The steam is discharged as condensate after transferring its available energy to the sludge. Dryers that use hot water or oil as the heat transfer medium are constructed internally in a manner different from those required for steam. Dewatered sludge is fed into the vessel continuously, with or without mixing with any recycled dried product. The transfer of heat from the heat transfer medium raises the temperature of the sludge and evaporates the water from the sludge solids surface. The water evaporated is transported out of the dryer by low-volume sweep gases or exhaust vapors.

If dried product is mixed with dewatered sludge, the moisture of the feed sludge can be reduced by 40 to 50%. The blending prevents agglomeration and fouling of the heat transfer surface. Dryers that dry unblended feed sludge should have internal breaker bars and must provide enough horse-

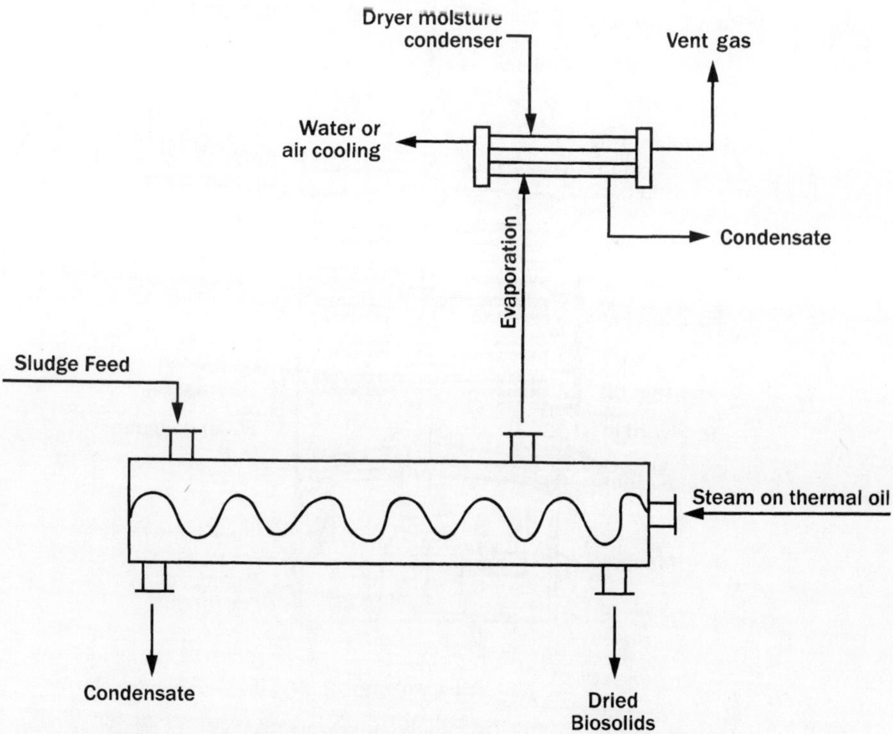

Figure 8.5 Flow schematic of horizontal indirect dryer system.

power to turn the agitator shafts to break up the clumps. Horizontal indirect dryers are capable of drying sludge with less than 10% moisture.

Vertical Indirect Dryer A vertical indirect dryer, such as the Pelletech dryer shown in Figure 8.6, dries and pelletizes sludge simultaneously. It is a vertically oriented multistage unit that uses steam or thermal oil in a closed loop as the heat transfer medium to achieve a dry solids content of 90% or more.

Dewatered sludge cake blended with dried product is fed at the top inlet of the dryer. The dryer is equipped with several trays heated by the heat transfer medium. The dryer has a central shaft with attached rotating arms. The rotating arms are equipped with adjustable scrapers that move and tumble the sludge in thin layers from one tray to another in a rotating zigzag motion until it exists at the bottom as a dried pelletized product. The process minimizes the formation of dust and oversized chunks.

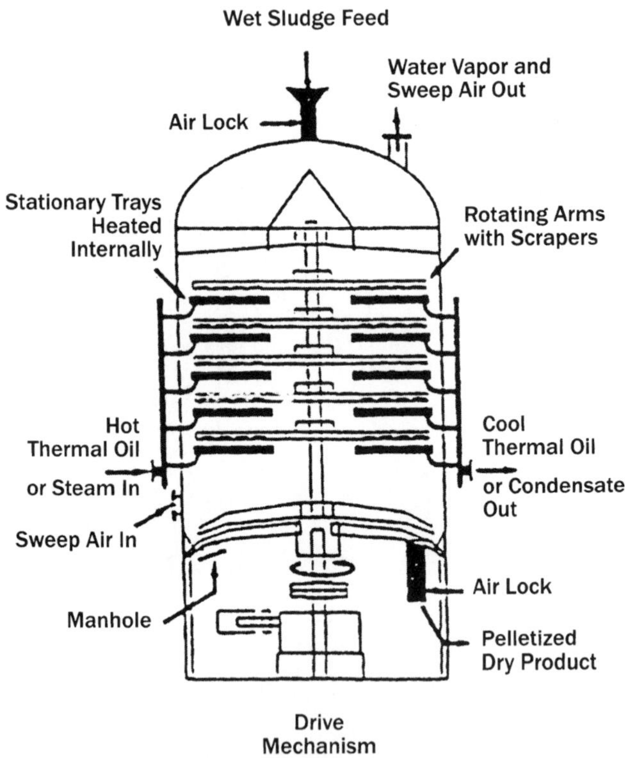

Figure 8.6 Vertical indirect dryer by Pelletech. (Reprinted with permission from WEF, 1998.)

The dryer's exhaust consists of water vapor, air, and some pollutants. After the water vapor is condensed, only a small amount of gases, mainly moist air, remains to be treated. These gases are vented from the dryer to an odor control unit for thermal destruction of odor-causing compounds.

The wastewater treatment plant in Largo, Florida, produces pelletized biosolids. The system includes aerobic digestion of primary and thickened waste activated sludge, dewatering on belt filter presses, a parallel two-train heat drying/palletizing system, product storage silos, energy recovery after-burners for process air, and a wet scrubber to control odor from a covered gravity thickener and drying building air. The thermal drying process reduces the moisture content of dewatered sludge cake from 85% to about 8% in the pelletized biosolids. In 2001, the facility produced dried biosolids at an average rate of 0.53 dry ton per hour at a net operating cost of $254.50 per dry ton. The dried product was sold for $38 per dry ton (Craven et al., 2004).

8.2.2 Design Considerations

The most important aspects of the design of a thermal drying system are discussed in the following sections.

Moisture Content of Feed Sludge Thermal evaporation of water from sludge requires considerable energy. The amount of fuel required to dry sludge depends on the amount of water evaporated. As mechanical dewatering methods are more efficient than thermal drying methods per volume of water removed, it is imperative that a dewatering step precede thermal drying so that overall energy requirements can be minimized.

Often, drying requires mixing of feed sludge with a portion of the dried sludge to achieve a solids content exceeding the plastic stage to prevent agglomeration and the fouling of internal dryer surfaces. Paddle mixers, pug mills, and hammer mills are devices for the thorough mixing of sludge cake and recycled product.

Storage Storage requirements for both the dewatered sludge feed and the dried product should be considered in the design of a drying system. Sufficient dewatered sludge storage should be provided to allow orderly shutdown of the drying system and to attenuate variations in production. A storage capacity of a minimum of three days of dewatered sludge production is recommended. Storage for the dried product depends on the final disposal arrangement. Sales of the product are likely to be seasonal, and storage for 90 days may be necessary unless bulk buyers provide off-site storage. If the dried product is burned as a fuel or it undergoes further processing, storage requirements are dictated by subsequent steps in the sludge processing system.

Dust can become a problem if the dried product is stored in bulk and is not pelletized. Stored product with a moisture content below 10% may combust spontaneously, especially when it contains large quantities of dust. To minimize dust generation, long conveyors that create an abrasive action, such as screw conveyors and drag conveyors, should be avoided. Even the use of pneumatic conveyors may be too abrasive. Open or folded belt conveyors are the preferred choices. Dried product with moisture content below 20% may absorb water from the environment.

Fire and Explosion Hazards Drying systems exposed to heavy dusting have had problems with fires. The combination of combustible particles, warm temperatures, sufficient oxygen, and high gas velocities make these systems susceptible to fire and explosion. Any material that will burn in air when it is in solid form may explode when in the form of finely divided powder. Biosolids are composed primarily of carbohydrates, proteins, and fats and will burn readily in solid form. For an explosion to occur, dust must be present in sufficient explosive concentration, and there must be a source of ignition. In addition, for an explosion to occur, particles must be sufficiently close together so that the heat released from one particle will heat the surrounding particles. An explosive concentration of dust ($320 g/m^3$) in the presence of oxygen at a concentration greater than 15% will explode when exposed to a spark at a temperature exceeding 355°C (671°F) (Barrett and Herndon, 2005). Cyclone separators, wet scrubbers, baghouses, or a combination of these can be provided to remove dust and fine particles from the process airstream.

Maintaining a minimum oxygen level in the dehydration chamber is essential during active sludge drying. It is even more critical during startup and shutdown of the dryer (when wet sludge is not present) because, historically, this is when most adverse dryer events have occurred. It has been demonstrated that using an inert gas such as nitrogen to purge available oxygen is the most effective option. Well-trained operators, controls with interlocks, and monitoring equipment are important considerations in reducing fire hazards.

When designing a sludge drying system, the following dryer safety standard checklist should be followed to avoid fire and explosion hazards (adapted from Barrett and Herndon, 2005). The items of importance are:

- An inert-gas purge system following the National Fire Protection Association (NFPA) Standard on Explosion Prevention Systems (NFPA 69)
- Level sensors and flow sensors located on the in-feed hopper and communicating with the dryer's control system to confirm sludge feed to the dryer
- An independent spark-detection system in place at the dryer's out-feed that communicates with the dryer's control system
- Careful consideration of the conveyance system used to transport dried product

- Control hardware designed and built to meet the requirements of Underwriters' Laboratories and, at a minimum, the National Electrical Manufacturers Association (NEMA) NEMA 4 rating
- Redundant safety reporting systems and instrumentation
- A human–machine interface protected by a multilevel pass-code protocol
- Historic-trending information recorded and pass code–protected in the control system
- Building(s) for drying systems that meet all codes and requirements of the local jurisdiction and NFPA 820 as appropriate
- Hot oil systems and steam boilers isolated behind fire-resistant walls
- Hot oil systems designed, built, and code-stamped according to the American Society of Mechanical Engineers (ASME) code (this applies to the entire system, not simply the hot oil heater)
- Storage systems (confined) equipped with temperature and carbon monoxide sensors
- Rupture disks incorporated into the dryer design following the guidelines in the NFPA codes
- Safety systems for heat dryers designed around specific maximum deflagration pressure (P_{max}) and deflagration index (K_{st}) values

Emissions and Odor Control Dryer equipment and handling and storage areas should be contained and vented to air pollution control equipment. Cyclone separators, wet scrubbers, baghouses, or a combination of these remove particulates from airstreams. Afterburners (incinerators) and chemical scrubbers typically control odors. Thermal oxidizers have proved to be the most effective in removing aldehydes and various species of sulfides and disulfides (methyl, dimethyl, and carbonyl).

Sidestreams Liquid sidestreams are produced by certain ancillary equipment such as wet scrubbers. Odorous liquid sidestreams produced by condensation of water vapor from dryers contain both organic oils and ammonia. These sidestreams can frequently be recycled to the treatment plant influent but may sometimes require separate treatment.

Heat Source and Heat Recovery The large amounts of energy required for thermal drying of sludge dictate that close attention be given to the source used to heat the drying medium. Natural gas and fuel oil are most frequently used but are becoming more expensive. Dryers should be designed for heat recovery and reuse to reduce energy use. Recovered heat from dryer or furnace exhaust may be reused to preheat combustion air, preheat feed sludge, or supplement plant heating requirements. The dried sludge itself has a fuel value and may be used as a heat source for the drying medium.

8.3 INCINERATION

Incineration is complete combustion, which is the rapid exothermic oxidization of combustible elements in sludge. Dewatered sludge will ignite at temperatures of 420 to 500°C (788 to 932°F) in the presence of oxygen. Temperatures of 760 to 820°C (1400 to 1508°F) are required for complete combustion of organic solids. In the incineration of sludge, the organic solids are converted to the oxidized end products, primarily carbon dioxide, water vapor, and ash. Particulates and other gases will also be present in the exhaust, which determines the selection of the treatment scheme for the exhaust gases before venting them to the atmosphere.

The principal advantages and disadvantages of incineration over other methods of sludge stabilization are listed in Table 8.3. Sludge is incinerated if its utilization is impossible or economically infeasible, if storage area is limited or unavailable, and in cases where it is required for hygienic reasons.

One of the principal parameters of sludge incineration is the sludge moisture. Sludge cake with 30 to 50% solids (50 to 70% moisture) is autogenous; that is, it can be burned without auxiliary fuel. Sludge cake with 20 to 30% solids (70 to 80% moisture) may require an auxiliary fuel for combustion. Therefore, before incineration, the moisture content of the sludge should be reduced by mechanical dewatering or thermal drying.

TABLE 8.3 Advantages and Disadvantages of Incineration

Advantages	Disadvantages
1. Reduces the volume and weight of wet sludge cake by approximately 95%, thereby reducing disposal requirements. 2. Complete destruction of pathogens. 3. Destroys or reduces toxins. 4. Potentially recovers energy through the combustion of waste products, thereby reducing the overall expenditure of energy.	1. High capital and operating costs. 2. Reduces the potential beneficial use of biosolids. 3. Highly skilled and experienced operating and maintenance staffs are required. 4. If residuals (ash) exceeds the prescribed maximum pollutant concentrations, they may be classified as hazardous waste, which requires special disposal. 5. Discharges to atmosphere (particulates and other toxic or noxious emissions) require extensive treatment to assure protection of the environment.

TABLE 8.4 Heating Values of Sludge and Other Residuals

Type of Sludge/Residual	Dry Solids	
	MJ/kg	Btu/lb
Primary sludge	20–28	8600–12,000
Activated sludge	16–22	6,900–9,500
Digested sludge	10–15	4,300–6,500
Grease and scum	39	16,800
Screenings	21	9,000

Another important parameter of sludge incineration is the heating value of sludge. It represents the quantity of heat released per unit mass of solids. The amount of heat released from sludge is a function of the types and combustible elements present in sludge. The primary combustible elements in sludge (and in most available auxiliary fuels) are carbon, hydrogen, and sulfur. Carbon burned to carbon dioxide has a heating value of 34 MJ/kg (14.6×10^3 Btu/lb), hydrogen has a heating value of 144 MJ/kg (62×10^3 Btu/lb), and sulfur has a heating value of 10 MJ/kg (4.5×10^3 Btu/lb). Consequently, any changes in the carbon, hydrogen, or sulfur content of sludge will raise or lower its heating value. Table 8.4 shows the heating values of various types of sludge, grease and scum, and screenings.

8.3.1 Methods of Incineration

The process of sludge incineration in furnaces can be divided into the following stages: heating, drying, distillation of volatile matter, combustion of the organic fuel matter, and calcination to burn the residual carbon. Heating the sludge to 100°C (212°F) and then drying it at about 200°C (392°F) consume the principal quantity of heat and are generally required for the incineration process. These parameters also affect the selection of the size of the main and auxiliary equipment and consequently, determine the cost in general. In the course of moisture evaporation in the drying zone, volatile substances are liberated together with the moisture, which sometimes results in objectionable odors.

The combustion of the sludge takes place at temperatures between 200 and 500°C (392 and 932°F), due to the thermal radiation of the flame and the incandescent walls of the combustion chamber, as well as the convection heat transfer from the exhaust gases. The calcination of the ash fraction of the sludge is completed by its cooling to a temperature at which it can be removed from the site.

The design temperature in the furnace should not exceed the melting point of ash [usually, about 1050°C (1922°F)] and should not be below 700°C (1292°F), thus providing reliable deodorizing of the gases. Systems for sludge incineration should provide complete combustion of the organic fraction of the sludge and utilization of the heat of the exhaust gases.

Furnace selection for combustion of dewatered sludge is determined by sludge moisture content and the noncaking nature of the carbonized residue. The first property excludes the possibility of combusting the sludge directly in a flame or in cyclone furnaces without predrying, and the second property excludes the possibility of combusting the sludge on grates; thus, at the present time, multiple-hearth, fluidized-bed, and rotating drum furnaces are most often used. These furnaces are described in the following sections.

Multiple-Hearth Incineration The flowchart of a system with a multiple-hearth furnace is presented in Figure 8.7. The furnace shell is a vertical steel

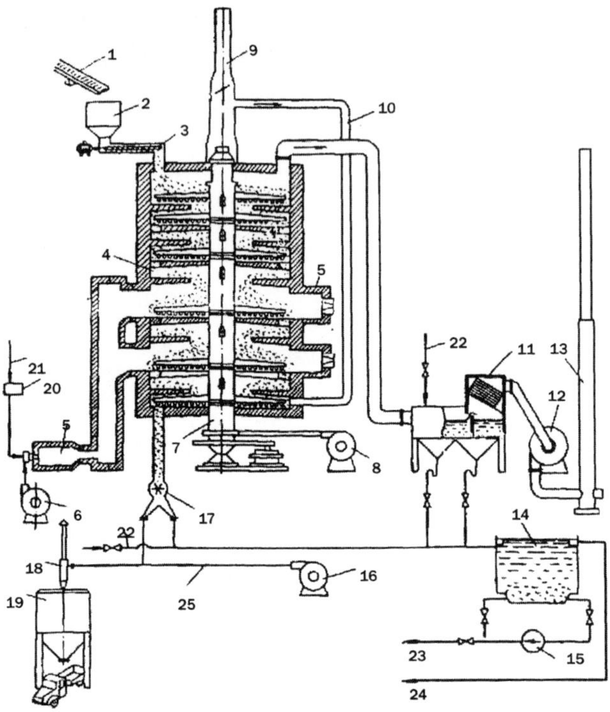

1 - Belt conveyor, 2 - Sludge charging hopper, 3 - Screw feeder, 4 - Multihearth furnace, 5 - External furnace, 6 - Draft blower, 7 - Furnace shaft, 8 - Cooling blower, 9 - Vent, 10 - Recirculation pipe, 11 - Scrubber, 12 - Blower,13 - Stack, 14 - Ash tank, 15 - Ash slurry pump, 16 - Pnuematic transport blower, 17 - Gate feeder,18 - Cyclone discharge device,19 - Ash discharge hopper, 20 - Gas regulator unit, 21 - Fuel gas pipe, 22 - Water pipe, 23 - Ash pipe, 24 - Wastewater pipe, 25 - Air pipe.

Figure 8.7 Flowchart of multiple-hearth incineration.

cylinder 6 to 8 m (20 to 26 ft) in diameter lined internally with refractory brick or heat-resistant concrete. The furnace is divided vertically into seven to nine refractory hearths. A vertical rotating shaft passes through the center of the furnace, to which the horizontal frames of the rake mechanisms, made of heat-resistant cast iron, are affixed. Each hearth has material transfer openings located alternatively on the periphery of one hearth and in the center section of the adjacent hearth.

The sludge moves by conveyors into the charging hopper and then onto the uppermost hearth of the furnace. The sludge is moved by rakes into the transfer openings, it drops to the next lower hearth, and continues its travel to the lower hearths. This provides continuous movement of the sludge mass in the opposite direction to the hot combustion air. The use of rake mechanisms to move and break up the clumps in the sludge intensifies the drying and combustion processes. The vertical shaft and the rake mechanism frames are hollow and are cooled by the air supplied by a blower. A fraction of this air enters the ash-cooling zone through a special air pipe and then goes to the sludge combustion zone. A multiple-hearth furnace operates with a supply of 50% excess air. The quantity of air is usually controlled automatically by monitoring the oxygen concentration in the exhaust gases by means of an oxygen analyzer. For ignition and also to maintain stable operation, the furnace is equipped with three or four burners and a forced-air supply.

The upper chambers of the furnace constitute the moist sludge drying zone, where the principal part of the moisture is evaporated. In the middle chambers, the organic sludge matter is combusted at a temperature of 700 to 900°C (1292 to 1652°F), and in the lower chambers, the ash is cooled before being discharged to the ash hopper. The ash from the hopper can be transported either in dry form (by pneumatic conveyance) to the ash hopper and then to the ash dump (by truck), or in wet form together with the dust collector ash to the sludge beds or ponds.

Countercurrent movement of the moist sludge and combustible furnace gases in the drying zone results in cooling of the gases to 250 to 300°C (482 to 572°F). The gases are discharged to the dust scrubber and vented to the atmosphere by a blower. Volatile substances are almost absent in the early stages of drying. Their intense liberation occurs in direct proximity to the zone of combustion of the principal mass of sludge, where it burns almost completely. Experience with operation of incinerators in Europe and Russia indicates that during normal operation of the furnace, the exhaust gases have no unpleasant odors. If necessary, moist gases discharged from the drying zone can be deodorized using a special afterburner, either separately or in the furnace proper.

The advantages of multiple-hearth furnaces include combusting both primary and secondary sludge, as well as trash from screens, scum from settling tanks and oil separators, dirty grit from grit chambers, and industrial wastes. They are characterized by their simplicity of service and by the reliability and stability of operation during significant variations in the quantity and quality of sludge treated. The furnaces can be installed in the open air. The drawbacks of multiple-hearth furnaces include high capital cost, large area required, presence of rotating mechanism in the high-temperature zone, and frequent failure of the rake devices.

Fluidized-Bed Incineration Fluidized-bed furnaces are well known in drying and roasting technology in various industrial fields. The furnace, a vertical steel cylinder lined internally with refractory brick or heat-resistant concrete, consists of a cylindrical furnace chamber, a lower conical section with an impermeable air distribution grate, and dome-shaped crown. Heat-resistant quartz sand 0.6 to 2.5 mm in size is placed on the grate to a depth of 0.8 to 1 m.

The turbulent (fluidized) bed in the furnace is formed when air is blown through the distribution grate at a rate at which the sand particles move in a turbulent manner and appear to boil in the flow of gas. The air is supplied by a blower to a recuperator, in which it is heated by the exhaust gases leaving the furnace to a temperature of 600 to 700°C (1112 to 1292°F). The heated air enters under the distribution grate at a pressure of 12 to 15 kPa (1.7 to 2.2 psi).

The design of the furnace amounts to determining the material and thermal balances of the sludge combustion process, establishing the geometric dimensions of the furnace elements, and the quantities of auxiliary fuel, air, and exhaust gases. The dimensions of the furnace are determined from the volume of the sludge combusted and the air velocity in the distribution grate. This velocity depends on the hydrodynamic regime of the furnace operation and size of the sand bed as well as the properties of the sludge (moisture, noncombustible solids, and ash size distribution). The quantity of air required for complete oxidation of the organic matter in sludge is determined from the sludge's ultimate composition.

Figure 8.8 presents a flowchart of an incineration system with a fluidized-bed furnace. The dewatered sludge with moisture of 60 to 75% is fed by conveyors into the charging hopper and then into the furnace at the top. Passing through the furnace chamber, in which the temperature is 900 to 1000°C (1652 to 1832°F), the sludge is dried and dispersed in the fluidized bed, where it is thoroughly mixed with the incandescent quartz sand. This results in the breakup of sludge agglomerates, instantaneous moisture evaporation, separation of volatile organic matter, combustion of carbonized residue, and calcinations of the mineral fraction. Due to the intense mass and heat transfer, the entire process occurs in less than 1 to 2 minutes. The volatile fraction of the sludge combustible matter is completely incinerated above the fluidized bed, and as a result the gas temperature is raised.

The initial charging and the subsequent maintenance of the designed quantity of sand on the grate are accomplished through the gate feeder. The sand (replaced when necessary) can be discharged at the bottom of the grate through a special slide gate. The fine ash and dust are discharged from the

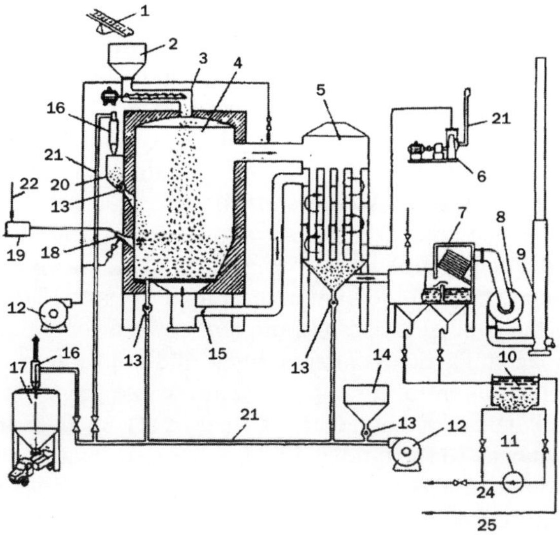

1 - Belt conveyor, 2 - Sludge charging hopper, 3 - Screw feeder, 4 - FB furnace, 5 - Recuperator, 6 - Air blower, 7 - Wet dust trap, 8 - Blower, 9 - Stack, 10 -Ash tank, 11 - Ash water pump, 12 - Ventilator, 13 - Gate feeder, 14 - Sand hopper, 15 - Gate valve, 16 - Cyclone discharge device, 17 - Ash discharge hopper, 18 - Gas burner, 19 - Gas control unit, 20 - Hopper-feeder, 21 - Air duct, 22 - Fuel gas pipe, 23 - Water pipe, 24 - Ash duct, 25 - Wastewater pipe.

Figure 8.8 Flowchart of fluidized-bed incineration.

furnace with the flow of exhaust gases, which go to the recuperator (air preheater). To decrease the temperature of the exhaust gases to 900°C (1652°F) before the recuperator intake, there is a forced supply of cold air, the quantity of which is controlled by an automatic system. The air, forced in under pressure, enters the space between the recuperator tubes and is directed in a crosscurrent–countercurrent manner. To provide such flow patterns, the space between the recuperator tubes is separated by horizontal partitions. The dust-containing exhaust gases cooled in the recuperator go to a wet scrubber, where ash and dust are removed and the gases are vented to the atmosphere. Ash handling is similar to the system described for multiple-hearth incineration.

The heated air, passing through the grate of the fluidized-bed furnace at the design velocity, fluidizes the bed and maintains the required combustion temperature. When the calorific value of the organic sludge fraction is not adequate to maintain the combustion process, auxiliary fuel is injected into the furnace through side burners.

Pyrofluid, a modified technology of a fluidized-bed furnace developed by Omnium de Treatment et de Valorisation (OTV, France), has been in operation since 1999 at the city of St. Petersburg, Russia, for the incineration of a mixture of dewatered primary and activated sludge. The incineration system consists of four lines and can treat a total of 250 tons/day of dry sludge solids. Each line has a Centrypress with polymer conditioning devices for sludge dewatering to 28 to 30% dry solids, an incinerator, a heat recovery and utilization system, an ash removing system, and a furnace gas wash and neutralization system. The furnace has two main chambers: a lower fluidized-bed chamber with quartz sand medium and an upper reactor. Dewatered sludge is pumped to the fluidized bed and air heated to a temperature of 500 to 600°C is introduced. Particles of dried sludge and sand at temperatures of 710 to 770°C are then taken to the upper reactor, where the temperatures reach 880 to 900°C. Particles of sludge are burned in this reactor, but the heavier sand goes down back to the fluidized bed. The exhaust gases contain the ash, which is separated in a boiler–electrofilter system. After acidity and alkali neutralization, the gases are cooled to 130°C and exhausted to the atmosphere. The concentration of oxygen in gases is kept at 6%. Ash at a temperature of 250°C is mixed with the water used for gas neutralization in a screw loader to concentrate it to 25 to 30% solids. Each incinerator uses 1000 to 1500 m^3/d of natural gas. In 2000, the cost of incineration was $50 to $60 per ton of dry solids.

Rotary Kiln Incineration Rotary kilns (or drum kilns) are often used in various fields of industry and are mass-produced. The kilns are used most often in the calcination of cement clinker and claydite and for incineration of sludge mixed with municipal solid waste (co-incineration).

Figure 8.9 is a diagram of a rotary kiln incineration system. The drum is installed with an inclination of 2 to 4° in the direction of the external furnace at the lower end. The furnace is cylindrical in shape, lined with refractory bricks, and equipped with gas–oil burners. The furnace is rolling on rails, which aids repair of the drum and replacement of the lining when required. The dewatered sludge (mixed with municipal solids waste for co-incineration) is charged at the upper end of the drum.

The sludge dries as it moves through the drying zone and burns in the incineration zone with the liberation of heat. The hot ash falls through an opening in the external furnace chamber and enters the air cooler, from which it goes by pneumatic transport into the receiving hopper and is then hauled to the ash dump. The ash is sometimes used as a conditioning agent in dewatering the sludge.

After cooling the ash to 100°C (212°F), the hot air goes to the furnace for use in combustion. The fine dust is carried out with the exhaust gases, as are the volatile organic substances liberated in the drying zone. When necessary, afterburning of the organic matter and deodorizing of the gases can be conducted in a special section of the charging chamber. In the drying zone, the

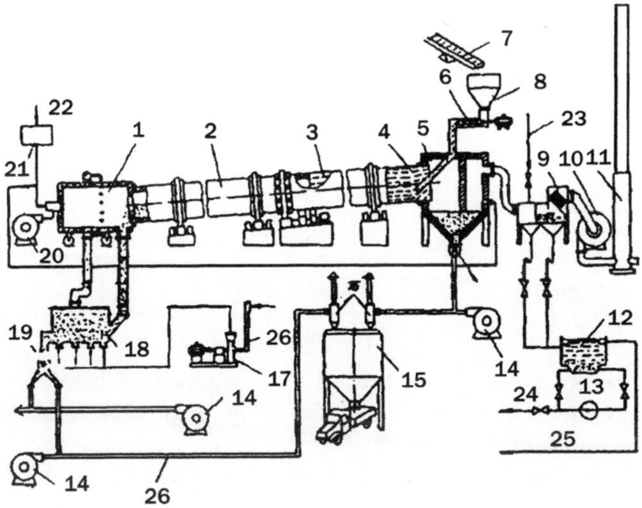

1 - External furnace, 2 - Rotating kiln, 3 - Vane fin, 4 - Spiral intake fin, 5 - Final combustion chamber, 6 - Screw feeder, 7 - Belt conveyor,8 - Sludge charging hopper, 9 - Scrubber, 10 - Blower, 11 - Stack, 12 - Ash tank, 13 - Ash slurry pump,14 - Pneumatic transport blower, 15 - Ash discharge hopper, 16 - Cyclone discharge device, 17 - Air blower, 18 - Air cooler, 19 - Gate feeder, 20 - Draft blower, 21 - Gas control unit, 22 - Gas pipe, 23 - Water pipe, 24 - Ash pipe, 25 - Wastewater pipe, 26 - Air pipe.

Figure 8.9 Flowchart of rotary kiln incineration.

temperature of the exhaust gases is 200 to 220°C (392 to 428°F), and moisture of the sludge decreases from 65 to 85% to 30 to 40%. In the combustion zone, the length of which usually does not exceed 8 to 12 m (26 to 39 ft), the temperature reaches 900 to 1000°C (1652 to 1832°F).

The advantages of rotary kiln incinerators include low emission of heat and low emission of particulates with the exhaust gases, the possibility of treating sludge with high ash and high moisture, and the possibility of installing the rotating section of the kiln in the open air (the furnace section and the charging chamber are usually in buildings). The drawbacks are their cumbersome size, substantial weight, high capital cost, and relative complexity of operation.

Emerging Technologies A new efficient technology for thermal drying and incineration of sludge has been studied in Russia. The technology included raw sludge thickening with polymers, centrifugal dewatering, and thermal drying and incineration by using low-temperature plasma without oxygen.

TABLE 8.5 Melting and Boiling Points of Heavy Metals

Metal	Melting Point (°C)	Boiling Point (°C)
Cadmium	321	767
Zinc	419.5	907
Lead	327.5	1749
Copper	1084.6	2562
Chromium	1907	2671
Nickel	1455	2913

Low-temperature plasma is also used for extraction of heavy metals from sludge. Heavy metals are generally insoluble; they usually are present at higher levels in sludge than in wastewater and concentrated even further by the dewatering of sludge. The experiments showed that heavy metals could be extracted from sludge by evaporation at increasing temperatures over time.

Different metals have different temperatures for melting and boiling. Table 8.5 lists the melting and boiling points of heavy metals of concern in sludge. These metals are extracted by using low-temperature plasma with catalysts. Various temperatures of the plasma allow extracting the different metals. The cooling of the vapor in reactors lets the metals separate as pure metals. This process is initiated by treating the dewatered sludge at low temperatures between 100 and 500°C. Once heated, the water evaporates, and then the organic substances burn off. The high-temperature process (500 to 3000°C) is followed by melting, boiling, and separating the metals by cooling the vapor from the reactor.

Another technology is a modified cyclone furnace for dewatered sludge cake incineration. Cyclone furnaces are used for combustion of liquid or finely dispersed dry materials in spray form. They have not been used for the combustion of dewatered sludge, due to the technical difficulties of pulverizing the sludge in the cylindrical furnace chamber. However, this does not exclude the possibility in principle of using cyclone furnaces for this purpose. Since the combustion of sludge in liquid form is economically infeasible, the sludge should be subjected to predrying and thorough grinding. Thermal drying can be conducted in opposing jet dryers or fluidized-bed dryers. Efficient systems for combustion of sludge make use of the furnace gases from the cyclone furnace to dry the sludge. The incineration system includes a dryer, sludge grinder, cyclone oven, gas deodorizing chamber, scrubber, blower, and exhaust stack.

8.3.2 Design Considerations

The most important design considerations for incineration systems are discussed in the following sections. Manufacturers of incinerators rely on empirical data obtained from extensive pilot plant tests. Such information is generally proprietary; therefore, the manufacturers will typically perform the design after receiving the input data, conditions, and functional specifications for the incineration system.

Moisture Content of Feed Sludge Traditionally, combined primary and secondary sludge is usually dewatered to 16 to 25% solids (75 to 84% moisture). Dewatering to this solids concentration means that for every kilogram of solids, 3 to 5.25 kg of water must be evaporated at high temperatures. Such systems require large amounts of auxiliary fuel to evaporate the water. The excess costs of such systems, from the increase in fuel price, have resulted in the shutdown of many incinerators.

Drier sludge cakes are produced at plants that use efficient belt filter presses, high-solids centrifuges, or recessed plate filter presses for dewatering. As a general rule, fluidized-bed furnaces are preferable to multiple-hearth furnaces when the solids content is greater than 30% (WEF, 1998). This is because fluidized-bed furnaces can operate with lower excess air levels while avoiding many of the operating complications of multiple-hearth furnaces. At less than 30% solids content, multiple-hearth furnaces merit consideration, provided that air emission regulations can be met without an afterburner.

Heat Recovery and Reuse Modern sludge incineration processes rely on waste heat recovery and recycle for economic operation. Heat recovery can be for internal reuse or for secondary use. Internal reuse includes direct recovery, in which the sludge is preheated or dried by flue gases, and indirect recovery, in which flue gases are used to preheat combustion air. Preheating the combustion air is the most common and economical approach to heat recovery in sludge incineration. Secondary uses can be in the form of space heating, power generation, or using in an indirect sludge dryer to increase solids concentration and thus to eliminate the need for an auxiliary fuel.

Ash Disposal Although incineration provides the greatest reduction in the volume of sludge, there is still a significant quantity of material to be disposed of. For a specific wastewater treatment plant, the quantity of ash produced will essentially be equal to the inert fraction in the sludge. Incineration of grit, screenings, and scum will also affect the quality and quantity of ash. The quantity of ash will vary seasonally in many facilities as a function of waste characteristics and where there are significant industrial discharges. However, most increases in the quantity of inerts result from infiltration or inflow to the sewer during wet weather periods. Chemicals also affect the quantity of ash and the removal of heavy metals. Thus, where chemicals are employed in the treatment processes, the effect on the quantity of ash and metals on the disposal facilities must be taken into consideration. The quantity of ash generally ranges from 200 to 400 g/kg (400 to 800 lb/ton) of raw dry solids combusted without consideration of grit. Digested sludge will produce proportionally more ash, about 350 to 500 g/kg (700 to 1000 lb/ton) of dry solids combusted because of the lower volatile content of digested sludge. Ash generally has a specific gravity of 2.4 to 3.0. Dry bulk densities range from 385 to 640 kg/m^3 (24 to 40 lb/ft^3), and wet bulk densities range from 1440 to 1920 kg/m^3 (90 to 120 lb/ft^3).

Ash from incinerators can be handled by two principal methods: dry ash handling or wet ash handling. Dry ash handling involves lifting the ash from the furnace with a bucket elevator or by a pneumatic method to a silo for later load-out. The bucket elevator method of lifting ash is generally applied to smaller installations. Lateral movement is usually provided by screw conveyors. This method is noisy, can be dusty, and is severe on the elevators and conveyors due to the abrasive nature of ash. The pneumatic method of lifting ash is generally applied to larger plants and where a lot of lateral movement is required, such as in multiple-unit installations. At the point of discharge, a baghouse or other type of air filter must be provided to capture the aerosol dust released when the ash drops into the silo. For ash load out from the silo, the ash is passed through a wetting device that provides enough water to control the dust that would be emitted during handling and transportation.

In wet ash handling, ash slurried with water is pumped to a lagoon for treatment and disposal. Therefore, this method is appropriate only where there is a feasible holding area (lagoon) on or near the plant site. Its principal merit is that it creates little dust. However, abrasion-resistant heavy-walled pipe and rubber-lined pumps are needed to minimize abrasion and wear.

Air Pollution Control Incineration of sludge results in the production of a sterile nonodorous ash, an approximately 90 to 96% volumetric reduction, and a large volume of combustion gases. Accompanying this process is the potential for significant degradation to air quality unless effective control technologies are employed to reduce emissions in the combustion gases.

If properly designed and operated, incineration can provide complete combustion of the organics in wastewater sludge to produce principally carbon dioxide, water, and sulfur dioxide. However, incomplete combustion can produce unacceptable intermediate products such as hydrocarbons, other volatile organics, and carbon monoxide. These products are often referred to as *products of incomplete combustion*. Some of these products can produce offensive odors; therefore, special attention is needed to minimize nuisance odor emissions.

Incineration of sludge has the potential for discharge of excessive particulates. These are the predominant air contaminants from thermal destruction, and they include both solid particles and liquid droplets (excluding uncombined water) that are swept along by the gas stream or formed through condensation of the flue gases. Particulates from incineration of sludge are enriched with volatile trace metals such as cadmium, lead, and zinc. Particles sizes are mostly smaller than 2 μm, and volatile elements are primarily in the submicrometer sizes (WEF, 1998). Technologies for controlling particulates from gas streams include (1) mechanical collectors, (2) wet scrubbers, (3) fabric filters, and (4) electrostatic precipitators. Selection of a particular collection system depends on the nature of the particulate matter, conditions of the gas streams, and emission limits.

Mechanical collectors exert inertia forces for particle separation. They have relatively low collection efficiency and are generally used as precollectors upstream of main particulate control devices. The three types of mechanical collectors in use include: (1) a settling chamber, which uses the low gas velocity through the chamber [less than 3 m/s (10 ft/s)] to settle the heavy particles; (2) an impingement separator, which directs the gas stream against collecting bodies, where particles lose momentum and drop out of the gas; and (3) a cyclone separator. In a cyclone separator, gas enters tangentially at the top of the cylindrical shell and is forced down in a spiral of decreasing diameter in a conical section. This action lets the particles spiral downward to the bottom through an airlock and lets the gas return back up at the center of the vortex and discharges from the top.

Wet scrubbers employ water to separate dusts or mists from gas streams. They are the most widely used emission control equipment for sludge incineration. They have the added advantage of removing water-soluble contaminants such as hydrogen chloride, sulfur dioxide, and ammonia. The four types of wet scrubbers are (1) spray towers, in which particles are captured by the droplets from a liquid spray at the top of the column against the rising gas stream; (2) cyclone scrubbers, in which centrifugal forces increase the momentum of the collision between the particles and the liquid droplets; (3) ejector-Venturi scrubbers, in which a high-pressure liquid jet scrubs the gas and provides the draft for moving the gas; and (4) Venturi scrubbers, in which the gas stream accelerates across the Venturi or orifice, and a scrubbing liquid is sprayed and mixed with the gas at the throat. The high turbulence causes collisions between liquid droplets and particulates; consequently, particulates are captured.

Fabric filters (baghouses) collect particulates in the gas stream by passing the gas through a filter medium or fabric. When the pressure drop across the filter increases as the dust accumulates, the filter is cleaned by mechanical shaking, pulse jet, or reverse airflow. Fabric filters are highly efficient (greater than 99% particulate removal efficiency); however, they can be used in an incineration system only if the gas temperature is reduced to 150 to 177°C (300 to 350°F).

There are two types of electrostatic precipitators: dry electrostatic precipitators and wet electrostatic precipitators. In a dry electrostatic precipitator, a negative charge is imparted to the particulates in the exhaust gases passing through a large chamber. The negatively charged particulates are then attracted to the positively charged collector plates in the chamber. The collected particles are removed by periodic vibration or rinsing. A wet electrostatic precipitator is similar to a dry electrostatic precipitator except that it contains a washing mechanism to counteract the buildup of volatile or particulate matter on the plates. Electrostatic precipitators can achieve removal efficiencies of 99% or greater with negligible pressure drop.

REFERENCES

Barrett, R., and Herndon, J. (2005), Safety First: Key Safety Considerations When Drying Biosolids to Class A Specifications, *Water Environment and Technology*, Vol. 17, No. 4, p. 35.

Craven, A., et al. (2004), Largo's Beneficial Use of Biosolids, *Florida Water Resources Journal*, April, pp. 42–44, 46, 48.

Currents, a U.S. Filter newsletter (2002), Vol. 1, No. 1, September.

Karmazinov, F. V. (Ed.) (2002), *Wastewater Treatment in Saint Petersburg*, New Magazine, Saint Petersburg, Russia.

Kowalski, D., et al. (2002), Benefits of a Progressive Centralized Biosolids Management System, *Florida Water Resources Journal*, April, pp. 31, 32, 39, 40.

Lide, D. R. (1995–1996), *Handbook of Chemistry and Physics*, 76th ed., CRC Press, Inc, Ann Arbor, MI.

Lue-Hing, C., Zeng, D. R., and Kucherither, R. (Eds.) (1992), *Water Quality Management Library*, Vol. 4, *Municipal Sewage Sludge Management: Processing, Utilization and Disposal*, Technomic Publishing Co., Lancaster, PA.

Metcalf & Eddy, Inc. (2003), *Wastewater Engineering: Treatment and Reuse*, 4th ed., Tchobanoglous, G., Burton, F. L., and Stensel, H. D. (Eds.), McGraw-Hill, New York.

Olson, R. T., and Bohman, R. T. (2001), Class A Biosolids: Simultaneous Dewatering and Pasteurization Using an FKC Screw Press, company brochure, FKC Co., Ltd, Port Angeles, WA.

Outwater, A. B. (1994), *Reuse of Sludge and Minor Wastewater Residuals* CRC Press/Lewis Publishers, Boca Raton, FL.

Pelletier, R. A., et al. (2001), To Lime or Not to Lime, That Is the Question, *Florida Water Resources Journal*, February, pp. 37, 38, 40.

Pennington, R. (2003), *Pollution Equipment News*, June, p. 37.

Samokhin, V. N. (1986), *Design Handbook of Wastewater Systems*, Vol. 2, Allerton Press, New York, pp. 587–606.

Shimp, G., and Childress, B. (2002), Improving the Process, *Civil Engineering*, Vol. 72, No. 9, September, pp. 74–77.

Spellman, F. R. (1997), *Incineration of Biosolids*, Technomic Publishing Co., Lancaster, PA.

The Wave (1998), *U.S. Filter* magazine, Municipal Water and Wastewater, February, Vol. 2, No. 1.

Turovskiy, I. S. (1988), *Wastewater Sludge Treatment*, Stroyizdat, Moscow.

U.S. EPA (1979), *Process Design Manual for Sludge Treatment and Disposal*, EPA 625/1-79/001.

——— (1999), *Biosolids Generation, Use, and Disposal in the United States*, EPA 530/R-99/009.

U.S. Filter, Prospect, Davis Process: Sludge Master, Indirect Rotating Chamber Sludge Dryer.

U.S. Filter, Prospect, Dewatering/Drying Systems.

WEF (1998), *Design of Municipal Wastewater Treatment Plants*, 4th ed., Manual of Practice 8 (ASCE 76), Water Environment Federation, Alexandria, VA.

——— (1992), *Sludge Incineration: Thermal Destruction of Residues*, Manual of Practice FD-19, Water Environment Federation, Alexandria, VA.

9

COMPARISON OF ENERGY CONSUMPTION

9.1 INTRODUCTION

The processes of anaerobic digestion, incineration, composting, and thermal drying of wastewater sludge require costly quantities of heat and electricity. Part of the cost of these can be offset by utilization of the organics in the sludge. In the following sections we compare the energy consumption of the four sludge treatment processes.

9.2 ANAEROBIC DIGESTION

The two types of anaerobic digestion processes are mesophilic and thermophilic. Mesophilic processes occur in the temperature range 32 to 35°C, whereas thermophilic process requires a temperature range of 50 to 55°C. Figure 9.1 shows the technological scheme for both mesophilic and thermophilic digestion in an activated sludge treatment plant.

Let us examine the energy required for each process by assuming the following:

Wastewater Sludge Processing, By Izrail S. Turovskiy and P. K. Mathai

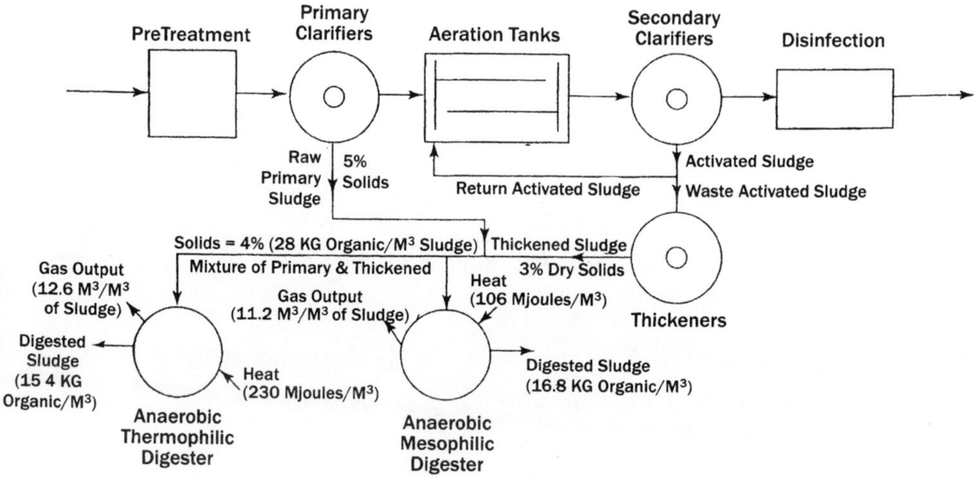

Figure 9.1 Technical scheme of anaerobic digestion.

- The quantity of gas obtained during digestion is approximately 1.0 m^3 for every 1.0 kg of volatile solids (VS) reduced.
- The ratio of raw sludge from primary clarifiers to the thickened waste activated sludge from secondary clarifiers, as calculated by the mass of dry solids, is 1:1.
- The solids content in the primary–secondary sludge mixture is 4%.
- Of the 4% solids, 70% is VS, resulting in 2.8% VS as a percentage of the mass of dry solids (70% × 4%).
- The disintegration of VS in the mesophilic process is 40%, and in the thermophilic process it is 50%.

One cubic meter of the primary–secondary sludge mixture contains 40 kg of dry solids (1.0 m^3 × 1000 kg/m^3 × 4%), and therefore 28 kg (40 kg × 70%) VS. The quantity of gas obtained from the digestion of this 1 m^3 of sludge will be 28 × 40% = 11.2 m^3 for the mesophilic process and 28 × 50% = 14.0 m^3 for the thermophilic process. The gas produced from digestion can be expected to be 60 to 70% methane, 16 to 34% carbon monoxide, and 0.4 to 6% nitrogen, hydrogen, and oxygen. The heat of combustion of the gas will average around 21 million joules/m^3 (MJ/m^3).

For digestion to occur, it is necessary to provide heat as follows:

$$\text{mesophilic process: } (1.1)(5.67)(33-16) = 106\ \text{MJ/m}^3 \text{ of sludge}$$

$$\text{thermophilic process: } (1.1)(5.67)(53-16) = 230\ \text{MJ/m}^3 \text{ of sludge}$$

where

1.1 = coefficient that takes into account heat loss
5.67 = quantity of heat (in MJ) required to raise the temperature of $1.0\,m^3$ of sludge with steam by 1°C
33 and 53 = temperatures (in °C) of mesophilic and thermophilic processes, respectively
16 = temperature (in °C) of the original sludge mixture prior to digestion

The excess heat released by the combustion of gas produced during the digestion of $1\,m^3$ of sludge is as follows:

$$\text{mesophilic process: } (11.2)(21) - 106 = 129\,\text{MJ/m}^3 \text{ of sludge}$$

$$\text{thermophilic process: } (14.0)(21) - 230 = 64\,\text{MJ/m}^3 \text{ of sludge}$$

where

11.2 and 14.0 = quantity of gas (in m^3) from the digestion of $1.0\,m^3$ of sludge in the mesophilic and thermophilic processes, respectively
21 = heat of combustion of $1.0\,m^3$ of gas (in MJ/m^3)
106 and 230 = energy required to heat $1.0\,m^3$ of sludge (in MJ) for the mesophilic and thermophilic processes, respectively (see above)

Since the digestion of sludge organics creates gas, there is a corresponding decrease in the amount of sludge organics remaining following digestion. The 28 kg of sludge organics (volatile solids) will be decreased by 11.2 kg, to 16.8 kg (28 – 11.2) in the mesophilic process. In the thermophilic process, the quantity of sludge organics will be decreased by 14.0 kg, to 14.0 kg (28 – 14.0).

9.3 INCINERATION

Incineration reduces sludge volume dramatically. Prior to incineration, sludge must be dewatered and thermally dried, with drying being the most energy-intensive step. Therefore, when considering the use of incineration, one must look for methods and techniques to reduce the amount of energy required and to provide some, if not all, of the required energy from another sludge process.

Dewatering can be accomplished by any number of mechanical processes, such as belt filter presses, pressure filter presses, and centrifuges. The less moisture in the dewatered sludge, the less the total consumption of energy (see Figure 9.2). Therefore, it is most energy efficient to remove as much moisture as possible prior to drying. The lower moisture content will reduce the energy requirements in the succeeding steps of thermal drying and incineration.

Let us examine the technology of treatment of a mesophilically digested primary and secondary sludge mixture using the thermal drying process. In

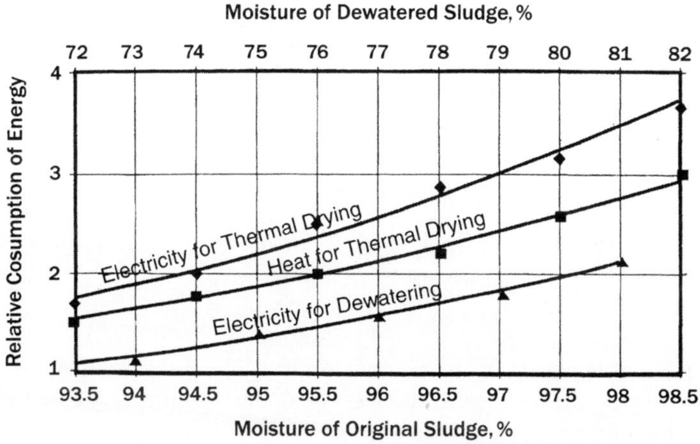

Figure 9.2 Moisture content versus energy consumption for dewatering and drying of sludge.

dryers with opposing jet streams of air, approximately 3.4 to 3.9 MJ of heat and 0.02 to 0.06 kWh of electricity are used for each 1.0 kg of evaporated moisture. Consider a wastewater treatment plant with a capacity of 100,000 m^3/d (26 mgd). The quantity of mesophilically digested primary–secondary sludge mixture with a solids content of 4% (96% moisture) is about 800 m^3/d. Assume that this 4% solid sludge is dewatered mechanically to a solids content of 18 to 24%, and compare the amount of energy required to dry the sludge with 24% solids versus one with 18% solids to a sludge with 60% solids.

The volume of dewatered sludge is calculated using the formula

$$V_2 = V_1\left(\frac{C_1}{C_2}\right)$$

where

V_2 = volume of dewatered sludge, m^3
V_1 = volume of sludge prior to dewatering, m^3 (800 m^3)
C_1 = dry solids concentration of sludge prior to dewatering, % (4%)
C_2 = dry solids concentration of dewatered sludge, %

$$\text{volume of dewatered sludge with 24\% solids} = (800\,\text{m}^3/\text{d})\left(\frac{4\%}{24\%}\right)$$
$$= 133\,\text{m}^3/\text{d}$$

$$\text{volume of thermally dried sludge with 60\% solids} = (800\,\text{m}^3/\text{d})\left(\frac{4\%}{60\%}\right)$$
$$= 53\,\text{m}^3/\text{d}$$

Thermal drying of dewatered sludge requires evaporation of 133 − 53 = 80 m^3/d of moisture. The energy required to evaporate the moisture is as follows:

$$\text{heat} = (80\,\text{m}^3/\text{d})(1000\,\text{kg}/\text{m}^3)(3.9\,\text{MJ}/\text{kg}) = 312{,}000\,\text{MJ}/\text{d}$$

$$\text{electricity} = (80\,\text{m}^3/\text{d})(1000\,\text{kg}/\text{m}^3)(0.03\,\text{kWh}/\text{kg}) = 2400\,\text{kWh}/\text{d}$$

$$\text{volume of dewatered sludge with 18\% solids} = (800\,\text{m}^3/\text{d})\left(\frac{4\%}{18\%}\right) = 178\,\text{m}^3$$

Therefore, it is necessary to evaporate 178 − 53 = 125 m^3/d of moisture for thermally drying the 18% solid sludge. The energy required to evaporate this moisture is as follows:

$$\text{heat} = (125\,\text{m}^3/\text{d})(1000\,\text{kg}/\text{m}^3)(3.9\,\text{MJ}/\text{kg}) = 487{,}500\,\text{MJ}/\text{d}$$

$$\text{electricity} = (125\,\text{m}^3/\text{d})(1000\,\text{kg}/\text{m}^3)(0.03\,\text{kWh}/\text{kg}) = 3750\,\text{kWh}/\text{d}$$

This is almost 1.6 times more heat and electricity than that required to dry a sludge that has 24% solids. Utilization of the methane gas generated by the mesophilic digestion process will reduce the required quantity of heat for thermal drying by (129 MJ/m^3)(800 m^3/d) = 103,200 MJ/d.

The heat value of municipal wastewater sludge (Q_b) in incineration normally ranges from 23.4 to 26.9 MJ/kg. This heat is obtained from the organics that are typically 65 to 72% of the sludge solids. The heat value is higher for raw sludge from primary clarifiers than for raw activated sludge from secondary clarifiers and digested sludge. When 1.0 m^3 of a 1 : 1 primary/secondary sludge mixture that contains 40 kg of dry solids (or 28 kg of organics) with a Q_b value of 25.5 MJ/kg is incinerated, approximately (28 kg)(25.5 MJ/kg) = 714 MJ of heat is obtained. Similarly, when 1.0 m^3 of mesophilically digested sludge with a Q_b value of 23.5 is incinerated, (16.8 kg)(23.5 MJ/kg) = 395 MJ of heat is obtained. For thermophylically digested sludge, the heat obtained is (14.0 kg)(23.5 MJ/kg) = 329 MJ.

The total energy from anaerobically digested sludge includes the energy from the methane gas produced from digestion plus the energy obtained from the combustion of the remaining sludge organics. For the mesophilically digested sludge, the total energy obtained is 129 + 395 = 524 MJ, and for thermophylically digested sludge it is 64 + 329 = 393 MJ. It follows from these calculations that it is reasonable to incinerate the raw sludge because of its higher heat value (714 MJ versus 524 or 393 MJ).

Combining the thermal drying process with the incineration of sludge may significantly reduce the energy expenditures for thermal drying. As pointed out before, a significant quantity of heat may be obtained by the incineration of sludge. However, most of the heat is spent for moisture evaporation and

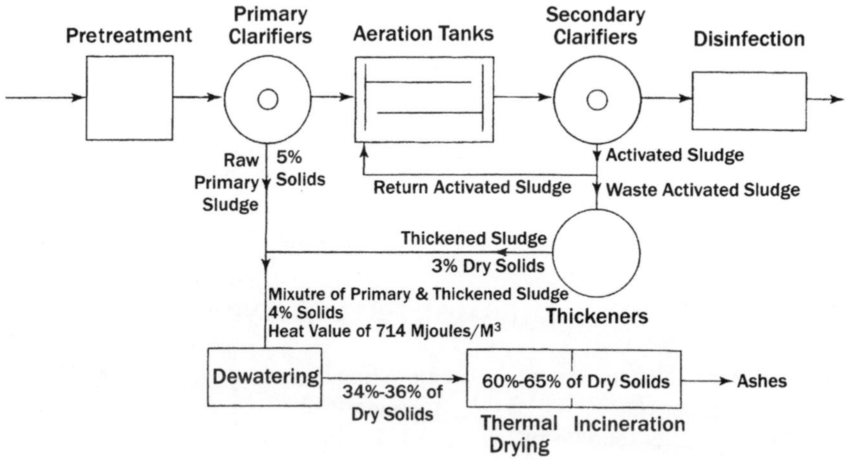

Figure 9.3 Schematic of autothermic incineration of sludge.

heating of the blast air, and there are some system loses. Therefore, incineration may cover only a part of the heat that is necessary for the thermal drying of sludge.

Incineration of a mechanically dewatered primary–secondary sludge mixture in an autothermic process (conducting the process of thermal drying and incineration without additional consumption of fuel) may be achieved when the moisture of the dewatered sludge mixture is 64 to 66%. An increase in the moisture content of the dewatered sludge requires using the appropriate quantity of energy for evaporation of moisture. An illustration of autothermic incineration of sludge is shown in Figure 9.3. It becomes reasonable to use incineration when toxic substances in the sludge prevent its use as a fertilizer or in a municipal landfill.

9.4 COMPOSTING

In sludge composting, a biothermal process takes place in which microorganisms reduce the sludge organics. This aerobic process is accompanied by a rise in temperature to 50 to 72°C and a decrease in the moisture content of the sludge. The quantity of sludge organics reduced during composting averages 25%. The reduction of 1.0kg of sludge organic creates an average 21MJ of heat. Taking into account the heat losses and the heating of compost material, it is necessary to spend approximately 4MJ of heat for the evaporation of 1.0kg of water. Thus, a reduction of 1.0kg of sludge organic allows the removal from the sludge of 5.0kg of water (21MJ/4MJ per 1.0kg of water).

Some moisture is removed from the sludge by natural evaporation. The total quantity of moisture removed from the sludge depends on such factors as climate, season of the year, dimensions of piles, duration of composting, and periods of turning over of piles. The moisture content of composted sludge is approximately 50%. Utilization of sludge organics as a plant fertilizer may bring greater economical benefit than their use as a heat source.

9.5 COMPARISON OF THERMAL DRYING AND COMPOSTING

In recent years, thermal drying and composting have become useful technologies for the preparation of class A biosolids. Let us examine three sludge-processing schemes, shown in Figure 9.4, to compare the heat consumption of these two technologies. Let the quantity of raw primary and thickened activated sludge be $800\,m^3/d$ with 4% dry solids.

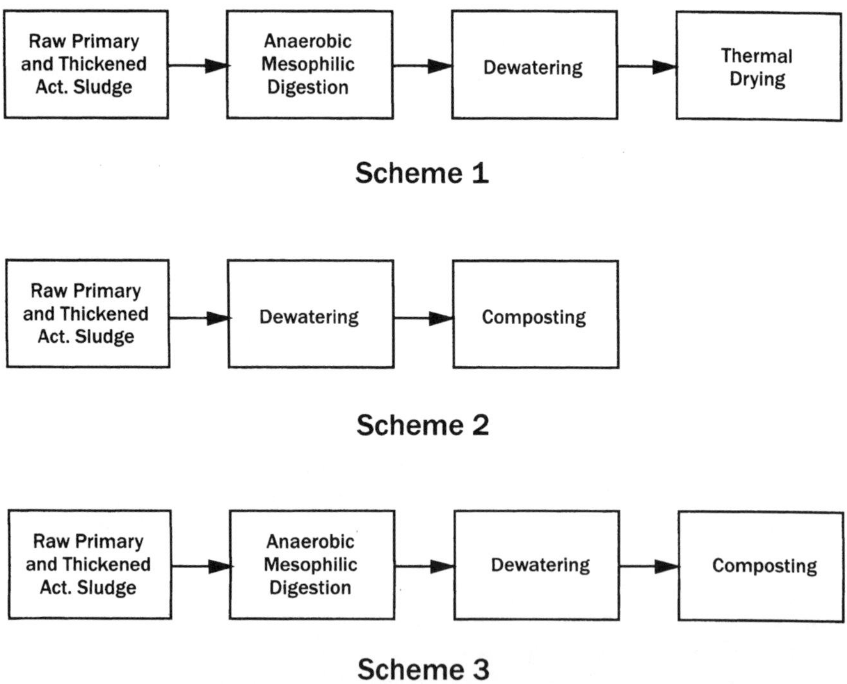

Figure 9.4 Sludge processing schemes.

Scheme 1

$$\text{heat necessary for sludge digestion} = (800\,\text{m}^3/\text{d})(106\,\text{MJ/m}^3) = 84{,}800\,\text{MJ/d}$$

$$\text{heat obtained by sludge digestion} = (800\,\text{m}^3/\text{d})(235\,\text{MJ/m}^3) = 188{,}000\,\text{MJ/d}$$

$$\text{excess heat from sludge digestion} = (800\,\text{m}^3/\text{d})[(235-106)\,\text{MJ/m}^3] = 103{,}200\,\text{MJ/d}$$

$$\text{volume of dewatered sludge with 24\% solides} = (800\,\text{m}^3/\text{d})\left(\frac{4\%}{24\%}\right) = 133\,\text{m}^3/\text{d}$$

$$\text{volume of thermally dried sludge with 90\% solids} = (800\,\text{m}^3/\text{d})\left(\frac{4\%}{90\%}\right) = 35.6\,\text{m}^3/\text{d}$$

$$\text{moisture to be evaporated for drying} = (133-35.6)\,\text{m}^3/\text{d} = 97.4\,\text{m}^3/\text{d}$$

$$\text{heat required to evaporated moisture} = (97.4\,\text{m}^3/\text{d})(1000\,\text{kg/m}^3)(3.9\,\text{MJ/kg}) = 379{,}860\,\text{MJ/d}$$

Utilization of the digester's excess heat (103,200 MJ/d) will reduce the energy consumption for thermal drying.

A heat exchanger following the thermal dryer can recover the heat required for sludge digestion, which is 84,800 MJ/d. Therefore,

$$\begin{aligned}&\text{net heat required for thermal drying of sludge}\\&= (379{,}860 - 103{,}200 + 84{,}800)\,\text{MJ/d}\\&= 361{,}460\,\text{MJ/d}\end{aligned}$$

The process allows the reduction of 188,000 MJ/d of heat.

Scheme 2 The quantity of organics reduced during composting process averages about 25%, but for digested sludge, with the reduced quantity of organics, it can be just 10%. For primary sludge, the reduction is more than 45%, and for activated sludge it is about 30 to 35%. Let us assume that the reduction of organics during composting of the mixture of raw primary and activated sludge is 35%. Then the quantity of reduced organics is $(28\,kg/m^3)(35\%) = 9.8\,kg/m^3$, which is $(800\,m^3/d)(9.8\,kg/m^3) = 7840\,kg/d$.

$$\text{heat released from composting} = (7840\,kg/d)(21\,MJ/kg) = 164{,}640\,MJ/d$$

Reduction of 1 kg of organics results in the removal of 5 kg of moisture. Therefore,

$$\text{moisture removed} = (7840\,kg/d)(5\,kg/kg) = 39{,}200\,kg/d = 39.2\,m^3/d$$

Scheme 3 The quantity of organics reduced during composting of digested sludge is 10%, which is $(28\,kg/m^3)(40\%)(10\%) = 1.1\,kg/m^3$, where 40% is the organics reduced in digestion.

$$\text{organics reduced in composting} = (800\,m^3/d)(1.1\,kg/m^3) = 880\,kg/d$$

$$\text{heat released during composting} = (880\,kg/d)(21\,MJ/kg) = 18{,}480\,MJ/d$$

$$\text{excess heat from sludge digestion} = 103{,}200\,MJ/d \quad \text{(see Scheme 1)}$$

$$\text{total heat obtained} = (18{,}480 + 103{,}200)\,MJ/d = 121{,}680\,MJ/d$$

$$\text{mositure evaporated from sludge} = (880\,kg/d)(5\,kg/kg) + \frac{103{,}200\,MJ/d}{3.9\,MJ/kg} = 30{,}900\,kg/d\ (30.9\,m^3/d)$$

Table 9.1 compares energy use in the three process schemes.

TABLE 9.1 Comparison of Energy Use (MJ/d)

Parameter	Scheme 1 (Thermal Drying)	Scheme 2 (Raw Sludge Composting)	Scheme 3 (Digested Sludge Composting)
Heat required for process	361,460	—	—
Heat reduced	188,000	—	—
Heat obtained	—	164,640	121,680

9.6 CONCLUSION

The calculation of expenses for the heat and electricity for the complete cycle of treatment and utilization of sludge should take into account the quantity of energy that may be obtained from sludge. Utilization of that energy can significantly reduce expenses for sludge treatment.

A comparison of sludge treatment processes shows that in terms of the energy required, composting is the most economical method. It is reasonable to prepare compost from dewatered raw sludge because it contains more organics than does digested sludge. At the same time, the scheme of sludge treatment should provide the possibility of the maximum reduction of moisture with the minimum expenditures.

REFERENCES

Turovskiy, I. S. (2001), Reduction of Energy Consumption in Wastewater Sludge Treatment, *Florida Water Resources Journal*, March, pp. 34–36.

———, and Westbrook, J. D. (2002), Recent Advances in Wastewater Sludge Composting, *Water Engineering and Management*, October, pp. 29–32.

———, Goldfarb, L. L., Pavlovskiy, L. L., and Minz, M. S. (1972), Method of Sludge Incineration, USSR patent 361983.

———, Goldfarb, L. L., Zamoschin, L. V., and Zhukov, Z. A. (1978), Method of Effluent Sludge Treatment, U.S. patent 4,125,465.

10

BENEFICIAL USE OF BIOSOLIDS

Wastewater Sludge Processing, By Izrail S. Turovskiy and P. K. Mathai

10.1 INTRODUCTION

The beneficial qualities of biosolids as a soil amendment are generally recognized. When added to soil, biosolids contribute nutrients and improve soil properties. Depending on agricultural needs, these benefits are even greater with composted biosolids, which enhance the physical, chemical, and biological properties of soil. Noncomposted biosolids have a high nutrient availability and decompose and mineralize quickly and easily in soil. This rapid decomposition can provide large amounts of nitrogen and phosphorus for immediate use by crops. Composted biosolids, on the other hand, retain highly stable organic materials that decompose at a slow rate, therefore releasing nutrients at a slower and steadier rate than do noncomposted biosolids. Composted biosolids thus provide a long-term source of slow-release nutrients.

Compost is just one form of biosolids that can be used beneficially. The combined potential of using either composted or noncomposted biosolids is great. By addressing the environmental and public health issues related to biosolids, the Part 503 rule has greatly encouraged land application of many of the other types of biosolids, including advanced alkaline stabilized, heat-treated, and pelletized biosolids, as well as less highly processed liquid biosolids and biosolids cake. The biosolids may be considered as lime fertilizer when lime is used as a reagent for sludge processing. In addition to nitrogen and phosphorus, biosolids contain minerals and micronutrients necessary for plant growth. The minerals contributed by biosolids include calcium, silica, aluminum, and iron. Micronutrients increase the speed of metabolism during plant growth, and their deficiency brings a breach in plant growth. One of the micronutrients in biosolids, copper, increases the harvest of wheat on marsh and sandy soils. Manganese contributes to an increase in sugar, beet, and corn harvests. Deficiencies in iron and zinc bring about a serious breach in the growth of cotton plants and fruit plants such as grape vines. Boron is vital for crops such as flax, sugar beets, cotton plants, feed legumes, peas, clover, alfalfa, and several other fruits, berries, and vegetables. Micronutrients also contribute to the assimilation of organics in biosolids by plants. However, excess concentrations of micronutrients such as boron, molybdenum, nickel, selenium, copper, and zinc exercise a negative influence on the growth and quality of plants.

It has been reported (U.S. EPA, 1999) that approximately 6.9 million dry tons of wastewater sludge solids were generated in 1998, of which 60% were used beneficially (e.g., land applied, composted, used as landfill cover), and 40% disposed of with no attempt to recover nutrients or other valuables. The U.S. EPA has projected that sludge solids generated will increase to 8.2 million dry tons in 2010. It is also projected that the percentage of beneficial use (rather than nonbeneficial disposal) will grow from 60% in 1998 to 70% in 2010. Figure 10.1 shows projections and trends in biosolids use and disposal methods. (Remember that sludge becomes a biosolid when it is treated to

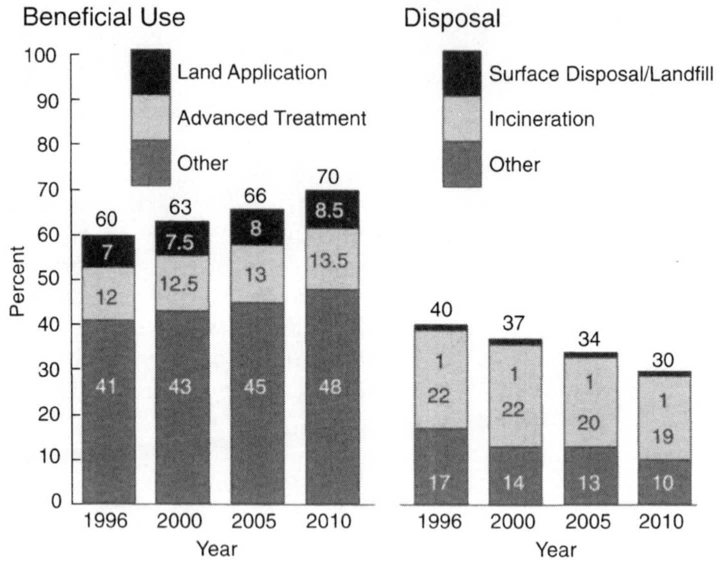

Figure 10.1 Beneficial use and disposal of biosolids. (From U.S. EPA, 1999.)

achieve compliance with federal regulations.) The expected increase in the use of biosolids is due to improvements in the quality of biosolids, regulatory influences, cost considerations, and public perception issues associated with disposal.

Over the past 25 years, industrial pretreatment and pollution prevention programs have substantially reduced levels of metals and other pollutants going into municipal wastewater treatment plants, which in turn have improved the quality of the sludge produced. The marked improvements in sludge quality have encouraged treatment plants to process their sludge further, such as composting or thermally drying it. When biosolids achieve the low levels of pollutants that make the widest distribution of biosolids products possible, processes such as composting and thermal drying become more attractive.

Regulatory influences on biosolids use include the indirect effects from the Clean Water Act of 1977 and 1987 and the Ocean Dumping Ban Act of 1988, and the more direct effects of the 40 CFR Part 503 rule. Part 503, codified in 1993, clearly defines biosolids requirements for use or disposal and has become a useful tool for biosolids managers in marketing efforts. The Part 503 rule helps biosolids managers identify *exceptional quality biosolids* (i.e., biosolids that meet the most stringent metals limits and class A pathogen and vector attraction reduction requirements). Exceptional quality biosolids meet the same regulations as those for any other fertilizer product. Therefore, the rule provides a useful public relations tool that has opened the door for greater use of biosolids as a fertilizer and soil conditioner and has also expanded the potential markets for products such as compost made from biosolids.

The increase in biosolids use associated with implementation of the Part 503 rule is measured according to a study performed by *BioCycle* magazine. This study notes that 37 states have regulations in place that are the same or more stringent than Part 503 (Henry and Brown, 1997). Thirty-four states regulate exceptional quality biosolids in the same way that they do fertilizers. Both of these regulatory conditions provide incentives for biosolids recovery and reuse by many communities. Composting programs on the community level, such as in Portland, Oregon; Palm Beach County, Florida; Hampton, New Hampshire; and St. Peters, Missouri, have lowered the cost of composting wastewater sludge by combining sludge with yard trimmings in their composting operations.

The Part 503 biosolids rule, Part 258 landfill rule, and various state requirements have driven up the cost of sludge disposal. Part 503 requires relatively expensive pollution control equipment and management practices, such as groundwater monitoring, thus encouraging biosolids recycling rather than disposal. Furthermore, public concern about the environmental and health impacts of incineration has made this disposal option even more costly and difficult to undertake. Any increased costs of biosolids disposal are expected to promote beneficial use.

Another factor influencing biosolids beneficial use is public perception. The ability to market biosolids will be determined by the public's perception of the safety and value of biosolids recovery. With the evolution of regulatory standards for biosolids, marketing and promotion of biosolids or products made from biosolids have improved public acceptance of the beneficial use of biosolids.

10.2 REQUIREMENTS FOR BENEFICIAL USE

Under the authorities of the Clean Water Act as amended, the U.S. EPA promulgated in 1993 at 40 Code of Federal Regulation (CFR) Part 503, *Standards for the Use or Disposal of Sewage Sludge*. The intent of this regulation (commonly known as the *Part 503 rule*) is to ensure that sewage sludge is used or disposed of in a way that protects human health and the environment. The regulation is described in detail in Chapter 1, and the relevant portions relating to beneficial use are summarized below.

10.2.1 Pollutant Limits

The first of the three parameters that must be assessed to determine the overall quality of biosolids is the level of pollutants (metals), the other two being two levels of quality with respect to pathogen densities (classes A and B), and vector attraction reduction. To allow land application of biosolids of variable quality, Part 503 provides four sets of pollutant limits: pollutant ceiling concentration limits, pollutant concentration limits, cumulative pollutant loading rates, and annual pollutant loading rates. These pollutant limits are listed in Table 10.1. All limits in the table are on a dry weight basis.

The *ceiling concentration limits* have to be met by all biosolids that are land applied and by both bulk biosolids and biosolids sold or given away in a bag or other container. These limits are absolute values, which means that all samples of biosolids have to meet the limits.

Biosolids meeting the *pollutant concentration limits* (also known as *exceptional quality pollutant concentration limits*) achieve one of the three levels of quality necessary for *exceptional quality biosolids* status, hence can be sold or given away in bags or other containers. These limits must also be met when bulk biosolids are land applied.

The *cumulative pollutant loading rates* apply to bulk biosolids that meet ceiling concentration limits but do not meet pollutant concentration limits. These rates limit the amount of a pollutant that can be applied to an area of land in bulk biosolids for the life of the application site.

The *annual pollutant loading rates* apply to biosolids that meet ceiling concentration limits but do not meet pollutant concentration limits and are to be sold or given away in a bag or other container—commonly, to homeowners.

10.2.2 Pathogen Reduction

The second parameter in determining biosolids quality is the presence or absence of pathogens such as *Salmonella* bacteria, enteric viruses, and viable helminth ova. Biosolids are classified as class A or B based on the level of pathogen present. Biosolids meet class A designation if pathogens are practically below detectable levels. All biosolids that are sold or given away in a bag or other container for application to land, lawns, or home gardens must meet class A status. Part 503 lists six alternatives for treating sludge to meet class A requirements. These alternatives are summarized in Table 1.6. One of the alternatives is to treat sludge in one of the processes to further reduce pathogens (PFRP), which are described in Table 1.8.

Biosolids are designated class B if pathogens have been reduced to levels that do not pose a threat to public health and the environment as long as actions are taken to prevent exposure to the biosolids after their use or disposal. Alternatives for meeting class B pathogen requirements, including processes to significantly reduce pathogens (PSRP), are described in Tables 1.9 and 1.10.

10.2.3 Vector Attraction Reduction

Attractiveness of biosolids to vectors, which are animals and insects such as rodents, flies, and birds, is the third parameter of biosolids quality. Part 503

TABLE 10.1 Land Application Pollutant Limits

Pollutant	Ceiling Concentration Limits[a,b]		Pollutant Concentration Limits[a,c,d]		Cumulative Pollutant Loading Rates		Annual Pollutant Loading Rates	
	mg/kg	lb/ton	mg/kg	lb/ton	kg/ha	lb/ac	kg/ha · yr	lb/ac-yr
Arsenic	75	0.15	41	0.08	41	37	2.0	1.78
Cadmium	85	0.17	39	0.08	39	35	1.9	1.70
Chromium	3000	6.00	1200	2.40	3000	2680	150	133.90
Copper	4300	8.60	1500	3.00	1500	1338	75	66.91
Lead	840	1.68	300	0.60	300	268	15	13.38
Mercury	57	0.11	17	0.03	17	15	0.85	0.76
Molybdenum	75	0.15	—	—	—	—	—	—
Nickel	420	0.84	420	0.84	420	374	21	18.74
Selenium	100	0.20	36	0.07	100	89	5	4.46
Zinc	7500	15.00	2800	15.00	2800	2498	140	124.91
Applies to:	All biosolids that are land applied		Bulk biosolids and biosolids sold or given away in a bag or other container		Bulk biosolids		Biosolids sold or given away in a bag or other container	

[a] Dry weight basis.
[b] Absolute values.
[c] Monthly averages.
[d] Exceptional quality biosolids.

allows 10 options to reduce vector attractiveness, which are described in Table 1.11. (Options 11 and 12 in the table are vector attraction reduction options for surface disposal of biosolids and treatment of domestic septage, respectively.) Options in Table 1.11 include some stabilization processes that reduce pathogens; therefore, vector attraction reduction alternatives can be combined with pathogen reduction alternatives for an acceptable land application project using class B biosolids.

10.2.4 Management Practices

Four management practices are specified for bulk *nonexceptional quality biosolids* that are applied to the land:

- Application of bulk non-EQ biosolids is prohibited if it is likely to affect threatened or endangered species or their designated critical habitat.
- Bulk non-EQ biosolids must not be applied to flooded, frozen, or snow-covered land, which allows the material to enter wetlands or other waters of the United States unless authorized by the permitting authority.
- Application of bulk non-EQ biosolids on agricultural land, forest, or reclamation site that is within 10m of any waters of the United States is prohibited unless authorized by the permitting authority.
- Bulk non-EQ biosolids should be applied to a site only at a rate equal to or less than the agronomic rate, which is the application rate designed to provide the amount of nitrogen needed by the crop or vegetation.

For biosolids sold or given away in a bag or other container, the management practice that applies is that a label or information sheet must be provided with the biosolids, indicating the appropriate application rate for the quality of the biosolids.

10.2.5 Surface Disposal

Surface disposal is placing biosolids on an area of land for final disposal. Examples of surface disposal practices include:

- Sludge-only landfills (monofills)
- Sludge piles or mounds
- Sludge lagoon used for final disposal
- Surface application sites where biosolids are applied at rates in excess of the agronomic rate

Surface disposal of biosolids differs from land application in that it principally uses the land for final disposal instead of using the biosolids to enhance the productivity of the land.

The Part 503 rule allows for storage of biosolids up to two years without any restrictions or control. However, biosolids remaining on the land for longer than two years is considered final disposal.

If a sludge disposal site is equipped with a liner and a leachate collection system, there are no pollutant limits because the liner retards the movement of pollutants in sludge into the groundwater. Sludge placed in an active unit without a liner and a leachate collection system must meet pollutant limits for arsenic, chromium, and nickel. The limits are listed in Table 1.4.

Disposal of sewage sludge in a municipal solids waste landfill (MSWLF) is not regulated by Part 503 other than requiring compliance with 40 CFR Part 258. To meet Part 258 requirements, the preparer must ensure that the sludge is nonhazardous and does not contain free liquid as defined by the paint filter test. If sludge is used as an alternative cover material, it may have to be treated for vector attraction reduction prior to its use.

10.3 LAND APPLICATION

Land application of biosolids refers to the spreading of biosolids on or just below the soil surface for beneficial use. The practice includes applying to:

- Agricultural land, such as fields used for the production of food, feed, and fiber crops
- Pasture- and rangeland
- Public contact sites, such as parks and golf courses
- Nonagricultural land, such as forests
- Disturbed lands, such as mine spoils, construction sites, and gravel pits
- Superfund sites for reclamation (Henry and Brown, 1997)
- Dedicated land disposal sites
- Home lawns and gardens

Biosolids are applied to the land at agronomic rates, that is, rates designed to provide the amount of such macronutrients as nitrogen, phosphorus, and potassium needed by the crop or vegetation grown on the land while minimizing the amount that passes below the root zone. Land-applied biosolids are subjected to further destruction of pathogens and destruction of many toxic organic substances from the combined effect of sunlight, soil microorganisms, and desiccation.

To qualify for land application, biosolids or materials derived from biosolids must meet at least the pollutant ceiling concentrations, class B pathogen requirements, and vector attraction reduction requirements. Biosolids that are sold or given away in bags or other containers for application to lawns and home gardens must meet the exceptional quality pollution concentration limits, class A pathogen requirements, and vector attraction reduction requirements. Because of the continuing concern with wastewater sludge disposal, many farmers accept only class A sludge, forcing wastewater treatment plant operators to modify their sludge treatment schemes (Vesilind, 2003).

10.3.1 Site Evaluation and Selection

A critical step in land application of biosolids is finding a suitable site. The suitability of a site considered depends on the land application options being considered. Site characteristics of importance are topography, soil characteristics, soil depth to groundwater, and accessibility to critical areas. For the initial screening of potential sites, areas with following characteristics should be avoided (WEF, 1998):

- Steep areas with sharp relief
- Shallow soils
- Environmentally sensitive areas such as intermittent streams and ponds
- Rocky, nonarable land
- Wetland and marshes
- Areas bounded by surface water bodies without appropriate set backs
- Culturally sensitive lands, such as old cemeteries and burial grounds

Topography is important, as it affects the potential for erosion and runoff. Slopes up to 30% can be accommodated, depending on the types of biosolids and the method of application. Table 10.2 shows the slope limitations for land application of biosolids. Loamy soil of slow to moderate permeability with alkaline or neutral pH and a soil depth of 0.6 m (2 ft) or more is preferred. Because groundwater fluctuates on a seasonal basis in many areas, an acceptable minimum depth to groundwater is difficult to establish. At least 1 m (3 ft) to groundwater is preferred.

Ideal sites for land application of biosolids have deep silty loam to sandy loam soils; groundwater deeper than 3 m (10 ft); slopes at 0 to 3%; no wells, wetlands, or streams; and few neighbors (Metcalf & Eddy, 2003). The location of the treatment plant also determines whether suitable sites are available within a feasible hauling distance, a decision typically based on economic considerations. Agricultural practices and inclement weather in the area also have a considerable effect on site availability throughout the year. Because sites may be available only at certain times, substantially more acreage than is needed in a given year should be permitted to receive biosolids. Facilities for storing the biosolids for later use should also be considered.

10.3.2 Design Application Rates

Application rates for biosolids can be limited by nitrogen or by pollutants (heavy metals listed in Table 10.1). For most biosolids that are land applied

TABLE 10.2 Typical Slope Limitation for Land Application

Slope (%)	Comment
0–3	Ideal, no concern for runoff or erosion of liquid or dewatered biosolids.
3–6	Acceptable, slight risk of erosion; surface application of liquid or dewatered biosolids is acceptable.
6–12	Injection of liquid biosolids required for general cases, except in closed drainage basin and/or when extensive runoff control is provided; surface application of dewatered biosolids is generally acceptable.
12–15	No application of liquid biosolids should be made without extensive runoff control; surface application of dewatered biosolids is acceptable, but immediate incorporation into the soil is recommeded.
>15	Suitable only for sites with good permeability where the length of slope is short and where areas with a steep slope constitute only a minor part of the total application area.

Source: Metcalf & Eddy, 2003.

in the United States, the nitrogen content is the limiting factor. Thus, the design application rate for agricultural use is determined by the crops to be grown and the nitrogen uptake values for those crops. Table 10.3 lists the nitrogen uptake rates for selected crops. The table also includes phosphorus and potassium uptake rates for forage and field crops.

Application Rates Based on Nitrogen Regular nutrient analyses of biosolids provide the database for the nitrogen-based application rate for a particular source of biosolids. Biosolids application rates are typically calculated to supply the available nitrogen provided by commercial fertilizers. Plant-available nitrogen in biosolids can be estimated using the equation

$$N_{PA} = [(NH_4^+)k_v + NO_3 + (N_o)f_m]F \qquad (10.1)$$

where

N_{PA} = plant available nitrogen in the application year, g/dry kg (lb/dry ton)

NH_4^+ = ammonia nitrogen in biosolids, %

k_v = volatilization factor for ammonia

= 0.5 for surface-applied liquid biosolids

= 0.75 for surface-applied dewatered biosolids

= 1.0 for injected liquid or dewatered biosolids

NO_3 = nitrate nitrogen in biosolids, %

N_o = organic nitrogen in biosolids, %

TABLE 10.3 Nutrient Uptake Rates for Selected Crops

	Nitrogen		Phosphorus		Potassium	
Crop	kg/ha · yr	lb/ac-yr	kg/ha · yr	lb/ac-yr	kg/ha · yr	lb/ac-yr
Forage crops						
Alfafa[a]	225–540	200–480	22–35	20–30	175–225	155–200
Bromegrass	130–225	115–200	40–55	35–50	245	220
Coastal bermudagrass	400–675	355–600	35–45	30–40	225	200
Kentucky bluegrass	200–270	180–240	45	40	200	180
Quackgrass	235–280	210–250	30–45	25–40	275	245
Reed canarygrass	335–450	300–400	40–45	35–40	315	280
Ryegrass	200–280	180–250	60–85	55–75	270–325	240–290
Sweet clover[a]	175	155	20	18	100	90
Tall fescue	150–325	135–290	30	25	300	270
Orchardgrass	250–350	225–310	20–50	18–45	225–315	200–280
Field crops						
Barley	125	110	15	13	20	18
Corn	175–200	155–180	20–30	18–25	110	100
Cotton	75–110	65–100	15	13	40	35
Grain sorghum	135	120	15	13	70	60
Potatoes	230	205	20	18	245–325	220–290
Soybeans[a]	250	225	10–20	9–18	30–55	25–50
Wheat	160	145	15	13	20–45	18–40
Trees						
Eastern forests						
Mixed hardwoods	220	195				
Red pine	110	100				
White spruce	280	250				
Pioneer succession	280	250				
Southern forests						
Mixed hardwoods	340	305				
Southern pine	220–320	195–285				
Lake state forests						
Mixed hardwoods	110	100				
Hybrid poplar	155	140				
Western forest						
Hybrid poplar	300–400	270–355				
Douglas fir	150–250	135–225				

Source: U.S. EPA, 1981.

[a] Legumes will also take nitrogen from the atmosphere.

f_m = mineralization factor for organic nitrogen
= 0.5 for warm climates and digested biosolids
= 0.4 for cool climates and digested biosolids
= 0.3 for cold climates and composted biosolids

F = conversion factor, 1000 g/kg (2000 lb/ton) of dry solids

The biosolids application rate based on available nitrogen is then calculated from the equation

$$A_{\mathrm{N}} = \frac{\mathrm{N}_U}{\mathrm{N}_{\mathrm{PA}}} \tag{10.2}$$

where

A_{N} = biosolids application rate based on nitrogen, dry metric tons/ha·yr (tons/ac-yr)

N_U = crop uptake of nitrogen, kg/ha·yr (lb/ac-yr)

N_{PA} = plant available nitrogen in biosolids, g/kg (2000 lb/ton)

Application Rates Based on Pollutants The pollutants (heavy metal) of concern are listed in Table 10.1. The biosolids application rates based on pollutant loadings can be calculated using the equation

$$A_P = \frac{L_P}{C_P F} \tag{10.3}$$

where

A_P = maximum amount of biosolids that can be applied, metric tons/ha·yr (tons/ac-yr)

L_P = maximum amount of pollutant that can be applied, metric tons/ha·yr (tons/ac-yr)

C_P = pollutant concentration in biosolids, mg/kg

F = conversion factor, 0.001 kg/metric ton (2000×10^{-6} lb/ton)

Land Requirements By comparing the values calculated using equations (10.2) and (10.3), the minimum biosolids application rate is determined, and then the field area required can be calculated using the equation

$$A = \frac{B}{L} \tag{10.4}$$

where

A = application area required, ha (ac)

B = biosolids production, metric tons/yr (tons/yr) of dry solids

L = nitrogen or pollutant application rate metric tons/ha·yr (tons/ac-yr) of dry solids

Design Example 10.1 A wastewater treatment plant produces 4 dry metric tons per day (4.4 tons/d) of dewatered digested biosolids. Following are the average concentrations of nitrogen and pollutants in the biosolids:

ammonia nitrogen :	0.5%
nitrate nitrogen :	0.02%
organic nitrogen :	3%
arsenic (As) :	35 mg/kg
cadmium (Cd) :	54 mg/kg
copper (Cu) :	2200 mg/kg
lead (Pb) :	700 mg/kg
mercury (Hg) :	15 mg/kg
nickel (Ni) :	400 mg/kg
selenium (Se) :	15 mg/kg
zinc (Zn) :	2500 mg/kg

Determine the application area required based on the nitrogen uptake rate for ryegrass and the pollutant loading rates when the biosolids are applied on the surface of the land.

1. Plant-available nitrogen in biosolids: From equation (10.1),

$$\begin{aligned} \mathrm{N_{PA}} &= (0.005)(0.75) + 0.0002 + (0.03)(0.4)(1000\,\mathrm{g/kg}) \\ &= 16\,\mathrm{g/kg}\,(32\,\mathrm{lb/ton}) \end{aligned}$$

2. Biosolids application rate based on nitrogen: From Table 10.3, the average nitrogen uptake rate for ryegrass is 230 kg/ha·yr (205 lb/ac-yr). From equation (10.2),

$$A_{\mathrm{N}} = \frac{230\,\mathrm{kg/ha \cdot yr}}{16\,\mathrm{g/kg}} = 14.4 \text{ dry metric tons/ha} \cdot \text{yr (6.4 dry tons/ac-yr)}$$

3. Land area required based on nitrogen application:

$$\begin{aligned} \text{yearly biosolids production} &= (4 \text{ metric tons/d})(365\,\text{d/yr}) \\ &= 1460 \text{ metric tons/yr (1610 tons/yr)} \end{aligned}$$

From equation (10.4),

$$A = \frac{1460 \text{ metric tons/yr}}{14.4 \text{ metric tons/ha} \cdot \text{yr}} = 101\,\text{ha (250 ac)}$$

Note: The initial nitrogen content of the soil is assumed to be zero. After the initial application, the mineralization rate of organic nitrogen will decrease in subsequent years; therefore, the plant-available nitrogen will also decrease, thus requiring a larger area for biosolids application.

4. Suitability of biosolids based on pollutant concentrations: Compare the concentrations of metals in the biosolids to the ceiling concentration and the exceptional quality pollutant concentration values in Table 10.1. All metal concentrations are below the ceiling concentration limits. Therefore, the biosolids are suitable for land application.

   The concentrations of cadmium, copper, and lead are above the values for exceptional quality biosolids. Therefore, annual application rates for those three metals need to be calculated.

5. Biosolids application rates based on pollutant loading rates:

   Cadmium: annual pollutant loading rate = 1.9 kg/ha·yr (Table 10.1) From equation (10.3), the maximum application rate

$$L_S = \frac{1.9\,\text{kg/ha}\cdot\text{yr}}{(55\,\text{mg/kg})(0.001\,\text{kg/metric ton})/(\text{mg/kg})}$$
$$= 34.5 \text{ metric tons/ha}\cdot\text{yr (15.3 tons/ac-yr)}$$

   Copper:

$$L_S = \frac{75\,\text{kg/ha}\cdot\text{yr}}{(2200\,\text{mg/kg})(0.001\,\text{kg/metric ton})/(\text{mg/kg})}$$
$$= 34.1 \text{ metric tons/ha}\cdot\text{yr (15.2 tons/ac-yr)}$$

   Lead:

$$L_S = \frac{15\,\text{kg/ha}\cdot\text{yr}}{(700\,\text{mg/kg})(0.001\,\text{kg/metric ton})/(\text{mg/kg})}$$
$$= 21.4 \text{ metric tons/ha}\cdot\text{yr (9.6 tons/ac-yr)}$$

   Based on the application rates above, the limiting rate is 21.4 metric tons/ha·yr of bisolids based on lead.

6. Land area required based on pollutant loading: From equation (10.4),

$$A = \frac{1460 \text{ metrictour/yr}}{21.4 \text{ metrictour/ha}\cdot\text{yr}} = 68\,\text{ha (165 ac)}$$

7. Comparing the area required for biosolids application based on nitrogen loading (101 ha) and lead loading (68 ha), the nitrogen loading is more limiting. Therefore, 101 ha (250 ac) of land is initially required for the application of biosolids.

10.3.3 Application Methods

Biosolids can be applied in liquid or dewatered form. The method of application depends on whether the biosolids are in liquid, dewatered cake, composted, or pelletized form; the topography of the site; and the type of vegetation, such as annual field crops, existing forage crops, trees, or preplanted land.

Liquid Biosolids Application The principal advantage of the application of biosolids in the liquid form is its simplicity. It has the added advantages of not requiring dewatering processes at the sludge processing facilities and that the liquid biosolids can be transferred by pumping. However, large quantities of biosolids have to be transported to the application site.

Liquid biosolids have solids concentrations in the range 1 to 7%. Application methods include simple surface spreading and subsurface incorporation by injection or plowing. For small facilities, tank trucks equipped with rear-mounted spreading manifolds or high-capacity spray nozzles are commonly used. If the application sites are near the treatment plant, spray irrigation with high-pressure pumps may be used, thus eliminating the need for vehicular transportation.

Liquid biosolids can be injected below the soil surface by using tank trucks equipped with injection shanks or incorporated immediately after surface application by using plows. A tractor equipped with liquid biosolids injection lines can also accomplish subsurface application. The liquid to be injected is supplied by a hose connected to the injection device. The tethered hose is dragged along by the tractor. Advantages of subsurface application include minimization of potential odor and vector attraction, reducing ammonia loss due to volatilization, and elimination of surface runoff.

Irrigation methods include sprinkling with large-diameter high-capacity sprinkler guns, and furrow irrigation. Sprinkling is used primarily for application to forested lands. The principal advantage of furrow irrigation is that liquid biosolids can be applied to row crops during the growing season without having the biosolids contact the crops themselves.

Dewatered Biosolids Application Dewatered biosolids application is generally used when the application site is some distance from the treatment plant or when the climatic conditions make it necessary to store the biosolids in a confined space for up to six months. Thirty-two to forty kilometers (20 to 25 miles) is usually the distance where liquid transport becomes more expensive than dewatered biosolids transport (Lue-Hing et al., 1998).

Achieving consistent surface application can be difficult with dewatered cake, although dewatered material has the advantage of allowing higher rates of application because of less moisture addition. The main advantage of dewatered biosolids is that farmers can apply them on their lands with their own conventional manure spreaders. Typical solids concentrations of dewatered biosolids applied to land range from 15 to 35%. They are most commonly spread by tractor-mounted box spreaders or manure spreaders followed by plowing into the soil. Compost and thermally dried biosolids are more easier to spread.

10.3.4 Application to Dedicated Lands

Dedicated land disposal means the application of biosolids to land for disposal purposes. The rates of application are much higher than agronomic rates, and no crop is grown. Disturbed land reclamation and high-rate land disposal are two types of dedicated land applications.

Disturbed Land Reclamation The surface mining of coal, exploration for minerals, generation of spoils from underground mines, and tailings from mining operations have created over 1.5 million hectares (3.7 million acres) of disturbed lands in the United States (U.S. EPA, 1983). These lands are usually a harsh environment for vegetation because of a lack of nutrients and organic matter, low pH, low water-holding capacity, low rates of water infiltration and permeability, and the presence of toxic levels of trace metals. The major reason for land reclamation is to revegetate a site so that water and wind erosion will be reduced. A revegetated site has the potential for agricultural production, animal grazing, and reforestation for lumber and pulp production. These benefits relate to the fact that biosolids can increase agricultural and forest utilization of disturbed lands and reduce environmental contamination from these lands.

Disturbed land reclamation consists of a one-time application of 110 to 220 dry metric tons/ha (50 to 100 dry tons/ac). If adequate topsoil is present on the site, annual application is feasible, especially if the site is used for reforestation for lumber and pulp production.

Dedicated Land Disposal Dedicated land disposal is application of biosolids on a continuing basis on a site where high-rate application of biosolids is acceptable environmentally. Avoidance of nitrate contamination of groundwater is the major criterion in selecting a suitable site. Groundwater contamination can be avoided by locating the sites remote from useful aquifers, intercepting and collecting leachate, and by installing impervious liners. Criteria for dedicated land disposal sites are presented in Table 10.4.

Application rates range from 30 to 250 dry metric tons/ha (13 to 110 dry tons/ac). The higher rates have been associated with sites that receive dewatered biosolids, have relatively low precipitation, and have no leachate problems because of site conditions or project design. Assuming a 50-year site operating life, total loading would be 1500 to 12,500 dry metric tons/ha (670 to 5575 dry tons/ac). Biosolids can be applied in slurry or dewatered form much like other land application methods.

TABLE 10.4 Criteria for Dedicated Land Disposal Sites

Parameter	Ideal Condition	Unacceptable Condition
Slope	<3%	Deep gullies, slope >12%
Soil permeability	$<10^{-7}$ cm/s[a]	$>1 \times 10^{5}$ cm/s[b]
Soil depth	>3 m (10 ft)	<0.6 m (2 ft)
Distance to surface water	>300 m (1000 ft) from any surface water	<90 m (300 ft) to any pond or lake used for recreational or livestock purposes, or any surface water body officially so classified under state law
Depth to groundwater	>15 m (50 ft)	<3 m (10 ft) to groundwater table (wells tapping shallow aquifers)[c]
Supply wells	No wells within 600 m (2000 ft)	Within a radius of 300 m (1000 ft)

Source: U.S. EPA, 1983; Metcalf & Eddy, 2003.

[a] When low-permeability soils at or too close to the surface, liquid disposal operations can be hindered due to water ponding.

[b] Permeable soil can be used if appropriate engineering design preventing dedicated land disposal leachate from reaching the groundwater is feasible.

[c] If an exempted aquifer underlies the site, poor-quality leachate may be permitted to enter groundwater.

10.3.5 Conveyance and Storage of Biosolids

Biosolids land application always requires the biosolids to be conveyed to the ultimate location. Successful biosolids land application programs should also have provisions to deal with daily biosolids production at wastewater treatment plants in the event that biosolids cannot be land-applied immediately. This contingency plan generally involves storage.

Transportation Biosolids may be transported by pipeline, truck, barge, rail, or a combination of these. The method of transportation chosen and it cost depend on the consistency and quantity of biosolids to be transported and the distance from origin to destination. Small to medium-sized treatment plants use trucks as the mode of transportation, as the distance to the destination is usually less than 100 km (62 miles). To minimize the danger of spills, odors, and dissemination of pathogens to the air, liquid biosolids should be transported in closed vessels such as tank trucks. Dewatered biosolids can be transported in open trucks, but the trucks should be covered and should have rubber-sealed rear gates to minimize nuisance odors and spills. One way to reduce public exposure to odors is to choose a hauling route that avoids densely populated residential areas. Making sure that trucks used to haul biosolids are clean and well maintained is another effective way to keep the road surfaces clean and to control odors during biosolids transport.

Storage Storage is necessary during inclement weather when land application sites are not accessible and during winter months when land application to snow-covered and frozen soil is prohibited or restricted. Storage may also be needed to accommodate seasonal restrictions on land availability due to crop rotations. For small biosolids generators, storage allows accumulation of enough material to complete land application efficiently in a single spreading operation.

The term *storage* refers to temporary or seasonal storage. Storage operations involve an area of land or facilities to hold biosolids until material is land applied on designated and approved sites. More permanently constructed storage facilities can involve state or locally permitted areas of land or facilities used to store biosolids. These facilities may be used to store any given batch of biosolids up to two years. They are usually located at or near land application sites and are managed so that biosolids come and go on a relatively short cycle, based on weather conditions, crop rotations, and land and equipment availability. Storage options include field stockpiling for short-term storage, and constructed facilities that include concrete, asphalt, clay, or compacted earth pads; basins and lagoons; tanks; or other structures that can be used continually to store liquid, semisolid, or solid biosolids.

Field Stockpiling Field stockpiling is used for short-term storage of dewatered cake, dried, or composted class A or B biosolids at the land application site. It is generally limited to the amount of biosolids needed to meet agronomic or reclamation requirements at a field or site. Field stockpiles should be placed in the best physical location possible in or adjacent to the fields that will receive the biosolids. For sites with a significant slope, provisions need to be made to manage up- and down-slope water. Forming windrows across slopes should be avoided to reduce the potential for piles to become anaerobic at the base where overland flow accumulates.

To the extent possible, piles should be shaped to shed water. Stockpiled biosolids form an air-dried crust that sheds precipitation and prevents significant percolation of water through the pile. Nonetheless, some states require that stockpiles be covered. For composted or dried (at least 50% solids) biosolids, tarps, wind barriers, or periodic wetting may be necessary to minimize blowing of dust, particularly in arid, windy climates when stockpiles are in close proximity to sensitive downwind areas. There have been some instances of tarps catching fire when used on compost materials. Incompletely composted materials have the potential to self-heat because microbial growth can still occur on the remaining nutrients. Heat-dried products that are rewetted or have not been sufficiently dried and cooled (<90% solids, >30°C) also can self-heat. In the presence of enough available water, microbes will utilize the nutrients in the biosolids and generate heat that cannot dissipate because of the mass of the stockpiles. Therefore, composted and heat-dried biosolids piles should be monitored so that a fire hazard does not develop. A noticeable increase in odor is a reliable indicator of microbial activity and the potential development of hot spots.

Storage Basins and Lagoons Storage basins for liquid or semisolid biosolids need to be large enough to provide adequate storage volume during worst-case weather conditions (long periods of inclement weather when field application is restricted and the basin cannot be emptied). Depth of storage basins may vary from 3 to 5m (10 to 16ft). The design volume must also include space for accumulation of precipitation expected over the storage period, plus the capacity to hold severe storm events. An impermeable liner is recommended to ensure against loss of biosolids constituents to groundwater by leaching. This type of design may negate the need for groundwater monitoring wells. Biosolids can be removed from the basins using a mud pump mounted on a floating platform or a mobile crane using a dragline.

Lagoons are used for long-term storage of sludge. They are usually part of the wastewater treatment plant. They are simple and economical if the treatment plant is located in a remote location. A lagoon is an earthen basin into which untreated sludge or digested biosolids are deposited. In lagoons with untreated sludge, the organic matter is stabilized by anaerobic and aerobic decomposition, which may give rise to objectionable odors. If the lagoons are used only for digested biosolids, nuisance odors are usually not a problem. The stabilized solids settle to the bottom of the lagoon and accumulate. Excess liquid from the lagoon may be decanted and returned to the plant for treatment. Solids concentrations as high as 35% have been reported in the bottom layer of lagoons used for treatment and long-term storage.

Storage Tanks Storage tanks for class A or B liquid biosolids are above- or belowground structures that are permanent and part of the sludge-processing facilities in a wastewater treatment plant. They are watertight and are generally concrete or steel structures which may be prefabricated or constructed entirely on site. Due to their impervious nature, these facilities do not warrant groundwater-monitoring wells.

Storage tanks may be open-topped or enclosed. The tank volumes should be large enough to contain daily biosolids produced during worst-case periods of inclement weather, or backup options must be part of the planning process. Enclosed tanks should be ventilated through passive vents or mechanical fans. Depending on the type of biosolids, tank design, climatic conditions, and airflow rates, a gas detection meter and alarm system tied to ventilation fans may be advisable to eliminate buildup of explosive levels of methane that might result from anaerobic biological activity in the tank.

Dewatered Biosolids Storage Dewatered biosolids storage facilities can be covered or uncovered and are designed to provide up to two years of storage for class A or B dewatered, air-dried, heat-dried, dry-lime stabilized, or composted biosolids. These facilities include open-sided or enclosed buildings and open-topped bunkers or pads. Storage facilities need to be large enough to provide adequate storage volumes during long periods of inclement weather when field application is restricted.

Unroofed facilities should have a durable hard pad with push walls and stormwater curbs, containment walls, and sumps. An impermeable floor is recommended to help control runoff, protect against loss of biosolids constituents to groundwater by leaching, and to accommodate vehicle traffic. Recommended materials include concrete or asphalt. For storing class A biosolids in arid areas, compacted soil or gravel with appropriate runoff controls may be satisfactory.

10.4 BENEFICIAL USE OF BIOSOLIDS IN RUSSIA

In Russia, long-term research and development of sludge treatment prior to land application was conducted with the principal author's participation and under his direct management (Turovskiy, 1999). The following organizations took part in that research: All-Union Research Institute of Water Supply and Sewage System, Russian Municipal Academy, Institute of Municipal Hygiene, Research Institute of Hygiene, Institute of Medical Parasitology and Tropical Medicine, Institute of Fertilizers and Agricultural Soil, Science Medicine Institutes, Agricultural Institutes of the Former Soviet Republics, and several wastewater treatment plants. The studies were performed with various types of sludge, such as raw sludge, aerobically digested sludge, anaerobic mesophilic and thermophilic digested sludge, mechanically dewatered sludge, and sludge dried on drying beds, from more than 100 wastewater treatment plants. Parameters analyzed included pathogenic bacteria, viruses, heavy metals, and helminth ova.

10.4.1 Pathogens

Large numbers of helminth eggs were found in both raw sludge and sludge digested under aerobic and anaerobic mesophilic conditions. Helminths are common parasitic worms that include cestodes (tapeworms), nematodes (roundworms or ascariasis, and whipworms or *Trichuris trichura*), and trematodes (flukes). Nematodes form the main mass of eggs.

The life cycle of nematodes is typical of helminths. Larvae bore through the intestinal wall of a host, either human or animal, enter the bloodstream, and are carried to the lungs, from which they are coughed up into the mouth and returned to the intestine, where they develop into adult worms. Humans are host to about 50 species of roundworms and, in fact, more than a third of the world's population suffers from diseases caused by roundworms (Goodman et al., 1986).

The number of nematode eggs in 1 kg of primary sludge, activated sludge, and the mixture digested in aerobic or anaerobic mesophilic conditions reached several hundreds; in mechanically dewatered sludge the number reached several thousands. A portion of the helminth eggs perishes during drying on drying beds and during storage, but a large part survives for a long time, and they can be transformed into larvae, which can survive for about five years in drying beds (Turovskiy, 1988). During the experiments, special little bags with helminth eggs were incorporated into the sludge, which allowed a precise determination of the quantity of damaged eggs resulting from the treatment. It was determined that helminth eggs are destroyed within 2 hours by heating at 50°C, within several minutes at 60°C, and within several seconds at 70°C.

After mechanically dewatering sludge and heating it to 60°C, inoculation on Wilson Bleaur or Ploskiryov medium, on media with different inhibitors, or on Miller, Kaufman, and other media revealed no presence of the intestinal typhoid group of bacteria. The studies showed that because of the extreme changeability of the colon bacillus that was revealed in the process of reactivation, there should be no fear of livability or virulence of pathogenic microbes during utilization of dewatered, heated sludge.

During aerobic anaerobic digestion, helminth eggs survive for a long time. In thermophilic digestion, additional equipment for disinfection may not be necessary. For predisinfection of mesophilic digestion, single-, two-, or multistage heat exchangers may be used for heating the sludge to 60°C. The temperature in the process that produces class A sludge that meets the fecal coliform requirements has to be maintained at either 53°C for 5 days, 55°C for 3 days, or 70°C for 30 minutes (Lue-Hing et al., 1998). These temperatures meet not only the fecal coliform requirements, but also density requirements for viruses and viable helminth ova.

Russian Construction Standards and Regulations for disinfecting sludge allow the following methods to be used:

- *Thermal*: heating, drying, incineration
- *Biothermal*: composting
- *Chemical treatment*
- *Biological*: extermination of microorganisms by unicellular fungi and by soil plants

For thermal drying, the standards require heating the entire mass of sludge to not less than 60°C. The moisture in the thermally dried sludge is recommended to be not less than 25% because evaporation of bound water requires excessive energy to achieve a moisture content below 25%, sludge with less than 25% moisture becomes hygroscopic and is able to absorb water from the air, and dust is formed when drying sludge to less than 25% moisture.

Raw sludge treated with hydrated lime to increase its pH to greater than 10 loses its odor, coliform and enterococcus are suppressed, but there is no significant effect on helminth eggs. Hydrated lime increases temperature during hydration; 1 mol (56 g) of calcium oxide generates 65 kJ of heat:

$$CaO + H_2O = Ca\,(OH)_2 + 65\,kJ \tag{10.5}$$

During the process of hydrating 1 kg of chemically pure lime (100% CaO), 1152 kJ of heat is produced, requiring 320 g of water.

The quantity of heat in kilojoules that is required for heating sludge by lime treatment may be determined by the formula

$$Q = (M_{SL}C_{SL} + M_{L}C_{L})\,\Delta T \tag{10.6}$$

where

Q = quantity of heat, kJ
M_{SL} = sludge mass, kg
C_{SL} = specific heat of sludge, kJ/kg · °C
M_{L} = lime mass, kg
C_{L} = specific heat of lime, kJ/kg · °C = 0.92
ΔT = difference in temperature that is necessary to heat the sludge and prime temperature of the sludge

The specific heat of sludge may be calculated as follows:

$$C_{SL} = 1.8\left(1 + 0.85 W_{SL}{}^{3}\right) \tag{10.7}$$

where

C_{SL} = specific heat of sludge, kJ/kg · °C
1.8 = specific heat dry sludge with a moisture of 5 to 10%, kJ/kg · °C
W_{SL} = sludge moisture, %

The heat in kilojoules produced by lime hydration, taking into consideration its activity by CaO, is

$$Q_{R} = 1152\,A M_{L} \tag{10.8}$$

where

Q_R = heat, kJ
A = lime activity, %
M_{L} = lime mass, kg

The increase in sludge temperature caused by a predetermined dose of lime may be calculated preliminarily by using the equation of material balance:

$$\Delta T_C = \frac{1152\,A M_{L}}{M_{SL}C_{SL} + M_{L}C_{L}} \tag{10.9}$$

Given the required difference in temperatures, the mass of lime necessary for sludge treatment may be calculated as follows:

$$M_{\mathrm{L}} = \frac{M_{\mathrm{SL}} C_{\mathrm{SL}}\, \Delta T}{1152A - C_{\mathrm{L}}\, \Delta T} \tag{10.10}$$

The formula for calculating sludge moisture after lime addition, assuming complete hydration of lime, is

$$W_K = \frac{1000 W_{\mathrm{SL}} - 0.32 A M_{\mathrm{L}}}{M_{\mathrm{SL}} + M_{\mathrm{L}}} \tag{10.11}$$

where W_K is the sludge moisture after lime addition, %.

The amount of lime (M_{L}) in kilograms required to decrease the moisture of 1 metric ton of sludge to the required level may be determined as

$$M_{\mathrm{L}} = \frac{(W_{\mathrm{SL}} - W_k)(1000)}{(0.32 + W_k)A} \tag{10.12}$$

The equations above may be used only for approximate calculations, because in reality, lime is not completely hydrated because of the internal structure of lime (which is not homogeneous by the size and quality of particles) and the presence of bound water in sludge with different forms of binding between solid particles and water. Lime consumption depends very strongly on the sludge moisture and therefore on the volume of the treated sludge. Therefore, dewatering of sludge is highly recommended.

10.4.2 Heavy Metals

Russian studies on the influence of biosolids on harvest, crop quality, and migration of heavy metals from biosolids into soil and from soil into plants (Goldfarb et al., 1983; Turovskiy, 1988) established that absorption of heavy metals by plants depends on the following:

- *Mobility of heavy metals in sludge.* Nickel, cadmium, and zinc are the most movable metals, but different methods of treatment may change the mobility of heavy metals in sludge. When treated by lime, for instance, the major part of the heavy metals does not migrate into plants.
- *Types of soils.* In acidic soils, the mobility of heavy metals is significantly higher than in alkaline soils. Organic and exchangeable cations in soils facilitate withholding of heavy metals. Until nitrogen is acting as a fertilizer, the influence of heavy metals on the growth of pants is insignificant.
- *Types of plants.* Crops have different absorptions of metals. For instance, during the study of the mineral composition of potatoes, carrots, and garden radishes grown on lands where sludge from Kiev (Ukraine) was

applied, researchers found that manganese was absorbed by all experimental crops; zinc and copper by only potatoes and carrots; iron by only carrots and garden radishes; and lead by only garden radishes. Iron and molybdenum decreased in potatoes, and nickel decreased in carrots. Other elements did not change significantly.

In the grain and straw of millet grown on black soils where turf–sludge fertilizers were applied, the content of copper and nickel increased. The content of chromium, lead, titanium and molybdenum in the ash of all plants were on the level of control.

Distribution of heavy metals in plants is uneven—the concentration is greater in plant's organs (stems, leaves) and smaller is the grain. The age of plants also affects the accumulation of heavy metals; there are more heavy metals in older tissues than in younger ones.

It was found that the conditions under which the sludge was applied affected the accumulation of heavy metals in plants. Applying sludge to soils in the spring tended to cause an increase in iron, barium, molybdenum, nickel, cobalt, and chromium in potato tubers.

Experiments to determine the toxicity and migration of heavy metals in plants were conducted because of the high content of heavy metals in the thermally dried sludge at the wastewater treatment at the city of Orechovo-Zuevo. The content of heavy metals (in mg/kg of dry substance) was as follows: lead, 200; chromium, 700; cadmium, 100; manganese, 500; copper, 200; zinc, 2500; and nickel, 100. The results of the experiments are shown in Table 10.5.

Because of the possibility of accumulating toxins in plants, it was decided to restrict the dosage of biosolids in the soil. Criteria for such restrictions were standard for limited concentration values in soil (PDS) for several heavy metals and toxins that were developed in Russia and approved by Russia's Ministry of Health. PDS values in dry soil were developed for the following metals (in mg/kg): lead, 20; cadmium, 9; arsenic, 20; nickel, 50; chromium, 100; mercury, 2.1; manganese, 1500; vanadium, 150; manganese + vanadium, 1000 + 100; and superphosphate (P_2O_5), 200.

The maximum rate of adding biosolids to soil is usually determined by a calculation that takes into account the possibility of embedding harmful impurities in the soil. The calculation is based on an assertion that after adding sludge to the soil, the total content of heavy metals in the soil (considering the dispersion in the plowed layer) should not exceed the permissible dose of heavy metals in soil (PDS) in mg/kg, defined by the equation

$$F + D < \text{PDS} \tag{10.13}$$

where F is the prime content of heavy metal in soil (mg/kg) and D is the additional supply of the same metal into the plowed layer of soil with fertilizer (mg/kg)

TABLE 10.5 Content of Heavy Metals in Plants Grown with the Use of Thermally Dried Sludge from Orechovo-Zuevo, Russia

Sample Analyzed	Content[a] (% of mass of dry substance) Cu	Zn	Mn	Ash Content (%)
Barley				
Control[b]	0.008	0.002	0.004	2.7
50 metric tons/ha TDS	0.001	0.003	0.0026	2.2
50 metric tons/ha TDS + N_{180}	0.002	0.004	0.005	5.4
50 metric tons/ha TDS + P_{120}	0.002	0.003	0.002	5.7
50 metric tons/ha TDS + K_{180}	0.002	0.0017	0.0036	2.2
50 metric tons/ha TDS + N_{180} + K_{180} + P_{120}	0.007	0.004	0.003	2.5
Oat				
Control[b]	0.0016	0.006	0.014	4.0
10 metric tons/ha TDS	0.0015	0.008	0.012	3.4
20 metric tons/ha TDS	0.0005	0.006	0.007	2.5
Corn				
Control[b]	0.01	0.016	0.005	10.5
10 metric tons/ha TDS	0.01	0.008	0.003	10.7
20 metric tons/ha TDS	0.007	0.007	0.005	10.2
Perennial herbs				
Control[b]	0.005	0.004	0.003	4.1
10 metric tons/ha TDS	0.004	0.003	0.003	5.0

[a] In addition to the elements listed, traces of Cr and Cd were found.
[b] Control = plant grown on soils without biosolids.

The value of permissible addition to soil of one or another toxin D_{tot} may be determined by the equation

$$D_{tot} = (\text{PDS} - F)(3000) \tag{10.14}$$

where 3000 is the mass of the plowed layer of soil in metric tons/ha recalculated for dry substance.

Depending on the type of soil, the magnitude of permissible supply of harmful impurities D_{tot} is usually decreased by the reducing factor K, which can be determined by the formula

$$K = K_1 K_2 K_3 \tag{10.15}$$

where

K_1 = factor based on the content of humus (H) in the soil (when H = 0.5 to 1%, $K_1 = 0.6$; when H = 1 to 2%, $K_1 = 0.8$; when H = 2 to 3%, $K_1 = 0.9$; when H > 3%, $K_1 = 1$)

K_2 = factor based on the mechanical composition of soil (for sandy and sandy-loam soils, $K_2 = 0.7$; for loamy soils, $K_2 = 0.9$; for all other soils, $K_2 = 1$)

K_3 = factor based on the concentration of hydrogen ion in the soil (for soil with pH < 5 and biosolids with pH < 6, $K_3 = 0.4$; for soil with pH = 6.5 to 7.0 and biosolids with pH < 6, $K_3 = 0.5$; for soil with pH = 7.0 to 7.5 and biosolids with pH > 6, $K_3 = 0.8$)

The average annual rate of adding biosolids to soil, D_{av} (in metric tons/ha), can be calculated using the equation

$$D_{av} = \frac{D_{tot}}{TC_N} \tag{10.16}$$

where T is the maximum time for adding sludge to the same site (years) and C_N is the concentration of the element in biosolids (g/ton of dry solids). The maximum rate of adding biosolids to soil (D_{max}) at a frequency of once in five years will be $5D_{av}$ ton/ha·yr dry solids. D_{max} is also restricted by the level of nitrogen in the soil, which should not exceed $N_P = 300$ kg/ha·yr.

Current specifications for thermally dried biosolids or compost used as a fertilizer limit the quality of the biosolids as follows:

moisture (%): <50

content as percentage of the mass of dry substance:

- organic: >40
- nitrogen (total N): >1.6
- phosphorus (as P_2O_5): >0.6
- potassium (as K_2O): >0.2

quantity of dry fractions D < 250 microns; % for sludge does not contain fraction more than 30 mm: <10

average density (kg/m^3): 500 to 700

Several articles have described how to meet 40 CFR Part 503 regulations (Foess and Siger, 1993; Siger, 1993; McDonald, 1995; U.S. EPA, 1995). The requirements for heavy metals content in biosolids become less restrictive when they are used for shrubs, flowers, fast-growing trees, development of low-productive soils, soil stabilization of ravines and hillsides, and planting trees and shrubs on former industrial waste sites.

Experiments performed in Russia, Belarus, and Ukraine showed that biosolids as fertilizers applied to the soil once continue to work for several years, although the effect decreased in subsequent years. Table 10.6 shows the data from those experiments for growing wheat and barley.

Compared to mineral fertilizers applied every year, biosolids applied once were effective for several years, as evidenced by the average harvest over

TABLE 10.6 Effect of Biosolids on Future Harvests

Condition	First-Year Harvest (kg/ha)	Second-Year Harvest (kg/ha)	Third-Year Harvest (kg/ha)	Average 3-Year Harvest (kg/ha)	Comments
Wheat					
Without fertilizers	3100–4800	3000–3250	2310–2700	3200	Average of experiments with different soils
Mineral fertilizers, $N_{90}P_{90}K_{60}$	4200–5400	3780–4970	3450–4300	4360	Used every year
Compost, 40 tons/ha	5510–5900	3980–5000	3500–3710	4600	Applied only in the first year
Thermally dried, 30 tons/ha	4980–5190	3860–4370	3120–3270	4130	Applied only in the first year
Barley					
Without fertilizers	1850–2970	1150–2450	1200–2300	1990	Average of experiments with different soils
Mineral fertilizers, $N_{90}P_{90}K_{60}$	2440–3070	2250–3150	2100–2920	2660	Used every year
Compost, 40 tons/ha	5100–4190	1890–3970	1620–2980	3290	Applied only in the first year
Thermally dried, 30 tons/ha	3100–4010	1890–3970	1890–2900	2960	Applied only in the first year

several years. Experiments also showed that compost and thermally dried biosolids were more effective with crop rotation every other year, such as potatoes in the first year and wheat in the second year, or barley in the first year and corn in the second year. Thermally dried biosolids are very effective when they are used to grow silage corn. For example, when sludge from the city of Orechovo-Zuevo, Russia (lime and ferric chloride conditioned, dewatered by vacuum filters, and thermally dried) was used as fertilizer to grow silage corn, the harvest increased almost fourfold. It was also found that biosolids as fertilizers are most effective when applied in autumn. For growing barley and corn, the best results are obtained when biosolids are used in combination with mineral fertilizers.

As a rule, application of biosolids to a field is recommended only once every five years, in a dosage from 5 to 20 dry tons/ha. This method provided higher average yields than did those using commercial fertilizers.

REFERENCES

Bitton, G. (1994), *Wastewater Microbiology*, Wiley-Liss, New York.

Epstein, E. (1997), *The Science of Composting*, Technomic Publishing Co., Lancaster, PA.

Federal Register (1993), 40 CFR Part 503, *Standards for the Use or Disposal of Sewage Sludge*, 58 FR9248 to 9404.

Foess, G. M., and Siger, R. B. (1993), Pathogen/Vector Attraction Reduction Requirement of the Sludge Rules, *Water Engineering and Management*, June, p. 25.

Garvey, D., Guario, C., and Davis, R. (1993), Sludge Disposal Trends Around the Globe, *Water Engineering and Management*, December, p. 17.

Goldfarb, L., Turovskiy, I., and Belaeva, S. (1983), *The Practice of Sludge Utilization*, Stroyizdat, Moscow.

Golueke, C. G. (1983), Epidemiological Aspects of Sludge Handling and Management, Part 2, *BioCycle*, Vol. 24, No. 3, p. 52.

Goodman, H. D., et al. (1986), *Biology*, Harcourt Brace Jovanovich, Orlando, FL.

Henry, C., and Brown, S. (1997), Restoring a Superfund Site with Biosolids and Fly Ash, *BioCycle*, Vol. 38, No. 11.

Kulik, A. (1996), Europe Cultivates Organics Treatment, *World Wastes*, Vol. 39, No. 2, pp. 37–40.

Lue-Hing, C., Zery, D. R., and Kuchenither, R. (Eds.) (1998), *Water Quality Management Library*, Vol. 4, *Municipal Sewage Sludge Management: A Reference Text on Processing, Utilization and Disposal*, Technomic Publishing Co., Lancaster, PA.

Maurice, B. (2003), *Seminar: Residuals Beneficial Use and Regulations in Florida*, Florida Department of Environmental Protection, Tallahassee, FL.

McDonald, G. J. (1995), Applying Sludge to Agricultural Land: Within the Rules, *Water Engineering and Management*, February, p. 28.

Metcalf & Eddy, Inc. (2003), *Wastewater Engineering: Treatment and Reuse*, 4th ed., Tchobanoglous, G., Burton, F. L., and Stensel, H. D. (Eds.), McGraw-Hill, New York.

Russian Government Standard (2001), *Demand of Wastewater Sludge Use as Fertilizers*, GOST R 17.4.3.07-2001, Moscow.

Siger, R. B. (1993), Practical Guide to the New Sludge Standards, *Water Engineering and Management*, November, p. 26.

Turovskiy, I. S. (1988), *Wastewater Sludge Treatment*, Stroylizdat, Moscow.

——— (1999), Beneficial Use of Wastewater Sludge Biosolids in Russia, *Florida Water Resources Journal*, May, pp. 23–25.

U.S. EPA (1981), *Process Design Manual for Land Treatment of Municipal Wastewater*, EPA 625/1-81/013.

——— (1983), *Process Design Manual for Land Application of Municipal Sludge*, EPA 625/1-83/016.

——— (1994), *A Plain English Guide to the EPA Part 503 Biosolids Rule*, EPA 832/R-93/003.

——— (1995), *A Guide to Biosolids Risk Assessment for EPA Part 503 Rule*, EPA 832/B-93/005.

——— (1999), *Biosolids Generation, Use, and Disposal in the United States*, EPA 530/R-99/009.

——— (2000), *Guide to Field Storage of Biosolids*, EPA 832/B-00/007.

Vesilind, P. A. (Ed.) (2003), *Wastewater Treatment Plant Design*, Water Environment Federation, IWA Publishing, Alexandria, VA.

Ward, R. L., McFeters, G. A., and Yeager, J. G. (1984), *Pathogens in Sludge*, Sandia Report 83-0557, TTC-0428, Sandia National Laboratory, Albuquerque, NM.

WEF (1998), *Design of Municipal Wastewater Treatment Plants*, 4th ed., Manual of Practice 8 (ASCE 76), Water Environment Federation, Alexandria, VA.

APPENDIX

UNITS OF MEASURE

A.1 Abbreviations for SI Units

Abbreviation	SI Unit	Abbreviation	SI Unit
°C	degree Celsius	kWh	kilowatthour
cal	calorie	L	liter
cm	centimeter	m	meter
cm^2	square centimeter	m^2	square meter
cm^3	cubic centimeter	m^3	cubic meter
d	day	mg	milligram
g	gram	Mg	megagram (1000 kg or metric ton)
h	hour		
ha	hectare	MJ	megajoule
J	joule	mL	milliliter
K	kelvin	mm	millimeter
kg	kilogram	MPa	megapascal
kJ	kilojoule	N	newton
km	kilometer	Pa	pascal
kN	kilonewton	s	second
kPa	kilopascal	W	watt
kW	kilowatt		

Wastewater Sludge Processing, By Izrail S. Turovskiy and P. K. Mathai

A.2 Abbreviations for U.S. Customary Units

Abbreviation	U.S. Customary Unit	Abbreviation	U.S. Customary Unit
ac	acre	gpm	gallons per minute
Btu	British thermal unit	hr	hour
cfm	cubic foot per minute (ft^3/min)	hp	horsepower
		hp-hr	horsepower-hour
cfs	cubic foot per second (ft^3/sec)	in.	inch
		lb	pound
cy	cubic yard (yd^3)	MG	million gallons
d	day	mgd	million gallons per day
°F	degree Fahrenheit	min	minute
ft	foot	scfm	standard cubic foot per minute
ft^2	square foot		
ft^3	cubic foot	ton	2000lb mass (short ton)
gal	gallon	wk	week
gpd	gallons per day	yr	year

A.3 Conversion from SI Units[a] to U.S. Customary Units

Multiply SI Units			To Obtain U.S. Customary Units	
Name	Symbol	by	Symbol	Name
centimeter	cm	0.3937	in.	inch
cubic meter	m^3	35.31	ft^3	cubic foot
		1.308	cy	cubic yard
		264.2	gal	gallon
cubic meters per day	m^3/d	0.1835	gpm	gallons per minute
		264.2	gpd	gallons per day
		2.642×10^{-4}	mgd	million gallons per day
cubic meters per second	m^3/s	22.83	mgd	million gallons per day
degree Celsius	°C	°C × 1.82 + 32	°F	degree Fahrenheit
	K	K × 1.8 − 459.67	°F	degree Fahrenheit
gram	g	2.205×10^{-3}	lb	pound
grams per cubic meter	g/m^3	8.34	lb/MG	pounds per million gallons
grams per kilogram	g/kg	1.0×10^{-3}	lb/lb	pounds per pound
		2.0	lb/ton	pounds per ton
hectare	ha	2.471	ac	acre
joule	J	0.2388	cal	calorie
		1.0	W-sec	Watt-second
kelvin		2.777×10^{-4}	W-hr	Watt-hour
kilogram	kg	2.2.5	lb	pound
kilograms per cubic meter	kg/m^3	6.243×10^{-2}	lb/ft^3	pounds per cubic foot
		1.686	lb/cy	pound per cubic yard
kilograms per cubic meter per day	kg/m^3·d	6.242×10^{-2}	lb/ft^3-d	pounds per cubic foot per day

A.3 *Continued*

Multiply SI Units			To Obtain U.S. Customary Units	
Name	Symbol	by	Symbol	Name
		62.42	$lb/10^3 ft^3$-d	pounds per thousand cubic foot per day
kilograms per day	kg/d	2.205	lb/lb	pounds per pound
kilograms per hectare	kg/ha	0.8922	lb/ac	pounds per acre
kilograms per square meter	kg/m^2	0.2049	lb/ft^2	pounds per square foot
kilojoule	kJ	0.9478	Btu	British thermal unit
kilojoules per kilogram	kJ/kg	0.4303	Btu/lb	British thermal unit per pound
kilometer	km	0.6214	mile	mile
kilopascal (gage)	kPa	0.145	psi	pounds per square inch (gage)
kilowatt	kW	1.341	hp	horsepower
liter	L	3.532×10^{-2}	ft^3	cubic foot
		0.2642	gal	gallons
liters per second	L/s	15.85	gpm	gallons per minute
megagram (1000 kg or metric ton)	Mg	1.1.5	ton	ton (short)
megajoule	MJ	0.3725	hp-hr	horsepower-hour
		0.2777	kWh	kilowatthour
meter	m	3.281	ft	foot
		1.094	yd	yard
meters per second	m/s	3.281	fps	feet per second
milligrams per liter	mg/L	8.34	ll/MG	pounds per million gallons
milliliters per cubic meter	mL/m^3	0.1337	ft^3/MG	cubic feet per million gallons
millimeter	mm	3.937×10^{-2}	in.	inch
newton	N	0.2248	lb	pound force
pascal	Pa	1.450×10^{-4}	psi	pounds per square inch
square kilometer	km^2	247.1054	ac	acre
		0.3851	mi^2	square mile
square meter	m^2	10.764	ft^2	square foot
		1.196	yd^2	square yard
watt	W	3.412	Btu/h	British thermal units per hour
		0.738	ft-lb/sec	foot-pounds per second

[a] Some of the units listed are metric units commonly used in environmental engineering.

A.4 Conversion from U.S. Customary Units to SI Units[a]

Multiply U.S. Customary Units			To Obtain SI Units	
Name	Symbol	by	Symbol	Name
acre	ac	0.4047	ha	hectare
bar	bar	100.0	kPa	kilopascal
British thermal units	Btu	1.055	kJ	kilojoule
British thermal units per pound	Btu/lb	2.326	kJ/kg	kilojoules per kilogram
calorie	cal	4.187	J	joule
cubic feet per million gallons	ft^3/MG	7.482	mL/m^3	milliliters per cubic meter
cubic feet per second	cfs (ft^3/s)	2.832×10^{-2}	m^3/s	cubic meters per second
cubic foot	ft^3	2.832×10^{-2}	m^3	cubic meter
		28.32	L	liter
cubic yard	cy	0.7646	m^3	cubic meter
degree	°F	0.5556(°F − 32)	°C	degree Celsius
Fahrenheit		(°F + 459.67)	K	Kelvin
degree Rankin	°R	0.5556	K	Kelvin
feet per second	fps	0.3048	m/s	meters per second
foot	ft	0.3048	m	meter
gallon	gal	3.785×10^{-3}	m^3	cubic meter
		3.785	L	liter
gallons per day	gpd	3.785×10^{-3}	m^3/d	cubic meters per day
gallons per minute	gpm	5.451	m^3/d	cubic meters per day
horsepower	hp	7.457	kW	kilowatt
horsepower-hour	hp-hr	2.685	MJ	megajoule
kiowatthour	kWh	3.60	MJ	megajoule
mile	mile	1.609	km	kilometer
million gallons	MG	3.785×10^3	m^3	cubic meter
million gallons per day	mgd	3.785×10^3	m^3/d	cubic meters per day
		4.383	m^3/s	cubic meters per second
pound (force)	lbf	4.448	N	newton
pound (mass)	lb	0.4536	kg	kilogram
pounds per acre per day	lb/ac-d	1.121	kg/ha · d	kilograms per hectare per day
pounds per gallon	lb/gal	0.1198	kg/L	kilograms per liter
pounds per million gallons	lb/MG	0.1198	g/m^3	grams per cubic meter
		0.1198	mg/L	milligrams per liter

A.4 *Continued*

Multiply U.S. Customary Units			To Obtain SI Units	
Name	Symbol	by	Symbol	Name
pounds per square inch (force)	psi	6895	Pa	pascal
pounds per thousand cubic feet	$lb/10^3 ft^3$	16.02×10^{-3}	kg/m^3	kilograms per cubic meter
square foot	ft^2	9.290×10^{-2}	m^2	square meter
square inch	in^2	645.2	mm^2	square millimeter
square mile	mi^2	2.590	km^2	square kilometer
square yard	yd^2	0.8361	m^2	square meter
ton (short)	ton	907.2	kg	kilogram
tons per acre	tons/ac	0.2242	kg/m^2	kilograms per square meter
		2.242	Mg/ha	megagrams per hectare
yard	yd	0.9144	m	meter

[a] Some of the units listed are metric units commonly used in environmental engineering.

INDEX

Wastewater Sludge Processing, By Izrail S. Turovskiy and P. K. Mathai